Design of Molecular Materials

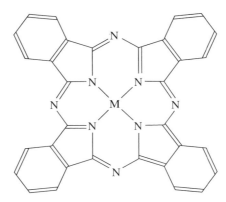
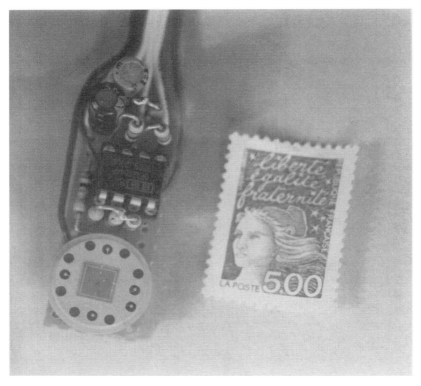

Photograph: F. G. Tournilhac

Design of Molecular Materials
Supramolecular Engineering

Jacques Simon
ESPCI-CNRS, Paris, France

Pierre Bassoul
ESPCI-CNRS, Paris, France

JOHN WILEY & SONS, LTD
Chichester • New York • Weinheim • Brisbane • Singapore • Toronto

Copyright © 2000 by John Wiley & Sons Ltd,
Baffins Lane, Chichester,
West Sussex PO19 1UD, England

National   01243 779777
International (+44) 1243 779777
e-mail (for orders and customer service enquiries): cs-books@wiley.co.uk
Visit our Home Page on http://www.wiley.co.uk
or http://www.wiley.com

All Rights Reserved. No part of this publication may be reproduced, stored in a
retrieval system, or transmitted, in any form or by any means, electronic, mechanical, photocopying,
recording, scanning or otherwise, except under the terms of the Copyright, Designs and Patents Act
1988 or under the terms of a licence issued by the Copyright Licensing Agency Ltd, 90
Tottenham Court Road, London, W1P OLP, UK, without the permission in writing of the Publisher

Other Wiley Editorial Offices

John Wiley & Sons, Inc., 605 Third Avenue,
New York, NY 10158-0012, USA

WILEY-VCH Verlag GmbH, Pappelallee 3,
D-69469 Weinheim, Germany

Jacaranda Wiley Ltd, 33 Park Road, Milton,
Queensland 4064, Australia

John Wiley & Sons (Asia) Pte Ltd, Clementi Loop #02-01,
Jin Xing Distripark, Singapore 129809

John Wiley & Sons (Canada) Ltd, 22 Worcester Road,
Rexdale, Ontario M9W 1L1, Canada

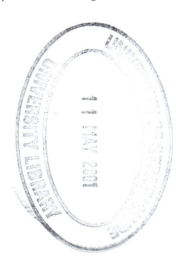

Library of Congress Cataloguing-in-Publication Data

Simon, J. (Jacques), 1947–
  Design of molecular materials : supramolecular engineering / by Jacques Simon, Pierre Bassoul.
    p. cm.
  Includes bibliographical references and index.
  ISBN 0-471-97371-8 (alk. paper)
    1. Organic solid state chemistry. 2. Macromolecules. I. Bassoul, Pierre. II. Title.

QD478.S52 2000
661′.805—dc21   99-089114

British Library Cataloguing in Publication Data

A catalogue record for this book is available from the British Library

ISBN 0 471 97371 8

Typeset in 10/12pt Times by Laser Words, Madras, India
Printed and bound in Great Britain by Biddles Ltd, Guildford, Surrey
This book is printed on acid-free paper responsibly manufactured from sustainable forestry,
in which at least two trees are planted for each one used for paper production

Contents

Preface ix

Introduction: Systemic Chemistry xi

1 Molecular Assemblies 1
   1.1 Generalities 1
   1.2 Solid State 5
   1.3 Thermotropic Liquid Crystals 9
   1.4 Lyotropic Molecular Assemblies 15
   1.5 Supramolecular Assemblies of Metal Complexes 23
   1.6 Conclusions 43
   1.7 References 44

2 A Few Notions of Symmetry 47
   2.1 Introduction 47
   2.2 Elements of Symmetry 47
   2.3 Relations between Symmetry Elements 52
   2.4 Symmetry of Molecular Units 54
   2.5 Determination of the Symmetry of a Molecular Unit 58
   2.6 Matrix Notation for Geometric Transformations 60
   2.7 Notion of Irreducible Representation 62
   2.8 Group–Subgroup Relationships 66
   2.9 References and Notes 67

3 Supramolecular Engineering: Symmetry Aspects 69
   3.1 Introduction 70
   3.2 One-Dimensional Space Groups 71
   3.3 Two-Dimensional Space Groups 79
   3.4 Notion of Two-Dimensional Molecular Shape 93
   3.5 Three-Dimensional Case 111
   3.6 References and Notes 128

4 Symmetry and Physicochemical Properties: The Curie Principle 131
   4.1 Electrical Potentials 131
   4.2 Electrical Multipoles in Chemistry 139
   4.3 Cause/Effect Symmetry Relationships 140

4.4	Generalization of the Curie Principle	153
4.5	References	155

5 Interactions and Organization in Molecular Media 156
5.1 Directional and Nondirectional Forces 156
5.2 Charge/Charge (Coulomb) Forces 159
5.3 Molecular Polarization 167
5.4 Induced Dipole/Induced Dipole Interactions 171
5.5 Polar Molecular Units 180
5.6 Directional Nonbonded Interactions 182
5.7 Segregation in Mesophases 183
5.8 Hydrogen Bonds 190
5.9 References 193

6 Molecular Semiconductors: Properties and Applications 196
6.1 Introduction 197
6.2 Collective and Individual Approaches to Electronic Levels in Molecular Materials 199
6.3 Molecular Semiconductors: Generalities 217
6.4 A Narrow-Band Molecular Semiconductor: Pc_2Lu 220
6.5 A Broad-Band Molecular Semiconductor: PcLi 229
6.6 Band Structure of Metallophthalocyanines 233
6.7 Liquid Crystalline Molecular Semiconductors 239
6.8 Junctions and Solar Cells 242
6.9 Conductivity-Based Gas Sensors 256
6.10 Field-Effect Transistors 269
6.11 References 291

7 Molecular Dielectrics 297
7.1 Cause, Effect, Cooperativity and Nonlinearity 298
7.2 Ferroelectricity 301
7.3 Pyroelectricity 308
7.4 Piezoelectricity 312
7.5 Polarizability and Hyperpolarizability in Optics 317
7.6 References 340

8 Industrial Applications of Molecular Materials 343
8.1 Introduction 344
8.2 Soaps 344
8.3 Organic Pigments and Dyes (colorants) 359
8.4 Photoconductors and Photocopying Machines 385
8.5 Liquid Crystal Displays 405
8.6 References 424

CONTENTS

Appendixes 428
1 Main Symmetry Point Groups: Notation and Symbols 429
2 Tables of Characters of the Main Point Symmetry Groups 433
3 Group–Subgroup Relationships for the Crystallographic and Infinite Groups 443
4 Two-Dimensional (Monocolor) Space Groups 444
5 Isohedral Tilings 450
6 Isohedral Tilings Derived from the Topological Class [3^6] 458
7 Isohedral Tilings Corresponding to a Given Site Symmetry 466
8 Piezoelectrical and Nonlinear Optical Tensor Coefficients 468
9 Irreducible Representations of the Group K_h 471
10 The Main Dyes and Pigments 475

Index 487

Preface

Les épreuves du livre relues, celui-ci ne nous appartient déjà plus. On se demande alors : à qui peut-il profiter? A quoi peut-il servir?

Il est probablement possible de le lire sans formation scientifique particulière. Il est cependant avisé, dans ce cas, de ne pas omettre trop de pages.

Si l'on n'essaie pas de démontrer mais seulement de montrer, la plupart des phénomènes physiques, peuvent être abordés simplement. Nous avons néanmoins vérifié que cette approche ne conduisait à aucun écart à la rigueur.

Je crains (sincèrement) que ce livre pose plus de problèmes qu'il n'en résoud. Le fait de cerner une difficulté, d'être conscient des efforts qu'il faudra déployer pour la résoudre, d' apercevoir des débuts de solution... tout cela suffit-il?

A Paris le 2 juin 2000

After having reread the book, it already no longer belongs to the authors. Is it too late to ask ourselves: who can take advantage of this book? What will it be used for?

It is probably possible to read this book without a deep knowledge of science. However, it is recommended, in this case, that not too many pages be left out. If we just try to show, and not to demonstrate, most of the scientific phenomena can be presented in a simple way. We have nevertheless verified that this approach does not bring about a gap in exactness.

I (sincerely) fear that this book might reveal more problems than it will solve. Outlining a difficulty, being aware of the efforts needed to solve it, perceiving the beginnings of a solution... is all that sufficient ???

Translation (preface): Jennifer York, Jean Le Bousse

Introduction: Systemic Chemistry

A major part of the job of the chemist, as I see it, is the prediction and control of the course of chemical reactions. For this, as for any other human attempt to master nature, two kinds of approach are possible. One is the construction of broad, far-reaching principles from which the detailed properties of matter may be deduced. The other involves the bit-by-bit development of empirical generalizations, aided by theories of approximate or limited validity whenever they seem either to rationalize a useful empirical conclusion or to suggest interesting lines of experimental investigation. By the nature of our material we chemists are forced to depend largely on the second pathway.

Louis P. Hammett

Systemic chemistry concerns the designing and fabricating of systems comprising individual units in interaction with each other. The behaviour of one unit depends on the numerous interrelations with other partners. The comportment of the whole system may be deduced from its earlier states. In this respect, this definition does not differ from L. van Bertalanffy's one [1]: 'Systemic qualifies a scientific approach of social, economical and political systems, which treats the problem as an ensemble of units in mutual interrelations'. This approach was developed mainly after discoveries made in cybernetics, information theory, biology and anthropology. This systemic approach of chemical problems, while sometimes perilous, is unavoidable whenever the fabrication of a system is targeted.

There is a long way and much work needed to obtain a system: (a) a molecular unit is first synthesized, (b) it is arranged into some condensed phase, (c) the reality of a suitable physicochemical property is demonstrated, (d) a device is made and (e) the interconnection of devices leads to a circuit or, in molecular terms, to a system, as shown in the figure.

At the very first step, *molecular engineering* is by now fairly well mastered. It is possible to design and synthesize almost any molecule with a given physicochemical property: polarity, polarizability, hyperpolarizability, redox potential, etc. Many choices may be made at every moment of the synthesis but only a few of them are taken on a purely rational basis. Consequently, the chemist is not bound to know *why* but must know *how*.

The second step is more difficult: is it possible to design molecular units in such a way that a condensed phase of predictable structure is obtained? This can be called *supramolecular engineering* (*supra* from the Latin meaning

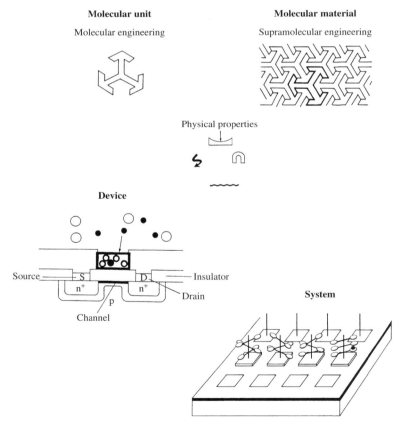

Figure The various steps from a molecular unit to systems (systemic chemistry). (A part is redrawn from M. C. Escher)

INTRODUCTION

'above', 'beyond'). Supramolecular chemistry has been defined in 1981 [2] as the chemistry of concave molecular units which can bind substrate molecules following the lock and key concept proposed by E. Fischer in 1894. At that time only processes involving enzyme/substrate-like reactions were given as examples of supramolecular chemistry. The word 'supermolecules' (*Ubermolekeln*) had already been proposed in 1937 by K. L. Wolf *et al.* for describing associations of molecules [3]. In 1977, a new type of surfactant possessing a crown-ether moiety as the polar head was described [4]. Structurally related amphiphilic complexes where said to form supermolecular transition metal assemblies in 1980 [5].

Supramolecular engineering requires, in order to solve a given problem, the gathering of a multitude of notions from many scientific fields: rudiments of geometry, physicochemistry, molecular chemistry, crystallogenesis, etc. All known relevant parameters must be taken into account. The risk is then high of using partly unmastered notions. It is obviously dangerous to recommend this approach to unexperienced scientists. However, a few guidelines can be found. The *Principe des Affinités Électives* is one of them; it is another statement of the well-known motto 'like-as-like goes together'.

Obtaining molecular materials is the next stage: the condensed phase must demonstrate a mechanical, optical, electrical or magnetic property. A combination of two or several of these *physical properties* may also be envisaged (electro-optical, piezoelectrical, etc.). A very reliable tool is furnished by the *Principe de Curie* [6] published in 1894: 'Lorsque certaines causes produisent certains effets, les éléments de symétrie des causes doivent se retrouver dans les effets produits'. This apparently simple principle can be misleading without careful examination. A simpler version resulting from group theory has been given in Ref. [7].

The notion of material is implicitly linked to practical uses. After characterizing the physical property, one must then think of the fabrication of *devices* and at that stage one reaches the borderline between science and technology (as long as the use of two different words is necessary). Devices may be classified depending on the nature of the active species used: electronics (electron), opto-electronics (photon), and iono-electronics (ion). In these cases, the word 'electronics' is taken in its general, commonly accepted, sense: 'domain of science dealing with the collection, the transmission, the treatment and the storage of information'. The term 'Molecular Electronics' has been used for related activities [8]. At that time (1982), it was said that 'simple extrapolation suggests that in approximately two decades electronic switches will be the size of large molecules'. It was, however, also added that 'communication with individual molecules is only one of several problems'. Subsequent works demonstrated the lucidity of this statement.

Beyond the device, one reaches the *system* (from the greek *sustêma*, meaning 'assembly', 'set'). A system is an assembly of elements, all in relation to each other, which then form a single entity. The system can be constituted of interconnected molecules or molecular assemblies. The system must be able to achieve a function during information processing. The system must also demonstrate new properties or functions impossible at the subsystem level. It can be shown that the relation between the various units must be nonlinear. In the case represented in the figure every unit (plug) must complex cations in a nonlinear process; this avoids a randomization of the information [9, 10]. Nonlinear complexation of the cation can occur whenever a positive cooperative effect takes place, i.e. the second and subsequent ions must be more easily bound to the ligand (or to the elementary unit) than the first one. Such systems are potentially able to perform the treatment and storage of information. Systemic chemistry is a generic name used to describe the various approaches in this field.

A systemic approach to a problem can be even more general. The final chapter of this book gives a few industrial applications of molecular materials: soaps, paints and inks, photoconductors and photocopying machines, liquid crystal displays. It will be seen that a huge number of physicochemical properties must be, at least partially, mastered in order to obtain a proper molecular material useful in a commercial device. In many, if not all, cases, the complexity of the problem is such that intuitive guidelines and empirical observations provide the only help that can be found.

The book has been written in such a way that only a very rudimentary background in chemistry and physics is necessary. The appropriate notions are recalled when necessary. As far as possible mathematical equations have been avoided and experimental data are not quantitatively fitted to theoretical models, while the chemical and physical processes important in the problem are exposed. The book could be appropriate for students or researchers.

REFERENCES

1. L. von Bertalanffy, *Théorie générale des systèmes*, Dunod, Paris (1973).
2. J. M. Lehn, *La Recherche*, **12** (127), 1213 (1981).
3. K. L. Wolf, F. Frahm and H. Harms, *Z. Phys. Chem. Abt.*, **B36**, 17 (1937); cited in H. Ringsdorf, B. Schlarb and J. Venzmer, *Angw. Chem. Int. ed.*, **27**, 113 (1988).
4. J. Le Moigne, Ph. Gramain and J. Simon, *J. of Coll. & Interface Sci.*, **60**, 567 (1977).
5. D. Markovitsi, A. Mathis, J. Simon, J. C. Wittmann and J. Le Moigne, *Mol. Cryst. Liq. Cryst.*, **64**, 121 (1980).
6. P. Curie, *J. de Phys.*, **3**, 26 (1894).
7. J. Simon, *C. R. Acad. Sci.*, **324**, Serie II, 47 (1997).
8. F. L. Carter (ed.), *Molecular Electronic Devices*, M. Dekker Inc., New York (1982).
9. J. Simon, M. K. Engel and C. Soulié, *New J. Chem.*, **16**, 287 (1992).
10. T. Toupance, V. Ahsen and J. Simon, *J. Am. Chem. Soc.*, **116**, 5352 (1994).

1 Molecular Assemblies

1.1 Generalities 1
1.2 Solid State 5
1.3 Thermotropic Liquid Crystals 9
1.4 Lyotropic Molecular Assemblies 15
1.5 Supramolecular Assemblies of Metal Complexes 23
1.6 Conclusions 43
1.7 References 44

Il est dangereux de s'exposer aux émotions dans l'art, lorsqu'on a résolu de s'en abstenir dans la vie.
M. Yourcenar (1903–1987)
Le Coup de Grâce

1.1 GENERALITIES

Molecular compounds form a huge number of condensed phases from perfectly organized single crystals to mesophases and isotropic liquids. The growth of single crystals from solutions is a very complex process in which very efficient self-recognition occurs. In 'modern' terms, single crystals could be called three-dimensional periodic supramolecular assemblies. In the subsequent sections less organized condensed phases will also be described.

Because of the fairly weak interaction energy between the molecular units, the properties of molecular materials may be derived from the characteristics of the molecular unit: shape, redox properties, polarity polarizability, etc. [1] (see Figure 1.1).

The energy needed to sublime a material (cohesive energy) is about 10 times higher for a covalent solid such as diamond, as compared to a typical van der Waals crystal such as benzene. This will be the origin of very different physical properties. In many examples, dynamic characteristics of the material (intra- and intermolecular vibrations) also play an important role.

DESIGN OF MOLECULAR MATERIALS

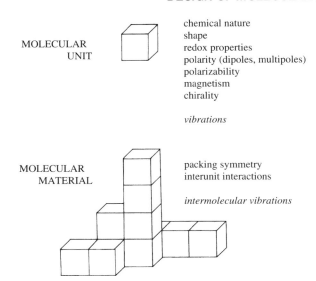

Figure 1.1 Schematic representation of a molecular material: a molecular unit is first synthesized and is then organized into a condensed phase

It is well known that to overcome a problem it is generally more cautious to proceed step by step: observation, belief, classification, comparison, understanding. It is noteworthy that this is not a linear but a cyclic procedure. In the case of molecular assemblies, the step of classification is already a difficult task. The notion of order breaking in a condensed phase may be taken as a first guide.

Three-dimensional periodic crystals offer an almost perfect state of organization. When some degree of disorder is introduced one can form *mesophases* (from the greek *mesos*: 'between', 'median').

Deviation from three-dimensional periodicity may be considered to arise from various types of rotational and/or translational breakings of order [2]. No difference is made between static and dynamic disorders since only time constants differentiate between the two processes. By combining translational and rotational breakings of order, a large number of mesophases may be identified.

In Figure 1.2, a few of the possibilities to generate disorder in a single crystal are shown. A more detailed description is given in the next chapter. In this figure the following hypotheses are taken. Disorder can be generated between layers while preserving the integrity of the layers. The same process may be envisaged for chains of molecules. In the last case, only the direction of the molecular units is not at random.

In further sections, the symmetry of molecular units or condensed phases will be considered. The symmetry of an object or an assembly of objects is not so obvious to find out. The position of a molecular unit, for example, may be averaged when the time of observation is longer than the time constant of the movement.

MOLECULAR ASSEMBLIES

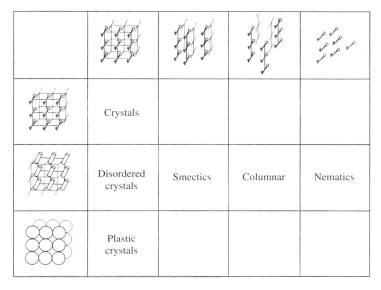

Figure 1.2 A few mesophases classified according to rotational/translational breakings of order [2]. The 'key' figures a molecular unit

Frequency (ν)	3×10^7	1×10^8	3×10^9	3×10^{10}	3×10^{11}	3×10^{12}	3×10^{13}	3×10^{14}	3×10^{15}	3×10^{16}	3×10^{17}	3×10^{18}	Hz
Wave number ($\bar{\nu}$)	10^{-3}	10^{-2}	10^{-1}	1	10	10^2	10^3	10^4	10^5	10^6	10^7	10^8	cm^{-1}
Wavelength (λ)		10	1										m
		10^3	10^2	10	1								mm
					10^3	10^2	10	1					µm
							10^3	10^2	10	1	10^{-1}	nm
	RADIO	RADAR		far INFRARED		near	VISI- -BLE	ULTRA- VIOLET		X-RAYS			
	NMR	EPR	Molecular Rotation	*Intermolecular vibration*	Intramolecular Vibration	*Peripheral*	Electronic transitions		*Deep*	X-ray diffraction			

Figure 1.3 The various domains of frequency of electromagnetic waves, with $\nu = c/\lambda$; $\bar{\nu} = 1/\lambda$

The means of characterization may be divided into two categories depending on whether periodic or nonperiodic condensed phases are under study. The various diffraction methods (X-ray, electron, neutron) can be used for periodic media. Other methods can characterize local order and/or movements, such as nuclear magnetic resonance (NMR) and electron paramagnetic resonance (EPR) (Figure 1.3). Developments of scanning tunneling microscopy (STM) and atomic force microscopy (AFM) allow information to be gained about nonperiodic media at a molecular scale.

Diffraction can be seen as the coherent interaction with a medium of a sinusoidal (or periodic) electromagnetic wave characterized by its wavelength and period (time constant). Diffraction patterns result from interferences arising from coherent sources in periodic media. A diffracted beam (Bragg's law) at an angle 2θ from the incident one can be observed whenever periodic planes separated by the distance d_r are present in the condensed phase:

$$2d_r \sin\theta = \lambda \tag{1.1}$$

where
λ = wavelength of the electromagnetic wave
d_r = interplanar distance (real space)

In the case of X-rays, the characteristic time is 10^{-17}–10^{-18} s, to be compared with vibronic relaxations (10^{-13}–10^{-14} s) or intermolecular vibrations (10^{-11}–10^{-12} s) [3].

A sinusoidal wave of wavelength λ_0 can be written as

$$\cos 2\pi \frac{c}{\lambda_0} t \tag{1.2}$$

where
c = velocity of the light

The wave can at first be considered as infinite and can be represented as shown in Figure 1.4 [4]. However, most sources emit coherently only over a limited period of time τ_c, called the *coherence time*, which corresponds to a coherence length in vacuum equal to $l_c = c\tau_c$. The sinusoidal but finite wave is this time the sum of an infinite number of sinusoids of *different wavelengths* (Fourier theorem) (Figure 1.5).

For ordinary light sources, l_c is about 50 cm; for lasers, l_c can reach several hundreds of kilometers. In X-ray diffraction, the coherence of the beam in the direction of propagation is about 1 μm limited by the spread of wavelengths [5]. However, the lateral coherence is only a few hundred Å due to the convergence of the beam. In order to illustrate how disorder or movements can influence the apparent symmetry of a condensed phase, one can consider successive snapshots over 10^{-17}–10^{-18} s of a columnar (discotic) mesophase investigated by X-ray diffraction. The time needed for the diffraction experiment is orders of magnitude longer than the successive X-ray snapshots. Consequently, the superimposition of a huge number of quasi-instantaneous diffraction patterns is experimentally observed. Moreover, if domains of limited size are present within the condensed phase, they will contribute differently to the X-ray pattern either because they possess the same structure but are oriented differently relatively to the incident X-ray beam or because, at a given time, they can have different structures due to molecular movements (Figure 1.6).

Rotation of the molecular units can occur around the column axis. Most of the time (cases A and B), there is no orientational order between the columns and only the periodicity associated with the overall shape-averaged mesogens can

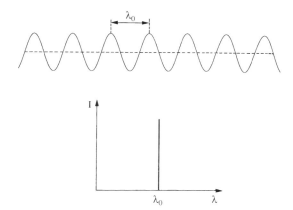

Figure 1.4 Amplitude as a function of time and intensity as a function of the wavelength for an infinite sinusoidal wave

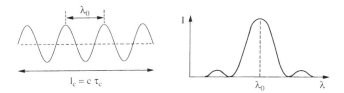

Figure 1.5 Representation of a sinusoid possessing a coherence length l_c and the approximate corresponding distribution of wavelengths. (After Ref. [4])

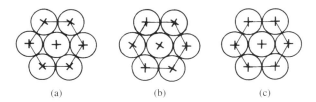

Figure 1.6 Three successive snapshots of a columnar mesophase (view along the column axis). The size of the domain to be considered is related to the coherence length of X-ray (usually 200–300 Å)

be observed. Rotationally oriented states are not in principle reached (case C) because of steric hindrance.

1.2 SOLID STATE

Single crystals, polycrystalline and amorphous materials may be encountered. The first thing to define in a crystal is the lattice. The origin of the lattice [6] (Figure 1.7) is arbitrary.

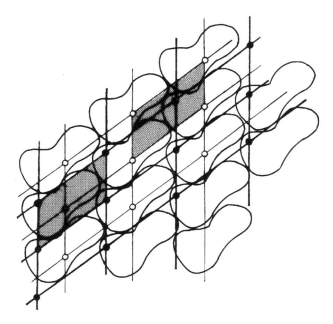

Figure 1.7 Illustration of the arbitrary choice of the origin of a lattice (two-dimensional)

The lattice can be superimposed to the crystalline structure in order to describe its translation properties. Any arbitrary point chosen in the molecular crystal keeps its environment unchanged when it is translated along the lattice vectors (Figure 1.7). A lattice can be constructed by the juxtaposition of identical unit cells which are parallelepipeds built on the three translation vectors with one node at each vertex. All the existing three-dimensional lattices may be described with seven different parallelepipeds. A. Bravais (1811–1863) proposed to take as the unit cell the parallelepiped of highest symmetry with the smallest volume. These cells can have nodes on their faces and at their center, and correspond to the 14 Bravais lattices (Figure 1.8).

The crystal is made of molecular units with their own symmetry and relative orientations. Thus 230 space groups are necessary to describe three-dimensional periodic arrangements. They derive from all the possible combinations of symmetry elements with the lattice translations.

In the neutral case, nonpolar molecular units are considered and nondirectional induced dipole/induced dipole interactions provide the main stabilization energy. In this case, a compactness factor k_p is a key parameter to determine the most stable molecular arrangement and

$$k_p = \frac{ZV}{V_0} \tag{1.3}$$

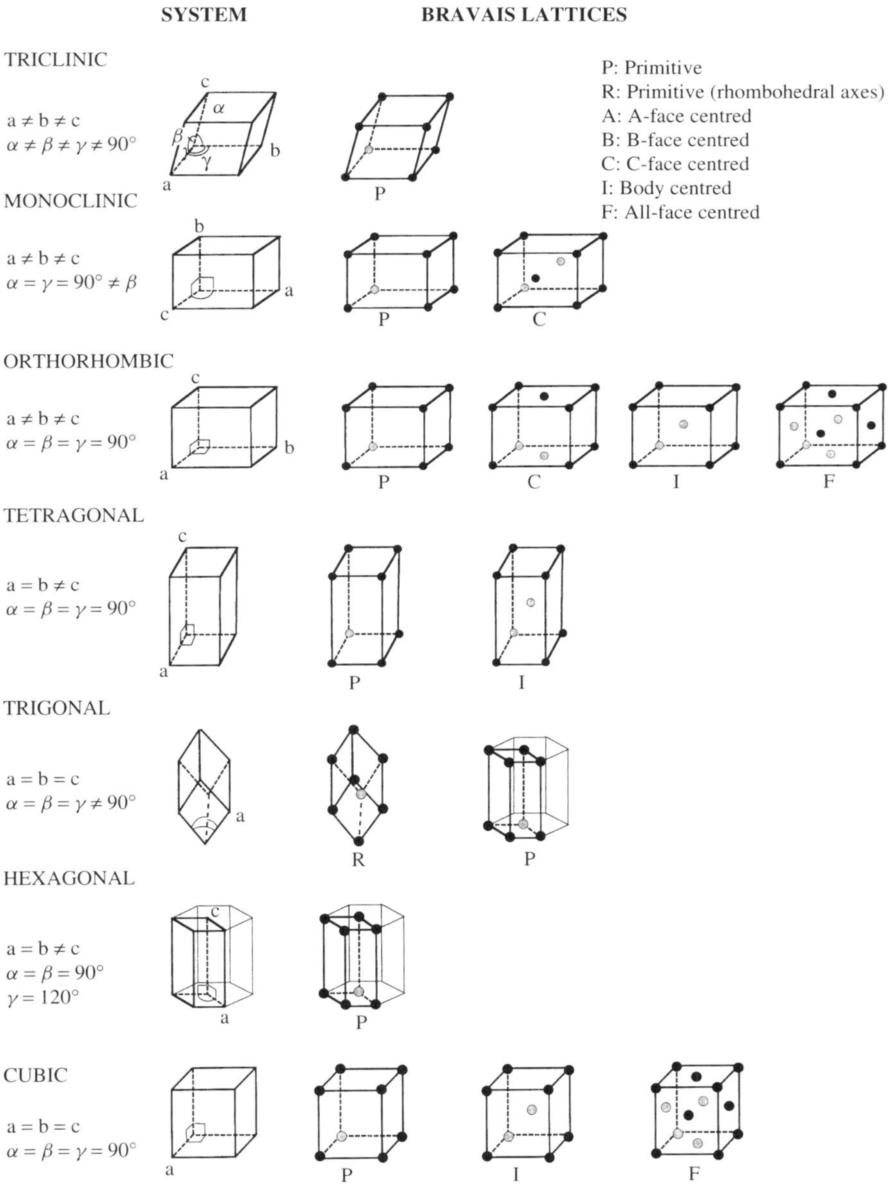
SYSTEM
BRAVAIS LATTICES
P: Primitive
R: Primitive (rhombohedral axes)
A: A-face centred
B: B-face centred
C: C-face centred
I: Body centred
F: All-face centred
TRICLINIC
a ≠ b ≠ c
α ≠ β ≠ γ ≠ 90°
c
α
β
γ
b
a
P
MONOCLINIC
a ≠ b ≠ c
α = γ = 90° ≠ β
b
a
c
P
C
ORTHORHOMBIC
a ≠ b ≠ c
α = β = γ = 90°
c
b
a
P
C
I
F
TETRAGONAL
a = b ≠ c
α = β = γ = 90°
c
a
P
I
TRIGONAL
a = b = c
α = β = γ ≠ 90°
a
R
P
HEXAGONAL
a = b ≠ c
α = β = 90°
γ = 120°
c
a
P
CUBIC
a = b = c
α = β = γ = 90°
a
P
I
F

Figure 1.8 The seven Laüe parallelepipeds and the corresponding 14 Bravais lattices. (Modified from Ref. [6])

where

Z = number of molecular units per unit cell
V_0 = unit cell volume
V = van der Waals volume of one molecular unit

The *closest packing principle* states that the compactness of the condensed phase must be the highest possible, as far as only the enthalpy part of the free energy is concerned (see Chapter 3). As an example, a few crystal packings encountered with metallophthalocyanines will be given. The chemical formula of phthalocyanines is shown in Figure 1.9.

The molecule is planar with a size of approximately $15 \times 15 \times 3.4\,\text{Å}^3$. Most of the metallophthalocyanines crystallize in a monoclinic lattice. A herringbone arrangement is found as for many other nonpolar aromatic derivatives (benzene, naphthalene, etc.). This type of structure ensures a high compactness (Figure 1.10).

Polycrystalline materials are made of small crystals, typically 500 Å to 1 μm in size. Crystallites joined by grain boundaries can be either randomly oriented or structurally correlated. Polycrystalline molecular films are easily obtained via

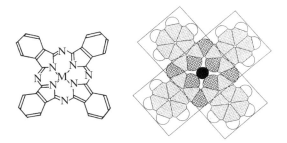

Figure 1.9 Chemical formula of a metallophthalocyanine, PcM, M = Zn, Ni, Cu, Co, etc.

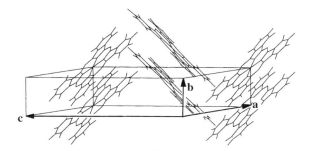

Figure 1.10 Structure of copper phthalocyanine (PcCu) ($a = 14.6\,\text{Å}$, $b = 4.8\,\text{Å}$, $c = 19.5\,\text{Å}$, $\beta = 120.9°$)

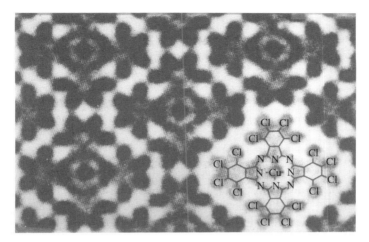

Figure 1.11 Electron microscopy image of chlorinated copper phthalocyanine (thickness of 50 Å with epitaxial growth on KCl). (Reproduced by permission of the International Union of Crystallography and T. Kobayashi from [7])

vacuum sublimation. Depending on the sublimation conditions, the size of the crystallites, their relative orientation and the overall compactness of the film may be varied. Thin crystals may also be prepared by epitaxial growth in a high vacuum chamber (Figure 1.11). The type of organization depends on the molecular characteristics and on the interactions with the substrate.

Amorphous solids have about the same density as crystalline solids but they do not have long-range periodicity. Standard organic and metallo-organic molecules have generally highly unsymmetrical shapes; consequently, thin films very rarely demonstrate a perfect amorphous state. Correlations in the position and in the relative orientations of the molecules still subsist even in disordered materials (quasi-amorphous state).

1.3 THERMOTROPIC LIQUID CRYSTALS

In 1888, F. Reinitzer prepared a derivative of cholesterol (cholesteryl benzoate) (Figure 1.12) and observed that, apparently, it possessed two different melting points: at 145 °C the crystals melt to give a viscous liquid, at 178 °C this latter transforms into an ordinary liquid.

H. Lehmann in 1889 concluded from optical microscopy observations that the intermediate phase was birefringent and anisotropic. It was shown, a long time afterwards, that 5% of all molecular compounds (mesogens) transform into liquid crystalline phases near or at their melting point. Under the microscope, entities of the order of 1 μm can be seen: from the geometrical figures observed, it has been possible since about 1920 to determine the correct structure at a molecular level [8].

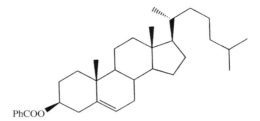

Figure 1.12 Chemical formula of cholesteryl benzoate (Ph = phenyl)

Many rod-like molecules yield lamellar structures. The smectic A phase is among the most commonly found layered structure. The term 'smectic' comes from the greek *smêgma*, meaning 'soap'. Smectic liquid crystals are usually denominated by a letter, which merely indicates the chronological order in which the mesophases have been first observed. This denomination does not recover any structural information. The observation by optical microscopy of liquid crystalline phases allows the vision of the *texture* of the mesophase: the image of the defects present in the structure. Different types of defects are compatible with a layered structure [9, 10] (Figure 1.13).

In smectic A, there is no measurable periodicity within the lamellar plane. The microscopic observations depend on the orientation of the sample. In the homeotropic arrangement the layers are parallel to the substrate, as shown in Figure 1.14.

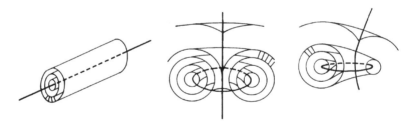

Figure 1.13 Three of the defects corresponding to folded layers. (Modified from Refs. [9] and [11] by permission of M. Kléman)

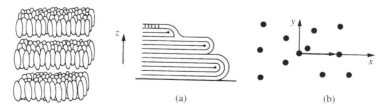

Figure 1.14 Homeotropic orientation of a smectic A phase: (a) side view, (b) top view (for microscopic observations, light propagates along the z axis)

MOLECULAR ASSEMBLIES

The long axis of the molecular units corresponds to the z direction, perpendicular to the substrate. In the plane (x, y), there is a statistical distribution of the molecules. If polarized light travels in the z direction with the electrical component of the electromagnetic wave in the (x, y) plane, the plane of the polarization is not rotated. The sample appears black between crossed polarizers. When the sample is not orientated, a texture is seen that can be correlated with the structure of the mesophase via the type of defect to which it can yield (Figure 1.15). The

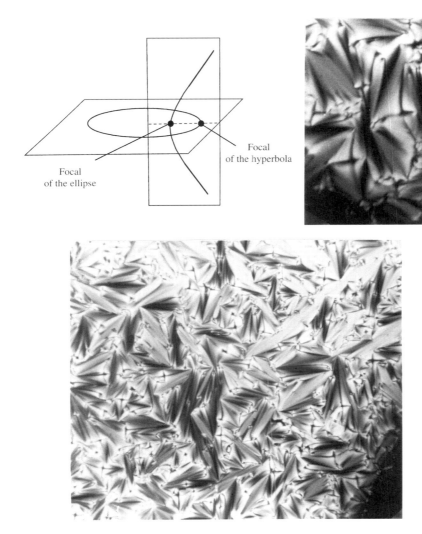

Figure 1.15 Experimental view by optical microscopy between crossed polarizers (texture) obtained with 1,4,5,8-tetradodecyloxyanthracene (at 102 °C) and associated geometrical features (by permission of Y. Bouligand and S. Norvez)

texture observed is called 'focal-conic' since the hyperpola (or straight lines) and the ellipses (or circles) are in a confocal relationship.

Various types of smectic mesophases may be encountered depending (a) on the degree of order within the layer and (b) on the tilting angle between the long axis of the mesogen and the lamellar plane. In disordered mesophases, the molecules within the layers are disposed at random, the molecules can more or less freely rotate around their long axes and they show a relatively high mobility within the planes. Depending on the tilting angle, smectic A or smectic C mesophases are distinguished (Figure 1.16).

In ordered smectic liquid crystals, some long-range positional order occurs within the layers. The two main types of positional order demonstrated within the layer planes correspond to hexagonal and orthorhombic arrangements. The S_B phase is hexagonal, the long axis of the molecular unit is approximately perpendicular to the layer plane and the molecular units rotate around the long axis. The phases S_F and S_I are also hexagonal but the molecules are tilted towards the sides (S_F) or the apices (S_I) of the hexagon. More ordered phases (E, G, H, J, K) derived from an orthorhombic ordering may also be obtained.

Nematic (from *nêmatos*, meaning 'threads') liquid crystals can be macroscopically almost as fluid as ordinary liquids (Figure 1.17).

The enthalpies of transition from the nematic phase to the isotropic phase, as measured by differential scanning calorimetry (DSC), are small compared to the transition solid → liquid crystal. The nematic phase, by optical microscopy, shows dark and mobile threads. The molecular units in these mesophases are highly disorganized and X-ray diffraction studies hardly distinguish nematic

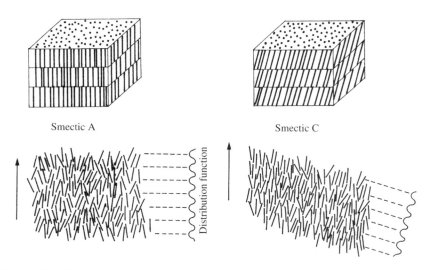

Figure 1.16 Schematic representations of the smectic A and smectic C mesophases. A more realistic view of the relative positions of the molecules is shown on the cross-sections. (Reproduced by permission of John Wiley & Sons Ltd from Ref. [12])

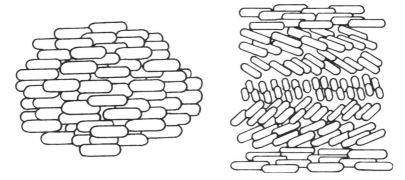

Figure 1.17 Simplified representations of nematic (left) and cholesteric (right) mesophases

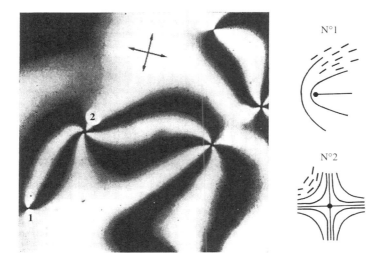

Figure 1.18 Schlieren texture (crossed polarizers) observed with 4-n-octyloxyphenyl-4-n-butyl-cyclohexanecarboxylate and molecular arrangements corresponding to defects 1 and 2. (Reproduced with permission from Demus and Richter [9])

phases from isotropic liquids. The long axis of the molecules are approximately all orientated in the same direction (director). In the case optically active molecules are used, *cholesteric* mesophases of related structures are formed, possessing an additional helical structure. The texture of nematic mesophases is said to be of the Schlieren type [9] (Figure 1.18). In the texture shown in Figure 1.18, there is extinction of light whenever the director is either parallel or perpendicular to the polarizer or the analyzer.

Smectics, nematics and cholesterics were, for almost one century, the only liquid crystalline phases characterized. In all three cases, rod-like molecular units (cigar-like) are employed to ensure a high anisotropy of the molecular polarizability. In 1977 a new type of molecular unit with a disc-like shape was used: the hexa-substituted ester of benzene [13]. In this mesophase, columns are formed by piling up the central aromatic cores. At higher temperatures nematic-like mesophases may be obtained (Figure 1.19).

The two-dimensional arrangement of the columns is either hexagonal (D_h), rectangular (D_r) or oblique (D_{ob}). The rigid cores of the discogen are either normal or tilted with respect to the axis of the columns. The intracolumnar intermacrocyclic distance may be approximately constant (ordered columnar) or not (disordered). The degree of order within the columns can be estimated from the diffuseness of the diffraction ring in the range 3.6–4.5 Å in the X-ray pattern.

Quasi-spherical (globular) molecular units can also be used to form plastic crystals [14] (Figure 1.20). The face centered cubic and hexagonal structures allow, among a few others, the packing of quasi-spherical molecular units. In

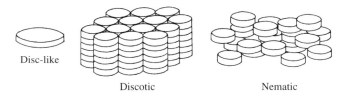

Figure 1.19 Columnar (or discotic) and nematic mesophases which can be obtained with disc-shaped molecular units

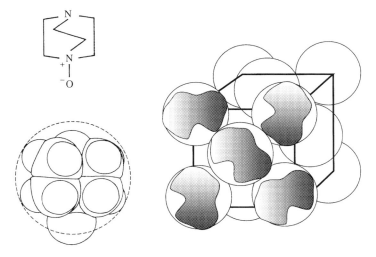

Figure 1.20 An example of globular molecule: the nitrogen oxide of triethylenediamine and a general cubic face-centred lattice often encountered for plastic crystals. (After Ref. [15])

these materials there is a more or less rapid molecular reorientation. Tetrahedral molecules — $C(SMe)_4$, $C(NO_2)_4$, SiF_4 — ditetrahedral molecules — Cl_3C-CCl_3, $Me_3Si-SiMe_3$ — Me_3C-CMe_2Cl, cyclopentane, cyclohexanol or derivatives of camphor, borneol, may form plastic crystalline phases. In all cases, the maximum diameter of the molecular unit is longer by approximately 1 Å than the average size of the effective sphere in which the rotational reorientation occurs. The rotation cannot therefore be considered as completely free. NMR analysis shows that the barrier to reorientation is of the order of a few kilocalories per mole [14].

1.4 LYOTROPIC MOLECULAR ASSEMBLIES

Amphiphilic molecules are constituted of two chemically antagonistic chemical fragments, very often one hydrophobic and the other hydrophilic. Lyotropic mesophases are made up of more than one component. One of these, generally an amphiphile, is mesogenic in nature; the others are plain liquids (solvents). Thermotropic liquid crystals are able to incorporate appreciable amounts of liquids without losing their mesomorphic character. Nevertheless, swollen thermotropic liquid crystals should not be confused with lyotropic mesophases. In the former, solvents are uniformly distributed throughout the volume of the sample playing the role of a diluent; in the latter, they participate in the ordering of the molecules in a very specific way [1].

Lyotropic liquid crystals are usually obtained with amphiphiles in the presence of water, and possibly, of other polar and nonpolar solvents, such as dimethylformamide [16], hydrazine [17], heptane or toluene [18–21]. However, they can also be produced with high molar mass (polymeric) mesogens such as polyamphiphiles (polysoaps) [22], macroamphiphiles (block copolymers) [23], natural rigid-chain polymers (DNA) [24], synthetic rigid or semi-rigid chain polymers (polypeptides, polyamides, aromatic polyesters) [25–27] and even viruses [28].

The ability of amphiphilic molecules to induce the formation of mesophases is due to the incompatibility of their constitutive parts. These fragments segregate in space and locate themselves within distinct domains separated from one another by interfaces of small thickness (~5 Å for soaps) [1].

The amphiphilic character is related to the nature of the interactions (van der Waals, dipole/dipole, ion/ion hydrogen bonding) involved. The hydrophobic moieties are nonpolar (aliphatic chains) or weakly polarizable (aromatic groups), while the hydrophilic ones are charged (ionic groups) or polar (oligoethylenoxide chains). The association of nonpolar molecular units is highly dependent upon the degree of order of the water molecules (structure of water) [29–33].

Amphiphilic molecules therefore form molecular materials whose structure can be partly predicted following two criteria: the closest packing principle and the segregation tendency of the hydrophilic and hydrophobic parts of the amphiphile (see later chapters). These two criteria are probably interdependent.

16   *DESIGN OF MOLECULAR MATERIALS*

Cationic	$-\overset{+}{N}R_3$
	$-\overset{+}{N}\text{(pyridinium)}$
Anionic	$-CO_2^-$
	$-OSO_3^-$
	$-OPO_3^{2-}$
Zwitterionic	$-Me_2N^+ \rightarrow O^-$
	$-CH(CO_2^-)\overset{+}{N}H_3$
	$-Me_2\overset{+}{N}(CH_2)_3SO_3^-$
	$-OPO_3^-(CH_2)_2\overset{+}{N}Me_3$
Non-ionic	$-(OCH_2CH_2)_nOMe$

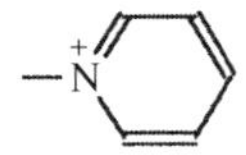

Figure 1.21 The main polar groups used in amphiphilic molecules

Numerous polar groups may be used as hydrophilic moieties (Figure 1.21): cationic (tetraalkylammonium, pyridinium, etc.), anionic (carboxylate, phosphate, etc.), zwitterionic (amino acids), non-ionic (polyoxyethylene, polyols).

Natural membranes are other examples of lyotropic molecular assemblies. They are mainly constituted of proteins (45%) triglycerides and paraffins (15%) and phospholipids (30%) (Figure 1.22).

Amphiphilic compounds constitute one-third of the constituents of the membranes. They play a major role in the type of organization obtained: paraffinic tails segregate to form the hydrophobic core of the membrane whereas the hydrophilic polar heads are solvated by the surrounding water molecules (Figure 1.23).

Soaps are obtained by treating grease or fat with sodium carbonate or another base. The most often encountered derivative is stearic acid, $C_{17}H_{35}CO_2H$, with eighteen carbon atoms (Figure 1.24).

A related subject started a long time ago when Babylonians spread oil droplets on water to foretell the future [36]. However, the history of this type of molecular assembly really started when B. Franklin (1706–1790) recovered the surface of a pond with a spoon of a greasy compound. A. Pockels (1862–1935) prepared the first monolayers at a water–air interface. Lord Rayleigh (1842–1919) reported on the nature of the layers. I. Langmuir (1881–1957) carried out the first systematic study of monolayers. The first multilayers were obtained by C. Blodgett using new techniques of transfer from water to solid substrates.

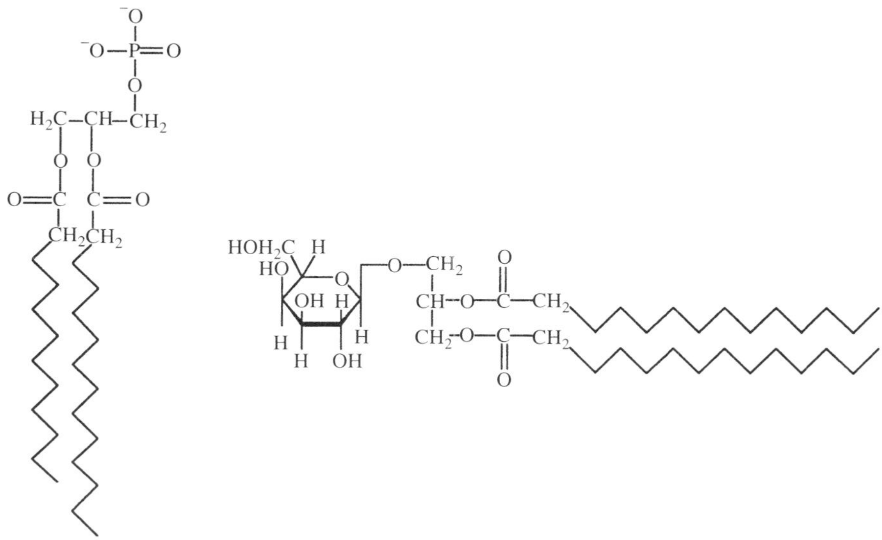

MOLECULAR ASSEMBLIES

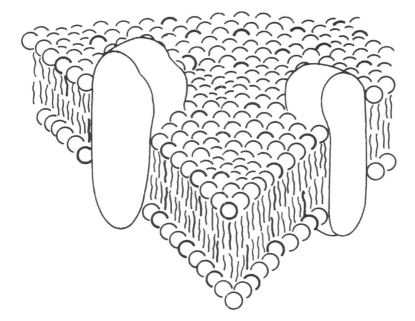

Figure 1.22 Two examples of amphiphilic compounds found in natural membranes: a phospholipid and a galactolipid

Figure 1.23 Schematic representation of a natural membrane constituted of phospholipids and proteins. The nature of the phospholipids is, for many of them, different for the two faces of the cellular membranes. (Redrawn from Ref. [34])

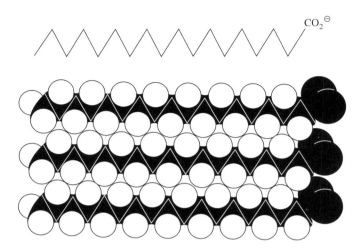

Figure 1.24 Molecular formula of the anion of stearic acid. (Redrawn from Refs. [34] and [35])

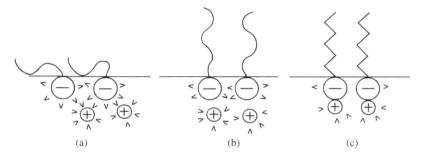

Figure 1.25 The three stages of the formation of a Langmuir–Blodgett thin film. The lateral compression increases from (a) to (c); correspondingly the area per polar head decreases

By spreading an organic solution of fatty acids on a water surface, the hydrophilic polar heads are directed towards the aqueous medium whereas the paraffinic tails remain at the surface (Figure 1.25, stage A). When the surface available for each polar head decreases, the repulsion between the negatively charged polar heads becomes important and the counterions come in closer proximity to RCO_2^-; stabilization is brought about by paraffin–paraffin interactions. The chains tend to crystallize and the polar heads form an ion pair with the counterions, while the strongly destabilizing charge–charge interactions are replaced by weaker and geometry-dependent dipole–dipole interactions (Figure 1.26). Langmuir–Blodgett films have been studied by Raman, IR spectroscopies and X-ray diffraction, allowing a fairly precise idea of their structures as a function of temperature to be found [37].

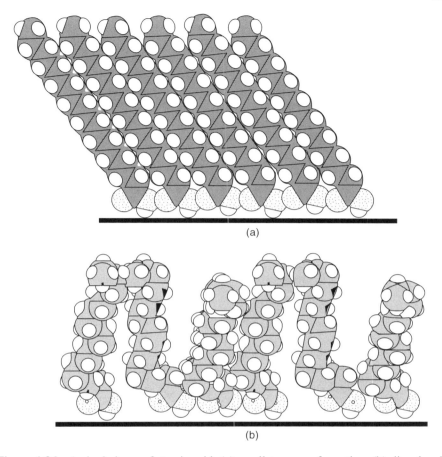

Figure 1.26 A single layer of stearic acid: (a) an all *trans* conformation, (b) disordered conformation. (Redrawn from Ref. [36])

Soaps (or phospholipids) can be dissolved in water at low concentrations. At some critical concentration, aggregates — micelles or related species — are formed. Micelles consist of approximately 60–100 amphiphilic molecules and they are roughly spherical in shape (Figure 1.27). By adding different amounts of soap in water, a phase diagram may be obtained indicating the presence of various types of phases (Figure 1.28).

Anhydrous soaps form crystals: both the hydrophilic and hydrophobic parts are in an ordered state (Figure 1.29). The cation associated with the anionic polar head forms a tightly bound ion pair. The area per polar head (S), i.e. the surface occupied by the polar head within the plane of the layers, is an important parameter for defining the degree of organization of the phase. In the case of crystalline anhydrous soaps, $S \sim 20 \text{Å}^2$: this value is imposed by the distance

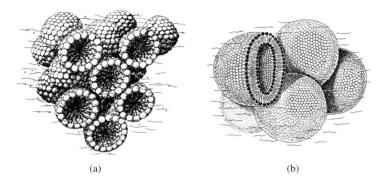

Figure 1.27 Two of the molecular assemblies that can be formed in aqueous media: (a) micelles, (b) vesicles. (Reproduced by permission of Springer-Verlag from [37])

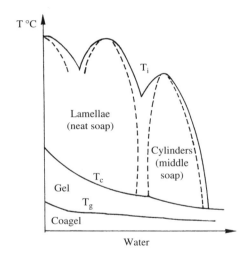

Figure 1.28 Phase diagram obtained for the system of soap in water. (After Ref. [1])

between the paraffinic chains in the crystalline state ($1 \sim 4.5$–4.7 Å). When the polar head is too large to be accommodated within 20 Å^2, a single layer structure is obtained (Figure 1.29b), in which two paraffinic chains correspond to one polar group.

In the presence of water, at intermediate temperatures, the gel phase is obtained [1, 38] (Figure 1.29c). Every soap layer is separated by water molecules which form a layer whose thickness can reach 100 Å. The polar heads are solvated and the ion pair between the polar head and its counterion is dissociated. Correspondingly, the electrostatic interactions between the polar heads increase and a single-layer geometry is adopted, the surface available for each polar head

MOLECULAR ASSEMBLIES

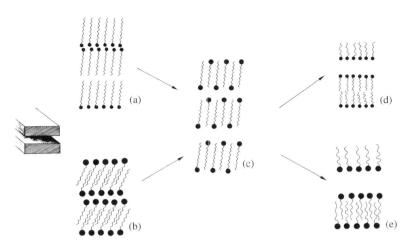

Figure 1.29 Schematic representation of the various lamellar phases which can be obtained as a function of temperature with amphiphilic molecules: (a, b) crystalline forms, $T < T_g$ (anhydrous soaps), (c) gel, $T > T_g$ (crystallized paraffinic chains, polar head ion pair dissociated), (d, e) $T > T_c$, both paraffinic tails and polar heads are in a disordered state. (After Ref. [1])

going from 20 Å^2 to approximately 40 Å^2. This permits the polar head/polar head repulsion energy to be minimized. The paraffinic chains are still in a crystalline or quasi-crystalline state and approximately fully extended. The thickness of the hydrophobic sublayer depends on the length of the hydrophobic moieties of the molecules, the amount of the solvent added and the degree of lateral spreading of the molecules. In the gel domain, when the concentration of water is varied, neither the thickness of the hydrophobic part nor the surface per polar head vary. Water molecules are incorporated between the layers of the amphiphiles without disrupting their organization.

When the amount of water present in the lyotropic mesophases increases lamellar and cylindrical organizations appear (Figure 1.30). In these structures, the paraffinic chains are in a quasi-liquid state. At the highest concentrations of surfactant in water, lamellar phases are formed whose structure presents some analogy with those found in natural membranes. The lamellar mesophases (neat soap) correspond to a periodic and alternate superimposition of layers of amphiphilic molecules. Solvent, if present, is located between the planes. The area per polar head is around 40 Å^2 [38]. At lower concentrations, infinite cylinders consisting of the amphiphilic molecules are formed, surrounded by the aqueous phase. Cylinders lead to hexagonal arrays (middle phase). As the area per polar group is of the order of 50 Å^2, the degree of organization is therefore lower than in lamellar mesophases.

In cylindrical phases as well as in lamellar ones the behavior of nonionic surfactants differs from usual ionic soaps. For these latter, the dissociation of

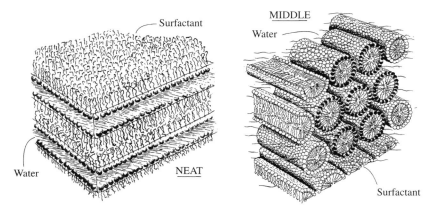

Figure 1.30 Structures of the lamellar (neat) and hexagonal (middle) phases. (Reproduced by permission of Springer-Verlag from [37])

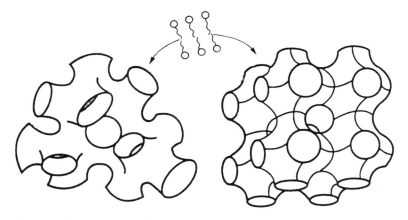

Figure 1.31 Cubic structure (right) and sponge phase (left): two types of organization found with amphiphilic compounds in water. The cubic structure is, however, rarely observed. The wall consisting of amphiphilic derivatives has been enlarged. (Reproduced with permission from [39])

the ion pair between the anionic polar head and its counterion leads to drastic differences in the structure of the mesophase. The concentration of electrolytes in water influences the structural characteristics of the mesophases. Nonionic amphiphiles are far less sensitive to this parameter.

Cubic mesophases can also be observed. They appear as clear, viscous gels that are optically isotropic. When produced in a homogeneous system (micellar solution, lamellar or hexagonal mesophase) by phase separation, due to a proper change of temperature or concentration, they occur in the form of small polyhedra of cubic symmetry. Their X-ray diffraction patterns clearly indicate their cubic crystal system (see Figure 1.31).

1.5 SUPRAMOLECULAR ASSEMBLIES OF METAL COMPLEXES

In this section, only the results coming from authors' laboratory are presented although many other publications should have been mentioned. Some of them can be found in Refs. [40] to [42] and in one book [43]. However, this chapter gave us the opportunity to publish in a coherent way our activities over approximately 20 years.

The chemistry of transition metal complexes presents many peculiarities. Electron transfer may, for example, occur in mixed valence complexes. This is not possible with closed shell organic derivatives. The use of d orbitals in addition to s and p orbitals offers a wider choice of synthetic pathways and physicochemical characteristics.

Coordination chemistry is mainly a study of single metal ions surrounded by ligands. Metal ion clusters are often considered to be situated between three-dimensional mineral solids and isolated metallic ions (Figure 1.32). The architecture of the assemblies is hardly predictable and there is no straightforward relationship between the characteristics of the bricks and the overall shape of the edifice.

In 1977 [44], an approach was proposed which allowed the formation of *metal ion complex assemblies*. An amphiphilic compound is synthesized in which the polar head is constituted of a ligand able to bind a given cation. The hydrophobic part, usually a paraffinic side chain, allows the formation of organized phases: micelles, cylinders, lamellae, etc. This type of aggregation forms *supramolecular transition metal assemblies* [45, 46] (Figure 1.33).

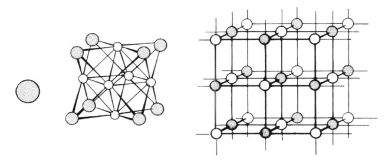

Figure 1.32 From individual to collective states: ions, clusters and three-dimensional solids

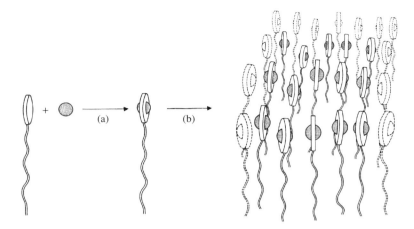

Figure 1.33 Supramolecular assemblies of transition metal complexes: (a) complexation, (b) organization. (After Ref. [46])

Following this approach, long-chain-substituted crown-ether macrocycles have been synthesized (Figure 1.34). The crown-ether macrocycles are known to selectively bind alkali and alkaline earth cations. The amphiphilic character of the complex can lead in a second stage, to ordered aggregates. The combination of their macrocyclic nature and their worm-like shape led the authors to propose the name 'annelides' for this new class of molecule [45, 46].

The rate of cation transfer between the macrocycles at a micellar subsurface was first studied with annelide-based assemblies [44] (Figure 1.35). It was demonstrated by ^{13}C NMR that the Ba^{2+} exchange between macrocycles was favoured at the micellar subsurface compared to the process in non-organized media. Physicochemical studies on the same type of compounds allowed the authors to determine the shape of the aggregates formed in aqueous solution (Figure 1.36).

The various geometrical parameters of the micelle may be deduced from light-scattering determinations and density measurements. The area offered to each

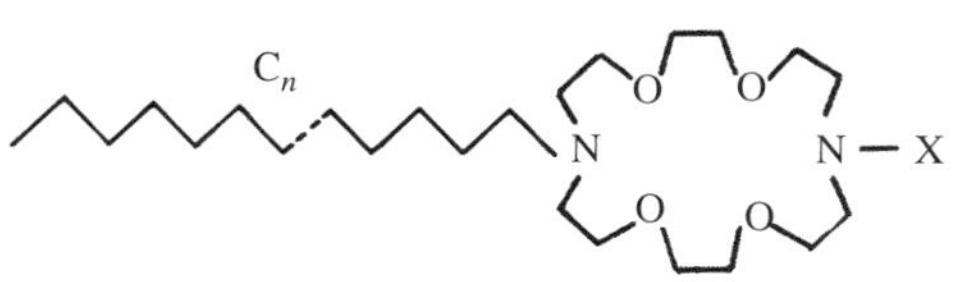

$n = 12 \quad X = -\text{Me}$
$n = 18 \quad X = -(CH_2)_2{}^+NMe_3$

Figure 1.34 The annelides: amphiphilic substituted crown-ethers. (After Ref. [46])

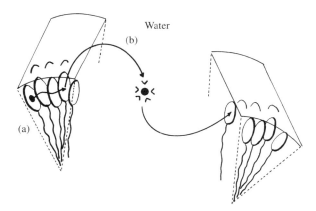

Figure 1.35 The two modes of cation exchange in a micellar solution of an amphiphilic macrocycle: (a) intramicellar hopping, (b) intermicellar exchange involving the dissociation of the complex

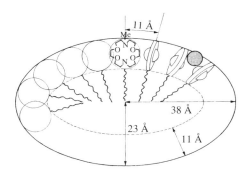

Figure 1.36 Geometrical parameters determined by light scattering of the micelle formed from a dodecyl-substituted crown-ether. (After Ref. [46])

macrocyclic polar head is not sufficient to allow a full free rotation of the rings at the micelle subsurface indicating some intersite organization. Calculations on previous micellar phases showed that one micelle contains almost as much water as amphiphilic ligand (volume/volume).

The local dielectric constant [47, 48] at the micellar subsurface has been measured by determining the pK difference between the first and the second protonation site of the two amino groups of the crown-ether macrocycle. The value found ($\varepsilon = 45$) is very different from the one measured in the absence of aggregation ($\varepsilon = 22$). Both values drastically differ from the dielectric constant of bulk water ($\varepsilon = 78$). Complexation phenomena may be perturbed at a hydrophobic–hydrophilic interface and in particular the magnitude of the constants may be drastically different from the values found in the bulk solution.

Crown-ether-substituted phthalocyanines have been synthesized [49–51] (see Figure 1.37). X-ray diffraction at small angles showed that the substituted phthalocyanines lead to a metastable columnar mesophase in which the phthalocyanine rings and the crown-ether macrocycles are approximately coplanar and form a two-dimensional square lattice ($a = b = 20.8$ Å). The phthalocyanine derivatives pile up with the crown-ether moieties in a staggered conformation (Figure 1.38). The distance between eclipsed crown-ether macrocycles is about 8.2 Å (one out of two molecular units); they form a channel-like structure in the condensed phase. It is, however, highly probable that ionic conduction is negligible in such highly structured and rigid materials.

Transition metal ions offer a wider choice of chemical reactions compared to closed shell diamagnetic alkali and alkaline earth cations. Triethylenetetramine (trien) and N,N'-bis (2-aminoethyl)-1,3-propanediamine (2,3,2-tet) subunits are among the most common complexing agents for most metallic ions. In order to induce the formation of supramolecular polymetallic assemblies [53], the ligands have been substituted with a hydrophobic paraffinic side chain [53, 54] (Figure 1.39).

By using the cobaltous complexes, it has been possible to study the complexation of dioxygen via the equilibria:

$$LCo^{II} + O_2 \rightleftharpoons LCo^{II}(O_2)$$
$$LCo^{II}(O_2) + LCo^{II} \rightleftharpoons LCo^{II}(O_2)Co^{II}L$$

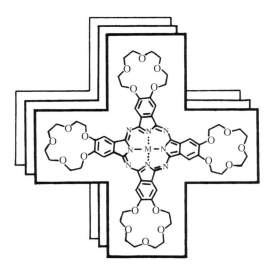

Figure 1.37 Molecular structure of benzo-15-crown5-substituted phthalocyanine. (After Ref. [52])

Figure 1.38 Staggered conformation found in the columnar stacking of benzo-15-crown5-substituted phthalocyanine (see Ref. [52])

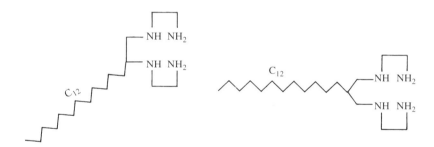

Figure 1.39 Side-chain-substituted tetraaza ligands. (From Ref. [53])

The species $LCo^{II}(O_2)$ is not usually observed due to the fact that the binuclear species are thermodynamically more stable. The dioxygen stability constants have been determined in micellar and nonmicellar phases. The micellar organization importantly influences the dioxygen–ligand binding constants. For the trien derivative, the dioxygen complex is 125 times more stable when the cobaltous complexes are preorganized at the micellar subsurface [53]. The various dioxygen complexes that can be formed are schematically shown in Figure 1.40.

The micellar effect may arise from two main contributions: (a) an effect of dimensionality change from three dimensions in an homogeneous medium to quasi two dimensions at the micellar subsurface and (b) an increase of local

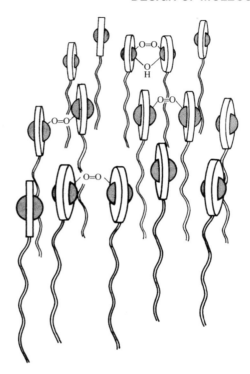

Figure 1.40 The various dioxygen complexes formed at the micellar subsurface of substituted trien and 2,3,2-tet amphiphilic ligands. (After Ref. [53])

concentration of the reactants. A simple model allowed us to show that the second factor alone is sufficient to take into account the stability constant increase due to the preorganization effect.

However, at that time, only a few transition metal complexes could be studied due to solubility problems. New ligands have then been designed to solve this difficulty (Figure 1.41). The amphiphilic balance of these ligands can be gradually adjusted by varying the length of the paraffinic chain or the degree of polymerization of the polyoxyethyleneglycol subunits. Several metal complexes (Co^{III}, Ru^{III}, Cu^{II}) have been isolated and characterized. Solutions of many other complexes (Cd^{II}, Mn^{II}, Ni^{II}, Zn^{II}, etc.) have been prepared by simply mixing aqueous solutions of the ligand and of the appropriate nitrate salt.

The metal complexes form thermotropic and lyotropic liquid crystalline phases [57, 58]. The anhydrous copper complex $LCu^{II}Cl_2$ shown in Figure 1.41 demonstrates four different mesophases before the transition to an isotropic liquid. By increasing the length of the polyoxyethylene side chains, the transition temperatures are lowered. Lyotropic mesophases can be formed in the presence of water, hexane, heptane, decane or benzene. Birefringence is observed up to 25% (v/v)

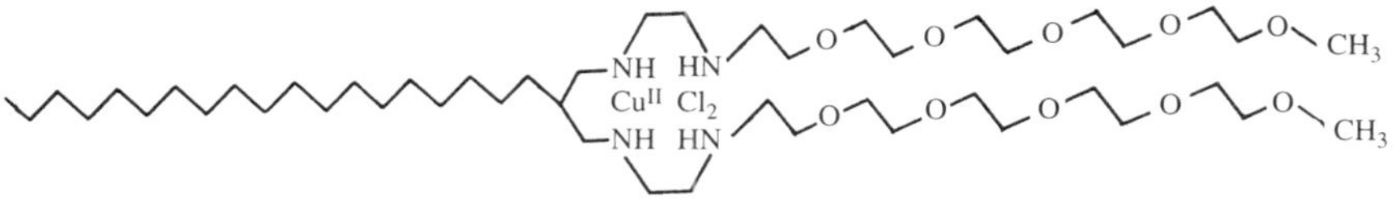

MOLECULAR ASSEMBLIES

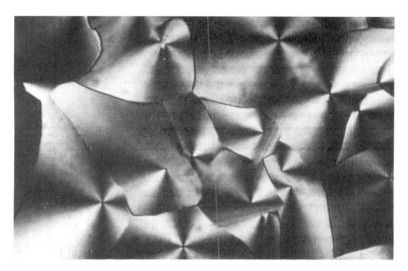

Figure 1.41 Formula of one of the synthesized paraffinic and polyethyleneoxide-substituted 2,3,2-tet ligands. (After Refs. [55] and [56])

Figure 1.42 Texture observed at 312 K for the mixture $LCuCl_2/9\%$ H_2O (L, ligand shown in Figure 1.41)

of solvent with the copper complex derivative. Optical microscopy shows well-developed textures when the sample is cooled down from the isotropic liquid (Figure 1.42). The liquid crystalline phase is metastable and crystallizes rapidly at 312 K.

The cobaltic complex $LCoCl_3$ has also been studied. In this case, the transition temperature to the isotropic liquid is near room temperature [57]. X-ray diffraction at small angles has been carried out, showing in all cases a lamellar structure. From the interlamellar distance and the characteristic lengths of the side chains in a molten state, a single-layer structure has been postulated (Figure 1.43).

The phase transitions of the amphiphilic copper complex have been studied by solid state ^1H and ^{13}C NMR as a function of temperature [59]. As the temperature is raised, the three different chemical fragments of the complex successively melt. At low temperature, both the polyethylene oxide and paraffinic chains show slow motions. The polyoxyethylene tails first melt at higher temperatures whereas the paraffinic chains are still rigid; these latter are then transformed into a quasi-liquid state, but the intralamellar organization of the copper complexes is preserved. Finally, an isotropic liquid forms (Figure 1.44) [59].

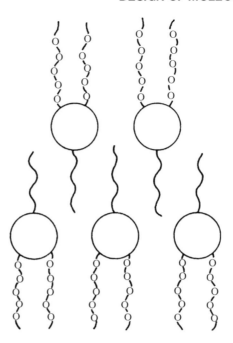

Figure 1.43 Single-layer structure obtained with cobaltic and cupric complexes of the amphiphilic 2,3,2-tet ligand (see the structure Figure 1.41). (After Ref. [56])

The effect of preorganizing the reactants within molecular assemblies has also been studied by synthesizing an amphiphilic photosensitizer (Figure 1.45) [60]. Ruthenium trisbipyridine salts, $Ru^{II}(bipy)_3X_2$, are exceptionally good photosensitizers, with an absorption maximum around 500 nm associated with a high extinction coefficient ($\varepsilon = 14\,000$). It is also very stable under irradiation. The quenching of the excited state has been shown to occur through an electron transfer mechanism:

$$Ru^{II}L_3 \xrightarrow{h\nu} (Ru^{II}L_3)^*$$

$$(Ru^{II}L_3)^* + Co^{III}L' \longrightarrow Ru^{III}L_3 + Co^{II}L'$$

$$Co^{II}L' + H^+ \longrightarrow Co^{II} + L'H^+$$

The effect of the preorganization of the reactants on the rate of electron transfer from the excited state of $Ru^{II}(bipy)_3$ towards the substituted cobaltic complex has been studied in micellar phases. Depending on the conditions, at least three pathways are possible, which differ by the dimensionality of the reaction (Figure 1.46).

An equation has been proposed that can be used in the three cases [60]:

$$v = k(d_{AB})^{-n} N_{AB} \qquad (1.4)$$

MOLECULAR ASSEMBLIES

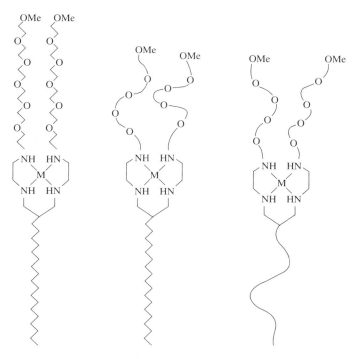

Figure 1.44 The three stages occurring in the melting process of an amphiphilic cupric complex (tegma crystals from Ref. [59])

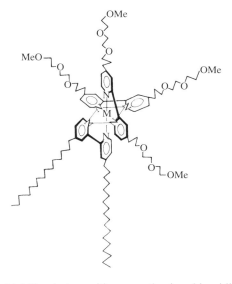

Figure 1.45 An amphiphilic photosensitizer: a ruthenium bipyridine derivative [60]

32 DESIGN OF MOLECULAR MATERIALS

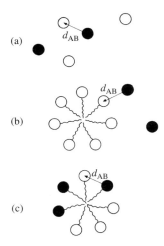

Figure 1.46 Three types of reactions differing by their dimensionality: (a) in an isotropic medium, (b) between an amphiphilic reactant belonging to a micelle and a molecule in the bulk, (c) between two amphiphilic reactants within a micelle. (After Ref. [60])

where

v = reaction rate
d_{AB} = mean distance between A and the closest B molecule
N_{AB} = number of couple AB per unit of volume
n = dimensionality of the reaction medium

A strong micellar effect has been experimentally found and an increase in the rate of electron transfer of 1800 has been measured when both reactants are amphiphilic and associated within micelles. This finding has been rationalized using equation (1.4). The luminescence properties of amphiphilic complexes of ruthenium in micellar phases have also been determined [61].

Phthalocyanines (from the Greek *naphtha*, meaning 'rockoil', and *cyanine*, meaning 'dark blue') are among the most widespread molecular colorants, with a world production of approximately 50 000 t per year. They provide a versatile chemical unit since elements from groups I_A to V_B can all combine with the phthalocyanine ring and more than 70 different metallic complexes are known [62].

Small divalent ion complexes of phthalocyanines (Pc) (Cu^{II}, Ni^{II}, Zn^{II}, Mg^{II}, etc.) form planar complexes whereas larger ions (Pb^{II}, Sn^{II}, etc.) lead to out-of-plane complexes. Proper macrocyclic substitution can afford chiral derivatives [63] (Figure 1.47). Rare earth ions (from Ce^{III} to Lu^{III}) yield sandwich bisphthalocyanine derivatives.

Substitution of the phthalocyanine ring with long-chain substituents allows the formation of columnar (discotic) mesophases [64] (Figure 1.48). The flat aromatic core is surrounded by flexible hydrocarbon chains: the molecular units

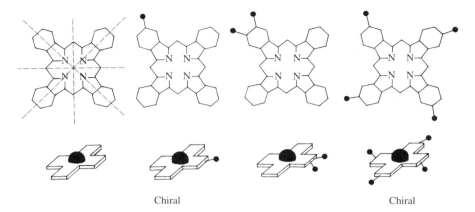

Figure 1.47 Chemical formulae of various types of phthalocyanines and a few of the geometries obtainable

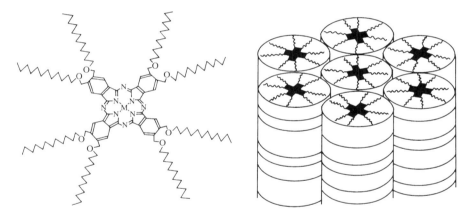

Figure 1.48 Liquid crystalline phase obtained from octasubstituted metallo-phthalocyanines. The macrocycles stack in columns isolated from each other by the disordered paraffinic chains [64]

stack in columns, the columns forming a two-dimensional hexagonal array. The spines formed by the superimposition of the aromatic cores are surrounded by the aliphatic tails in a quasi-molten state.

Optical microscopy shows characteristic flower-like textures for the substituted phthalocyanines over a very large range of temperatures: from approximately 50 °C to more than 300 °C (Figure 1.49, Plate 1). At first only a few transition metal complexes showing a mesomorphic behavior were described in the literature (see the references cited in Ref. [66]). Since then, a lot of work has been published concerning metal-containing liquid crystal phases [67], transition metal liquid crystals [40], calamitic metallomesogens [41] and metallomesogens [42]. In 1985, hydrophilic polyethyleneoxide side chains were used, instead of the hydrophobic paraffinic tails, to substitute the phthalocyanine core [66] (Figure 1.50).

The compound $[Me(OCH_2CH_2)_2OCH_2]_8PcH_2$ shows a birefringent phase from room temperature to more than 300 °C [68]. X-ray diffraction indicates an hexagonal structure as previously found with paraffinic substituents.

In-plane polarity necessitates the synthesis of unsymmetrically substituted phthalocyanines. Tetracyanobenzene has been used to form a dicyano-substituted phthalocyanine [69] (Figure 1.51). The electrical dipole moment of the molecular unit has been estimated to be $7D$ [69].

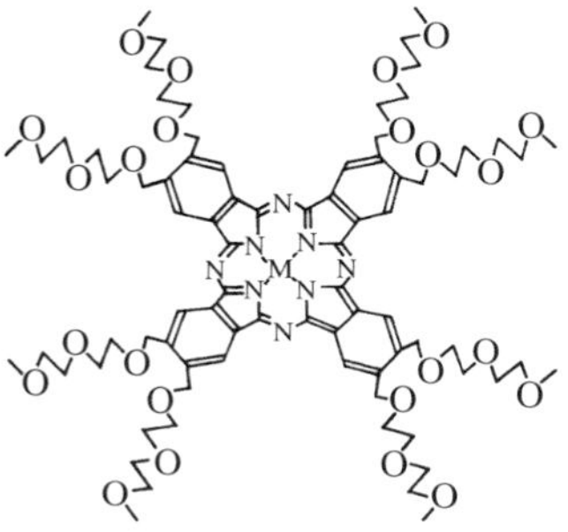

Figure 1.50 Discogens substituted with polyethyleneoxide side chains. Different lengths of polyethyleneoxide chains have been used (After Refs. [66] and [68])

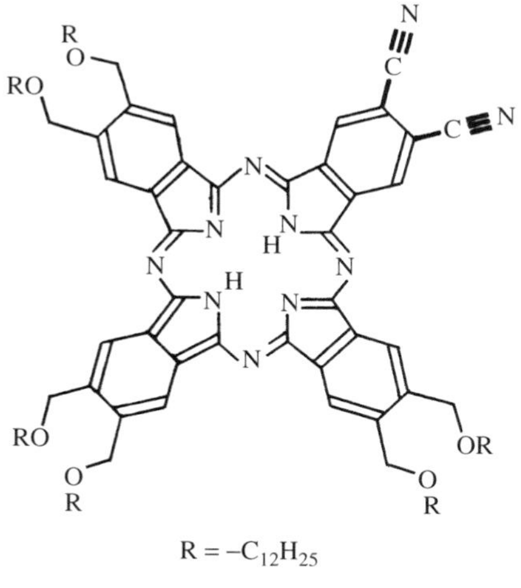

$R = -C_{12}H_{25}$

Figure 1.51 A polar discogen based on the phthalocyanine subunit. (After Ref. [69])

MOLECULAR ASSEMBLIES

X-ray diffraction studies of the mesophases show an hexagonal order of the columns. ^{1}H NMR on aggregates in solution demonstrates a slow rotation between adjacent macrocycles (Figure 1.52).

Out-of-plane polar molecular units have also been prepared [70]. Divalent lead and tin ions have too large radii (1.2 and 0.93 Å, respectively) to fit the central cavity of the phthalocyanine ring; in consequence, these ions form a protuberance above the molecular plane of the phthalocyanines (Figure 1.53).

The ground state dipole moment of PcSn has been calculated to be of the order of $5D$. The experimental value found with (2-ethyl-hexyl O)$_8$PcPb is $5.7D$. The stability domains of the mesophases have been determined for $(C_n OCH_2)_8 PcPb$ with $n = 8, 12, 18$. The first two compounds are liquid crystalline at room temperature with a transition to an isotropic liquid at 155 and 125 °C, respectively. X-ray

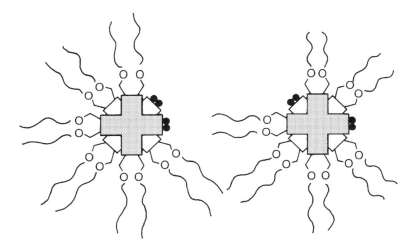

Figure 1.52 Two of the possible conformations for polar-substituted phthalocyanine discogens. (After Ref. [63])

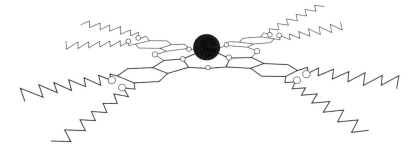

Figure 1.53 Representation of out-of-plane divalent complexes of $(C_n OCH_2)_8 PcM$, $M = Sn^{II}, Pb^{II}$. (After Ref. [70])

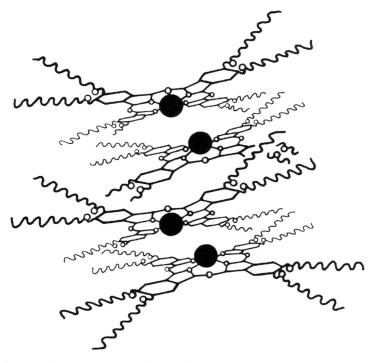

Figure 1.54 Schematic representation of the head-to-tail arrangement probably occurring in the liquid crystalline phase of $(C_8OCH_2)_8PcPb$. (After Refs. [70] and [71])

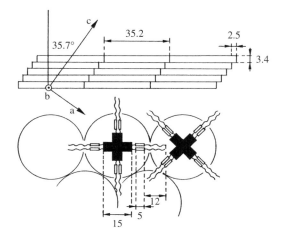

Figure 1.55 Schematic representation of the rectangular mesophase D_{rd} derived from $(C_{12}OPh)_8PcCu$: a hexagonal packing of columns in each layer can be seen; however, the axis of the column is tilted relative to the macrocyclic ring. The distances are in Å. (After Ref. [75])

diffraction measurements demonstrate the columnar structure of the mesophases. A band around 7.4 Å has been observed which probably corresponds to periodic pairs of molecules in a head-to-tail arrangement within the columns (Figure 1.54).

The nature of the connecting link between the side chains and the aromatic core influences the type of mesophase that can be obtained [72, 73]. $-CH_2OR$ and $-CH_2CH_2R$ side chains lead to columnar liquid crystals with no detectable order within the columns. On the contrary, $-OR$ side groups yield mesophases with some intracolumnar organization. When the length of the paraffinic chains is varied [74], both the crystal-to-mesophase and the mesophase-to-isotropic phase transition temperatures are lowered.

n-Alkoxyphenyl side chains have also been used to substitute the phthalocyanine moiety [75]: in this case, a rectangular disordered columnar mesophase (D_{rd}) is observed. Although of very different symmetry, this phase is structurally closely related to conventional hexagonal liquid crystals (Figure 1.55).

Tetrapyrazinoporphyrazine is a phthalocyanine analog with pyrazine (PZ) subunits instead of benzene rings (Figure 1.56). The corresponding long-chain-substituted derivatives yield liquid crystals [76]. Both hexagonal disordered and centred rectangular disordered columnar mesophases have been observed. Because of the pyrazine subunits, the half-wave reduction potentials of the molecular unit are shifted from -0.94 V for (2-ethyl-hexyl O)$_8$PcH$_2$ to -0.41 V for $(C_{12})_8PZH_2$ [76].

2-Ethyl-hexyl side chains have been used to promote the formation of nematic liquid crystalline phases with disc-like molecular units (Figure 1.57) [77]. Indeed, in the temperature range 223–270 °C, X-ray diffraction indicates the presence of a nematic phase for the corresponding phthalocyanine derivatives; a homeotropic orientation on untreated surfaces of a glass slide is observed [77].

The presence of heavy metals within the columns enhances the X-ray diffraction intensities. The intercolumnar distance of octa-substituted phthalocyanines is in the range 25–35 Å, in good correspondence with the wavelengths 10–100 Å associated with soft X-rays. It has therefore been proposed that the platinum

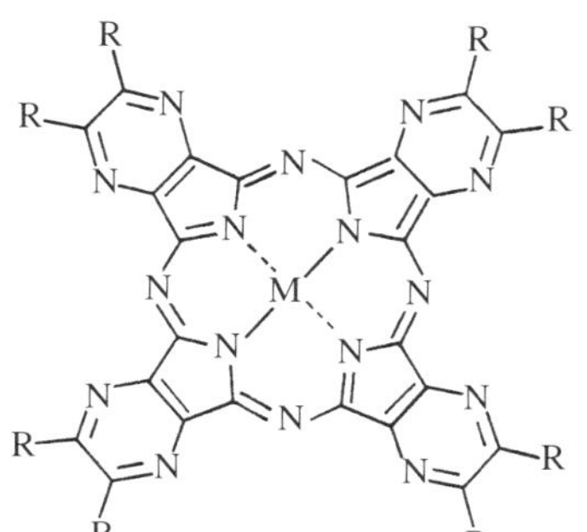

Figure 1.56 Various substituted tetrapyrazinoporphyrazines synthesized: $(C_n)_8$PZM [76]

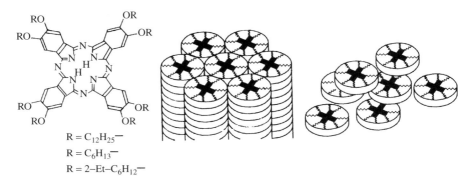

Figure 1.57 The 2-ethyl-hexyl-substituted phthalocyanine derivatives synthesized and a schematic representation of the molecular organizations that have been obtained (columnar and nematic phases). (After Ref. [77])

complexes of octasubstituted phthalocyanine macrocycles should be utilized to fabricate soft X-ray analyzers [78].

The electrical conductivity of molecular solids may be related to the redox molecular properties and to the structure of their condensed phases (see Chapter 6). Monophthalocyanines of divalent metals afford insulators when undoped. On the contrary, bisphthalocyanines Pc_2M (M = rare earth) give semiconducting materials. In 1985, it was demonstrated that Pc_2Lu led to the first *intrinsic molecular semiconductor* [79]. In thin films or single crystals of Pc_2Lu, disproportionation reaction occurs since this molecular unit can be both easily oxidized and reduced:

$$2Pc_2Lu \rightleftharpoons Pc_2Lu^+, Pc_2Lu^- \qquad \Delta E = 0.48 \text{ eV}$$

The ionic species Pc_2Lu^+ and Pc_2Lu^- may be considered as charge carriers. The intrinsic conductivity of thin films Pc_2Lu is of the order of $10^{-5} \, \Omega^{-1} \, \text{cm}^{-1}$. Pc_2Lu was known, on the other hand, to have remarkable electrochromic properties: blue for the reduced form and red for the oxidized form. The exceptional conduction and electrochromic properties of Pc_2Lu subunit led us to use it as the central core of discotic mesogens (Figure 1.58).

The neutral green $[(C_nOCH_2)_8Pc]_2Lu$ complexes ($n = 8, 12, 18$) show a mesomorphic behavior only for the octadecyl derivative and over a small temperature range (5 °C). The oxidized species $[(C_nOCH_2)_8Pc]_2Lu^+ \, SbCl_6^-$ yield liquid crystals for the three different chain lengths and over extended domains of temperature (100 °C) [80].

It is possible to extend the stability domain of the substituted lutetium complexes by changing the nature of the connecting link of the side chains. $[(C_{12}O)_8Pc]_2Lu$ forms a liquid crystalline phase from 85 to 189 °C. The structure is, as usual, hexagonal [81]. Oxidation to $[(C_{12}O)_8Pc]_2Lu^+$, BF_4^- further increases the domain of stability of the mesophase: this latter is observable from

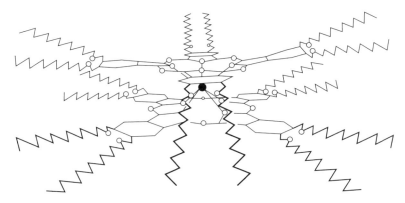

Figure 1.58 Schematic representation of [(C$_{12}$OCH$_2$)$_8$Pc]$_2$Lu. (After Ref. [80])

−3 to 253 °C. Columnar mesophases that are stable at room temperature may therefore be obtained with these compounds.

The redox potentials in solution of [(C$_n$OCH$_2$)$_8$Pc]$_2$Lu ($n = 8, 12$) are very close to those of the unsubstituted lutetium complex [81]. The oxidized and reduced species are formed via a fully reversible electrochemical process. Further studies have been made on thin films of [(C$_n$OCH$_2$)$_8$Pc]$_2$Lu [82]. The redox properties were studied by cyclic voltammetry in the presence of aqueous solutions of various salts. Long-term stability under cycling was observed: 10^6 redox cycles lead to only a 5% decrease of the electrochemical activity [82].

The electrical properties of the liquid crystalline phases as a function of frequency have been studied [83–85]. The overall electrical response depends on the electrical properties of the various domains (Figure 1.59).

The distance between the two electrodes is of the order of 20 µm. The molecular material is constituted of domains or microcrystals whose average size is of

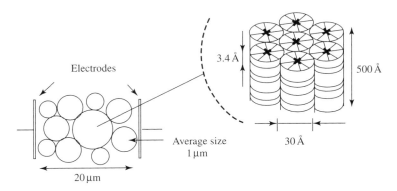

Figure 1.59 The various relevant parameters present in columnar mesophases which must be considered in electrical determinations

the order of 1 μm. The electrical properties will arise from the charge transport within and between the domains. Within a single domain, intracolumnar and intercolumnar charge-transfer processes must also be distinguished. Electrical determinations as a function of frequency allow the measurement of the electrical characteristics of the various domains.

The electrical properties of R_8PcH_2 (R = $-OC_{12}$, $-CH_2OC_{18}$), $[R_8Pc]_2Lu$ (R = $-OC_{12}$, $-CH_2OC_{18}$) and R_8PcLi (R = $-OC_{12}$) have been determined [83]. At low frequency (<100 Hz), all of these compounds are insulators ($\sigma < 10^{-11}\,\Omega^{-1}\,cm^{-1}$). A ω^0 dependency is observed. The rate-limiting step for charge carrier transport is probably related to intercrystallite hopping properties.

The number of charge carriers per unit of volume can be estimated from $\sigma = ne\mu$, assuming that the mobility, μ, is in the range 10^{-6}–$10^{-2}\,cm^2/V\,s$. By taking the highest value of the mobility, the concentration of charge carriers is calculated to be more than 10^{11} carriers/cm^3 for $[(C_{12}O)_8Pc]_2Lu$ and $(C_{12}O)_8PcH_2$ and more than 10^{10} carriers/cm^3 for $(C_{12}O)_8PcLi$.

At higher frequencies (>10^2–10^5 Hz), $\omega^{0.5}$ and $\omega^{0.8}$ dependencies are observed. The mechanism is then probably associated with intracolumnar or intercolumnar processes. The ionization potentials and electron affinities of paraffinic and aromatic moieties are significantly different and charge carriers must be localized on the aromatic subunits. The transport processes must therefore involve the aromatic cores of the columns. A columnar mesophase derived from $[(C_{12}O)_8Pc]_2Lu$ oriented in a magnetic field showed an anisotropy of about 10 for the conductivity (10 kHz) [85].

The substituted phthalocyanine subunit may also lead to polymeric materials via polycondensation of dihydroxystannyl- or dihydroxysilyl-phthalocyanine derivatives (Figure 1.60). In this way *spinal columnar* liquid crystals are formed [86–88]. The mesomorphic properties of both the monomeric and polymeric forms have been studied.

The dihydroxo-tin(IV) phthalocyanine derivative was obtained in two steps [70, 86] starting from $(C_{12}OCH_2)_8PcH_2$. This latter was treated with $SnCl_2$ in

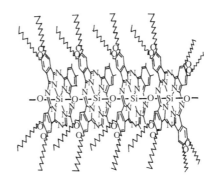

Figure 1.60 Spinal columnar liquid crystals formed by condensation of $(C_{12}OCH_2)_8$ $PcSi(OH)_2$. (After Ref. [88])

refluxing pentanol to give quantitatively $(C_{12}OCH_2)_8PcSn$, which, when reacted with H_2O_2, yielded the dihydroxo-tin(IV) derivative. The liquid crystalline properties of $(C_{12}OCH_2)_8PcSn(OH)_2 \cdot 2H_2O$ were studied by conventional techniques (DSC, optical microscopy, X-ray diffraction at small angles) [86]. A transition is observed at 59 °C from a highly ordered crystalline phase to a liquid crystal. At 95 °C a new texture slowly appears with a probable concomitant loss of water. The isotropic liquid forms at 114 °C. The solid phase, at room temperature, presents an X-ray pattern compatible with an orthorhombic structure; it transforms into a rectangular mesophase at 59 °C. The orthorhombic lattice (two units per cell) is transformed into a rectangular lattice (one unit per cell) with an increase of volume of approximately 10% (Figure 1.61). In the second case, the columns are all tilted in the same direction. The transformation between the two lattices does not involve a drastic displacement of the axes of the columns, as could be thought from the very different lattice parameters.

Above 100 °C, the fluid isotropic liquid slowly transforms into a highly viscous partially anisotropic mass with a change of color from green to yellow-green. This was interpreted as being due to the formation of oxystannyl polymers. ^{1}H NMR, UV–visible absorption spectra and IR spectroscopy substantiated this assignment [86].

A very similar behavior is obtained for silicon derivatives. The dihydroxysilicon (IV) phthalocyanine derivative $(C_{12}OCH_2)_8PcSi(OH)_2$ shows a birefringent viscous mass from room temperature to 300 °C with no detectable transition. DSC reveals a peak at -7 °C (15 kcal/mole). $(C_{12}OCH_2)_8PcSi(OH)_2$ is therefore a liquid crystal at room temperature. X-ray diffraction measurements at room temperature and at 150 °C both show a series of five Bragg reflections, in agreement with a hexagonal lattice (intercolumnar spacing of 30.8 Å).

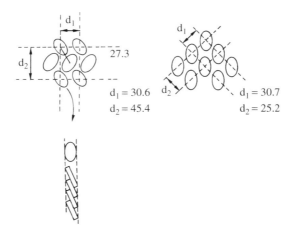

Figure 1.61 Two of the lattices that can be associated with $(C_{12}OCH_2)_8PcSn(OH)_2 \cdot 2H_2O$ (See Ref. [86])

Polycondensation is achieved by heating $(C_{12}OCH_2)_8PcSi(OH)_2$ in the liquid crystalline phase at 180 °C for 7 hours. Polymerization can be followed by optical absorption spectroscopy and gel permeation chromatography. The X-ray pattern of the heated samples is significantly modified as compared to the starting materials. A series of three Bragg reflections with reciprocal spacings in the ratio 1:2:3 is indicative of a lamellar order (interlamellar distance of 31 Å). This is not modified from room temperature to 60 °C. For temperatures higher than 60 °C, optical microscopy and X-ray diffraction both indicate the formation of an isotropic liquid [88].

Planar bisphthalocyanine may be employed as a central rigid core of mesogens (Figure 1.62) [89, 90]. An anisotropic phase is observed for the metal free derivative (M = H_2) by polarized light microscopy from room temperature to 300 °C. No transition is detected by DSC. X-ray diffraction at small angle indicates an orthorhombic lattice ($a = 42$ Å, $b = 28$ Å) at room temperature and a square symmetry ($a = b = 25$ Å) at 300 °C [90].

Globular metallomesogens have been prepared from a tetra-substituted tribenzosilatrane subunit [91, 92] (Figure 1.63). X-ray studies on different tetra-substituted tribenzosilatranes indicate the formation of various types of condensed phases showing columns and ordered layers perpendicular to the axis of the columns. These compounds do not form plastic crystals. Columnar structures with a hexagonal lattice have been characterized by X-ray diffraction (Figure 1.64)

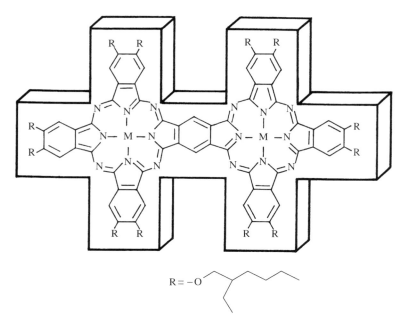

Figure 1.62 Dodeca-substituted planar bisphthalocyanines, with R = 2-ethyl-hexyl. (According to Ref. [89])

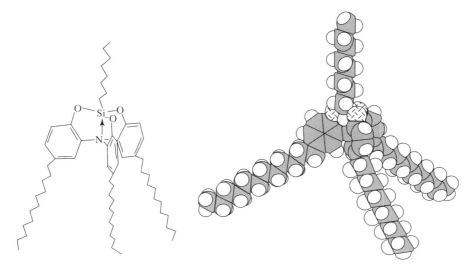

Figure 1.63 Chemical formula of tetra-substituted tribenzosilatranes and a molecular model of one of them [91, 92]

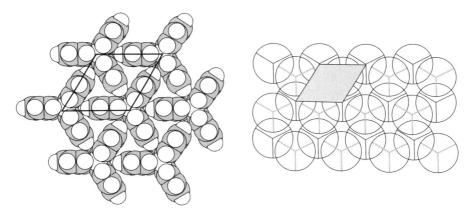

Figure 1.64 Cross-section of a molecular model of tribenzosilatrane and structure obtained from one of the derivatives. (Modified from Ref. [93])

[93]. By adding four different chains on the tetrabenzosilatrane subunit, a *macrochiral* centre can be formed [91].

1.6 CONCLUSIONS

There is therefore a long way to go from the design and the synthesis of a molecular unit to the formation of molecular assemblies. There is not such a large difference between chemical molecular assemblies and natural systems: the driving forces of the organization processes are the same. However, the synthetic assemblies are rudimentary as compared to those found in biology. In particular, the molecular systems, in most cases, do not lead to a function, they are not interconnected with each other, they are not part of a chemical chain necessary for a transformation and they are not, or only with difficulty, addressable. However, self-assembling processes permit the fabrication of submicronic devices to be envisaged [94].

1.7 REFERENCES

1. J. Simon, J.-J. André and A. Skoulios, *Nouv. J. de Chimie*, **10**, 295 (1986).
2. J. Simon, 'Mésomorphe (Etat)', in *Encyclopaedia Universalis*, Paris.
3. E. A. Silinsh, *Organic Molecular Crystals*, Springer Series in Solid-State Sciences, Vol. 16, Springer-Verlag, Berlin (1980).
4. M. Françon, *L'optique moderne et ses développements depuis l'apparition du laser*, Hachette, Paris (1986).
5. J. M. Cowley, *Diffraction Physics*, North Holland, Amsterdam (1975).
6. M. van Meersche and J. Feneau-Dupont, *Introduction à la cristallographie et à la chimie structurale*, Oyez, Leuven (1977).
7. N. Uyeda, T. Kobayashi, K. Ishizuka and Y. Fujiyoshi, *Jeol News*, **19E**, 2 (1981).
8. G. Friedel, *Ann. de Phys.*, **18**, 273 (1922).
9. D. Demus and L. Richter, *Textures of Liquid Crystals*, Verlag Chemie, Weinheim (1978).
10. G. W. Gray and J. W. G. Goodby, *Smectic Liquid Crystal*, Leonard Hill, London (1984).
11. C. E. Williams, M. Kléman, *J. de Phys. (Paris) Colloq.*, **36**, C1-315 (1975); see also J. B. Fournier and G. Durand, *J. Phys. II. France*, **1**, 845 (1991).
12. G. W. Gray (ed.), *Critical Reports on Applied Chemistry*, Vol. 22, *Thermotropic Liquid Crystals*, John Wiley & Sons, Chichester (1987).
13. S. Chandrasekhar, B. K. Sadashiva and K. A. Suresh, *Pramana*, **9**, 471 (1977).
14. J. N. Sherwood (ed.), *The Plastically Crystalline State*, John Wiley & Sons, Chichester (1979).
15. P. Mihailovic, P. Bassoul and J. Simon, *J. Phys. Chem.*, **94**, 2815 (1990).
16. C. McDonald, *J. Pharm. Pharmacol.*, **22**, 148 and 774 (1970).
17. M. S. Ramadan, D. F. Evans and R. Lumry, *J. Phys. Chem.*, **87**, 4538 (1983).
18. P. A. Winsor, *Chem. Rev.*, **68**, 1 (1968).
19. P. Ekwall, *Advances in Liquid Crystals*, Vol. 1 (ed. G. H. Brown), Academic Press, New York (1975).
20. I. Danielsson, *Adv. Chem. Ser.*, **152**, 13 (1976).
21. D. M. Small, *J. Colloid. Int. Sci.*, **58**, 581 (1977).
22. A. Mathis, A. Skoulios, R. Varoqui and A. Schmitt, *Eur. Polym. J.*, **10**, 1011 (1974).
23. A. Skoulios and G. Finoaz, *J. Chim. Phys.*, **59**, 473 (1962).
24. C. Robinson, *Tetrahedron*, **13**, 219 (1961).
25. G. D. Fasman, *Poly-alpha-aminoacids*, Marcel Dekker, New York (1967).
26. M. Arpin, C. Strazielle and A. Skoulios, *J. Phys.*, **38**, 307 (1977).
27. E. T. Samulski and D. B. Du Pre, *Adv. Liquid Cryst.*, **4**, 121 (1979).
28. J. D. Bernal, *Faraday Disc. Chem. Soc.*, **25**, 7 (1958).
29. Y. A. Ovchinikov, V. I. Ivanov and A. M. Shkrob, *Membrane Active Complexones*, Elsevier Scientific, New York (1974).

30. G. Eisenman, *Membrane*, Vol. II, *Lipid Bilayers and Antibiotics*, Marcel Dekker, New York (1973).
31. R. A. Horne, *Water and Aqueous Solutions*, Wiley-Interscience, New York (1972).
32. H. S. Frank and M. W. Evans, *J. Chem. Phys.*, **13**, 507 (1945).
33. G. Nemethy and H. A. Scheraga, *J. Chem. Phys.*, **36**, 3382 (1962).
34. L. Stryer, *La biochimie* (Trad. S. Weinman), Médecine-Sciences Flammarion, Paris (1995).
35. P. Atkins, *Molécules au quotidien*, Interédition, Paris (1989).
36. A. Ulman, *An Introduction to Ultrathin Organic Films from Langmuir-Blodgett to Self Assembly*, Academic Press, Boston (1991).
37. P. Diehl, E. Fluck and R. Kosfeld (eds.), *NMR: Basic Principles and Progress*, Vol. 9, *Lyotropic Liquid Crystals*, Springer-Verlag, Berlin (1975).
38. A. Skoulios, *Adv. Colloid. Int. Sci.*, **1**, 72 (1967).
39. J. Charvolin, *Images de la Recherche*, CNRS, Paris (1994).
40. P. Espinet, M. E. Esteruelas, L. A. Oro, J. L. Serrano and E. Sola, *Coord. Chem. Rev.*, **117**, 215 (1992).
41. S. A. Hudson and P. M. Maitlis, *Chem. Rev.*, **93**, 861 (1993).
42. A.-M. Giroud-Godquin and P. M. Maitlis, *Angew. Chem. Int. ed. Engl.*, **30**, 375 (1991).
43. J. L. Serrano (ed.), *Metallomesogens: Synthesis, Properties and Applications*, VCH, New York (1996).
44. J. Le Moigne, Ph. Gramain and J. Simon, *J. Colloid. Int. Sci.*, **60**, 565 (1977).
45. D. Markovitsi, A. Mathis, J. Simon, J. C. Wittmann and J. Le Moigne, *Mol. Cryst. Liquid Cryst.*, **64**, 121 (1980).
46. J. Le Moigne and J. Simon, *J. Phys. Chem.*, **84**, 170 (1980).
47. J. Bjerrum, *Z. Phys. Chem.*, **106**, 219 (1923).
48. J. G. Kirkwood and F. H. Westheimer, *J. Chem. Phys.*, **6**, 506 (1938).
49. A. R. Koray, V. Ahsen and Ö. Bekâroglu, *J. Chem. Soc. Chem. Commun.*, 932 (1986).
50. N. Kobayashi and Y. Nishiyama, *J. Chem. Soc. Chem. Commun.*, 1462 (1986).
51. R. Hendriks, Ot. E. Sielcken, W. Drenth and R. J. M. Nolte, *J. Chem. Soc. Chem. Commun.*, 1464 (1986).
52. C. Sirlin, L. Bosio, J. Simon, V. Ahsen, E. Yilmazer and Ö. Bekâroglu, *Chem. Phys. Lett.*, **139**, 362 (1987).
53. J. Simon, J. Le Moigne, D. Markovitsi and J. Dayantis, *J. Am. Chem. Soc.*, **102**, 7247 (1980).
54. J. Simon and J. Le Moigne, *J. Mol. Cat.*, **7**, 137 (1980).
55. D. Markovitsi, J. Simon and E. Kraeminger, *Nouv. J. de Chimie*, **5**, 141 (1981).
56. P. Bassoul and J. Simon, *New J. Chem.*, **20**, 1131 (1996).
57. D. Markovitsi, A. Mathis, J. Simon, J. C. Wittmann and J. Le Moigne, *Molec. Cryst. Liquid Cryst.*, **64**, 121 (1980).
58. D. Markovitsi, Thèse de Doctorat d'Etat (1983).
59. D. Markovitsi, J.-J. André, A. Mathis, J. Simon, P. Spegt, G. Weill and M. Ziliox, *Chem. Phys. Lett.*, **104**, 46 (1986).
60. D. Markovitsi, R. Knoesel and J. Simon, *Nouv. J. de Chimie*, **6**, 531 (1982).
61. R. Knoesel, D. Markovitsi, J. Simon and G. Duportail, *J. Photochem.*, **22**, 275 (1983).
62. J. Simon and J.-J. André, *Molecular Semiconductors*, Springer-Verlag, Berlin (1985).
63. J. Simon and P. Bassoul, in *Phthalocyanines*, Vol. 2 (eds. C. C. Leznoff and A. B. P. Lever), VCH, Weinheim (1993).
64. C. Piechocki, J. Simon, A. Skoulios, D. Guillon and P. Weber, *J. Am. Chem. Soc.*, **104**, 5245 (1982).
65. D. Guillon, A. Skoulios, C. Piechocki, J. Simon and P. Weber, *Molec. Cryst. Liquid Cryst.*, **100**, 275 (1983).
66. C. Piechocki and J. Simon, *Nouv. J. de Chimie*, **9**, 159 (1985).

67. A. P. Polishchuck and T. M. Timofeeva, *Russian Chem. Rev.*, **62**, 291 (1993).
68. D. Guillon, P. Weber, A. Skoulios, C. Piechocki and J. Simon, *Molec. Cryst. Liquid Cryst.*, **130**, 223 (1985).
69. C. Piechocki and J. Simon, *J. Chem. Soc. Chem. Commun.*, 259 (1985).
70. C. Piechocki, J.-C. Boulou and J. Simon, *Molec. Cryst. Liquid Cryst.*, **149**, 115 (1987).
71. P. Weber, D. Guillon and A. Skoulios, *J. Phys. Chem.*, **91**, 2242 (1987).
72. D. Masurel, C. Sirlin and J. Simon, *New J. Chem.*, **11**, 455 (1987).
73. K. Ohta, L. Jacquemin, C. Sirlin, L. Bosio and J. Simon, *New J. Chem.*, **12**, 751 (1988).
74. M. K. Engel, P. Bassoul, L. Bosio, H. Lehmann, M. Hanack and J. Simon, *Liquid Cryst.*, **15**, 709 (1993).
75. K. Ohta, T. Watanabe, S. Tanaka, T. Fujimoto, I. Yamamoto, P. Bassoul, N. Kucharczyk and J. Simon, *Liquid Cryst.*, **10**, 357 (1991).
76. K. Ohta, T. Watanabe, H. Hasebe, Y. Morizumi, T. Fujimoto, I. Yamamoto, D. Lelièvre and J. Simon, *Molec. Cryst. Liquid Cryst.*, **196**, 13 (1991).
77. D. Lelièvre, M. A. Petit and J. Simon, *Liquid Cryst.*, **4**, 707 (1989).
78. J. Vacus, P. Doppelt, J. Simon and G. Memetzidis, *J. Mater. Chem.*, **2**, 1065 (1992).
79. J.-J. André, K. Holczer, P. Petit, M.-T. Riou, C. Clarisse, R. Even, M. Fourmigué and J. Simon, *Chem. Phys. Lett.*, **115**, 463 (1985).
80. C. Piechocki, J. Simon, J.-J. André, D. Guillon, P. Petit, A. Skoulios and P. Weber, *Chem. Phys. Lett.*, **122**, 124 (1985).
81. F. Castaneda, C. Piechocki, V. Plichon, J. Simon and J. Vaxivière, *Electrochimica Acta*, **31**, 131 (1986).
82. S. Besbes, V. Plichon, J. Simon and J. Vaxivière, *J. Electroanal. Chem.*, **237**, 61 (1987).
83. Z. Belarbi, C. Sirlin, J. Simon and J.-J. André, *J. Phys. Chem.*, **93**, 8105 (1989).
84. Z. Belarbi, M. Maitrot, K. Ohta, J. Simon, J.-J. André and P. Petit, *Chem. Phys. Lett.*, **143**, 400 (1988).
85. Z. Belarbi, *J. Phys. Chem.*, **94**, 7334 (1990).
86. C. Sirlin, L. Bosio and J. Simon, *J. Chem. Soc. Chem. Commun.*, 379 (1987).
87. C. Sirlin, L. Bosio and J. Simon, *J. Chem. Soc. Chem. Commun.*, 236 (1988).
88. C. Sirlin, L. Bosio and J. Simon, *Molec. Cryst. Liquid Cryst.*, **155**, 231 (1988).
89. D. Lelièvre, L. Bosio, J. Simon, J.-J. André and F. Bensebaa, *J. Am. Chem. Soc.*, **114**, 4475 (1992).
90. D. Lelièvre, O. Damette and J. Simon, *J. Chem. Soc. Chem. Commun.*, 939 (1993).
91. C. Soulié, P. Bassoul and J. Simon, *J. Chem. Soc. Chem. Commun.*, 114 (1993).
92. C. Soulié and J. Simon, *New J. Chem.*, **17**, 267 (1993).
93. P. Bassoul, J. Simon and C. Soulié, *J. Phys. Chem.*, **100**, 3131 (1996).
94. J. Simon and C. Sirlin, *Pure Appl. Chem.*, **61**, 1625 (1989).

2 A Few Notions of Symmetry

2.1 Introduction 47
2.2 Elements of Symmetry 47
2.3 Relations between Symmetry Elements 52
2.4 Symmetry of Molecular Units 54
2.5 Determination of the Symmetry of a Molecular Unit 58
2.6 Matrix Notation for Geometric Transformations 60
2.7 Notion of Irreducible Representation 62
2.8 Group–Subgroup Relationships 66
2.9 References and Notes 67

Le monde de la Science est beau avant d'être vrai.
Ce monde est admiré avant d'être vérifié.
G. Bachelard (1884–1962)
La Formation de l'Esprit Scientifique

2.1 INTRODUCTION

Most rationalizations and theories (with a few exceptions) in reality follow careful phenomenological observations. At that stage, an aesthetic approach is often taken since no other method is available. Beauty is a personal notion and, as all things based on emotions, can lead to devastatingly false visions. In most cases, the sense of beauty alone is insufficient. In this respect, symmetry concepts are quite representative. In the next sections a mathematics-free approach will be tentatively given. However, since a powerful mathematical tool is on perusal (group theory), an incessant check of the validity of our proposals will be made.

2.2 ELEMENTS OF SYMMETRY

The chemist's sense of beauty is almost always associated with highly symmetrical figures (Figure 2.1). When the molecule is not beautiful, it can be either complicated or useful. We can classify the various symmetry elements than can be met. The apparently simplest (and probably the first discovered) symmetry element is the mirror plane (Figure 2.2, Plate 1). From a careful examination of the illustration, one can verify that the mirror image of the lady's right hand is superimposable with her left hand.

More conventional ways may be employed to represent this property. A stereographic projection can be used to represent a mirror plane (Figure 2.3).

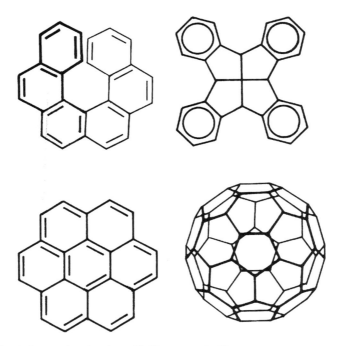

Figure 2.1 A few molecules: beautiful? symmetrical?

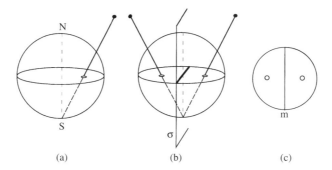

Figure 2.3 Stereographic projection of (a) one point, (b) two points related with a mirror plane, (c) conventional representation of a mirror plane

A FEW NOTIONS OF SYMMETRY

(a)   (b)   (c)

Figure 2.4 Three examples of symmetry axes: (a) flower, (approximate) C_5 axis [2], (b) snowflake, C_6 axis (after Ref. [3]), (c) diatom skeleton, C_{68} axis [4]

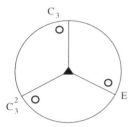

Figure 2.5 Stereographic representation of a symmetry axis $C_n (n = 3)$. The notation C_n^m means m successive rotations of $2\pi/n$

Figure 2.6 Yin-Yang symbol: man and woman, sunny and dark, dry and humidity are complementary approaches of the same thing. It can be seen as an inversion centre

Besides mirrors, the next symmetry element one can meet is the symmetry axis (Figure 2.4). The notation C_n means that a rotation $2\pi/n$ allows an identical figure to be obtained. The corresponding stereographic representation is given in Figure 2.5.

The inversion center (i) is more difficult to visualize in three dimensions. However, a satisfactory two-dimensional representation may be given (Figure 2.6).

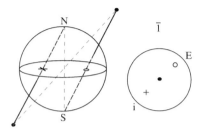

Figure 2.7 Stereographic representation of an inversion centre (notation: i or $\bar{1}$) ○, point above the plane; +, point below the plane (as shown)

The mathematically derived stereographic projection defines the inversion centre more clearly. At that stage, it is necessary to distinguish between elements and operations of symmetry. To each symmetry element corresponds one or several symmetry operations. It can be readily seen that the inversion centre can be considered as the result of two symmetry operations carried out successively: a C_2 (180°) rotation and a mirror axis perpendicular to the rotation axis (Figure 2.7):

$$i = C_2 \cdot \sigma_h$$

where

σ_h = mirror perpendicular to the C_n axis

It is possible to generalize this approach. An axis of rotoreflexion of order n is given by

$$S_n = C_n \cdot \sigma_h \tag{2.1}$$

Once again, naturally occurring examples can be found (Figure 2.8).

In most classifications, the following notation of the main symmetry elements is taken [5]:

Identity E
Symmetry axis C_n (including C_∞)
Mirror plane σ
  v = vertical (containing C_n)
  d = dihedral
  h = horizontal (perpendicular to C_n)
Rotoreflexion S_n: rotation of $2\pi/n$ followed by a symmetry plane perpendicular to the axis

An easy way to find all possible symmetry operations is to alternately use the product:

$$\bar{n} = C_n \cdot i \tag{2.2}$$

A FEW NOTIONS OF SYMMETRY

Figure 2.8 Copulation of two starfishes (*Archaster typicus*). An example of an S_{10} symmetry element

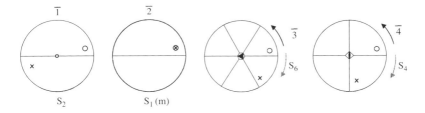

Figure 2.9 Correspondence between S_n and $\bar{n}$ symmetry operations

Symmetry elements can be separated into proper and improper rotations (involving i) and a correspondence between $\bar{n}$ and S_n symmetry elements may be found (Figure 2.9).

Whenever an improper rotation is applied to an object, it cannot be chiral. It is superimposable with its image in a mirror plane. Most textbooks use the rotoreflexion symmetry elements. The correspondence for the first rotation axes is found in Table 2.1.

Table 2.1 Correspondence between $\bar{n}$ and S_n symmetry elements

Direct		C_1	C_2	C_3	C_4	C_5	...
Inverse	$\bar{n}$	$\bar{1}$	$\bar{2}$	$\bar{3}$	$\bar{4}$	$\bar{5}$	...
	S_n	S_2	S_1	S_6	S_4	S_{10}	

2.3 RELATIONS BETWEEN SYMMETRY ELEMENTS

When an object possesses two different symmetry elements this, in most cases, leads to the presence of additional symmetry elements. In the example shown in Figure 2.10, the presence of two symmetry elements, a C_2 axis and a mirror plane, yields obligatorily to the presence of another symmetry plane.

This can also be seen by using the stereographical projections (Figure 2.11). The application of the identity E on 1 leaves it unchanged; the application of C_2 gives 2. The introduction of a mirror plane $(\sigma_v)_1$ leads to 3 and 4, where 3 can be considered to be given by the combination of two symmetry operations:

$$C_2 \cdot (\sigma_v)_1$$

and 4 is obtained directly from 1 via the symmetry plane $(\sigma_v)_1$. Consequently, a second mirror plane is present, $(\sigma_v)_2$. Alternatively, 3 may be obtained using the operation $(\sigma_v)_2$ from 1.

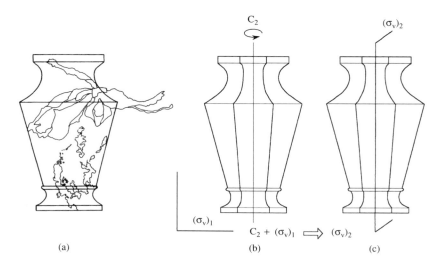

Figure 2.10 Absence or presence of symmetry elements: (a) no symmetry element (after [6]); (b) undecorated version; two symmetry elements, C_2 and $(\sigma_v)_1$, are easily seen; (c) the mirror plane $(\sigma_v)_2$ is also a symmetry element of the object

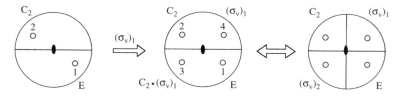

Figure 2.11 The presence of a C_2 axis and a mirror plane implies the presence of a second mirror plane

A FEW NOTIONS OF SYMMETRY

A rotation axis C_n in addition to a mirror plane containing it implies the presence of n mirror planes. In the same way, the presence of two C_2 axes (around x and y) implies a C_2 axis around z (Figure 2.12). The same thing can be written using stereographic projections (Figure 2.13).

The previous operation may be written as [7]

$$C_2(x) \cdot C_2(y) = C_2(z)$$

More generally, the product:

$$C_2(i) \cdot C_2(j)$$

about axes which intersect at an angle θ is equivalent to a rotation 2θ around an axis perpendicular to the two C_2 axes (Figure 2.14).

Two planes of symmetry, A and B, separated by an angle ϕ yield a $C_{2\phi}$ axis [7].

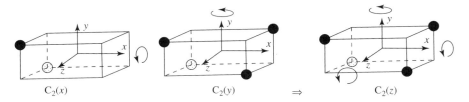

Figure 2.12 Two perpendicular C_2 axes lead to the presence of a third C_2 axis

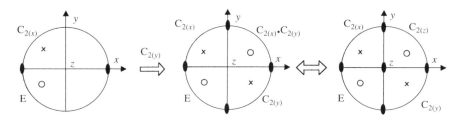

Figure 2.13 Same as Figure 2.12 using stereographic projections

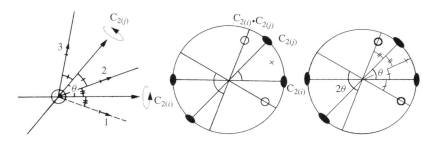

Figure 2.14 $C_2(i) \cdot C_2(j)$ is equivalent to a $C_{2\theta}$ rotation axis perpendicular to the two others: classical and stereographical representations

From these examples, one can verify that a set of symmetry operations, when combined, possesses the structure of a mathematical group. This means that the set of symmetry operations has the following properties:

(a) Any product $S_i \cdot S_j$ is an element of the set.
(b) In successive products, any association of elements can be achieved:

$$(S_i \cdot S_j) \cdot S_k = S_i \cdot (S_j \cdot S_k)$$

(c) An identity element is present: $S_i \cdot E = E \cdot S_i = S_i$
(d) An inverse element S_i^{-1} can be found for each of the operations:

$$S_i \cdot S_i^{-1} = S_i^{-1} \cdot S_i = E$$

2.4 SYMMETRY OF MOLECULAR UNITS

It is possible to determine the sets of symmetry operations compatible with each other. One can start from a molecular units with no symmetry element and then one can add symmetry operations step by step until the most symmetrical object is obtained: a sphere. As an example, one can consider the unit shown in Figure 2.15 which has no symmetry. The addition of a C_2 axis imposes a modification of the molecular unit. Two mirror planes containing the symmetry axis must be introduced at the same time.

It is now possible to postulate the presence of an additional mirror plane by further transforming the starting unit (Figure 2.16). At that stage numerous new symmetry operations are generated (Figure 2.17). The unit obtained is tetrahedral and the set of corresponding symmetry operations are:

$$E \quad 8C_3 \quad 3C_2 \quad 6S_4 \quad 6\sigma_d$$

Further transformation of the unit in order to introduce other symmetry operations necessitates an increase in the number of subunits. For example, consideration

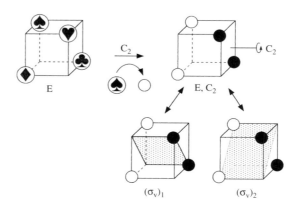

Figure 2.15 The addition of one symmetry operation (C_2) leads to a new set of symmetry operations: E, C_2, $2\sigma_v$

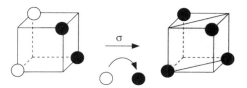

Figure 2.16 Transformation of the molecular unit to generate an additional mirror plane

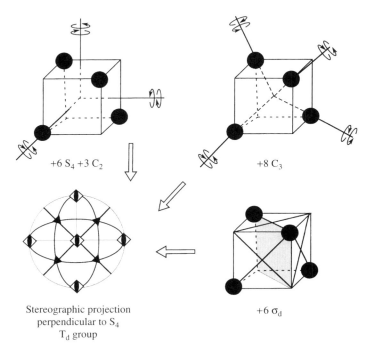

Figure 2.17 Further modification of the unit allowing the presence of an additional mirror plane induces the presence of the above symmetry operations, the overall number of symmetry operations is given. (Modified from Ref. [8])

of a C_4 symmetry axis leads to the modification shown in Figure 2.18. As expected, extra-symmetry elements appear (Figure 2.19). The cube which has been constructed has therefore the following symmetry elements (and no others):

E $8C_3$ $6C_2'$ $6C_4$ $3C_2$ i $6S_4$ $8S_6$ $3\sigma_h$ $6\sigma_d$

It is not possible to go further with conventional symmetry elements. It is, however, possible to consider infinite symmetry axes which are defined as

$$C_\infty \quad \text{with} \quad \frac{2\pi}{n}, \quad n \longrightarrow \infty \text{ (see Figure 2.20)}$$

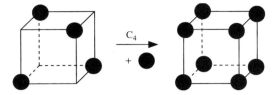

Figure 2.18 Further modifications of the unit allowing the presence of a C_4 axis

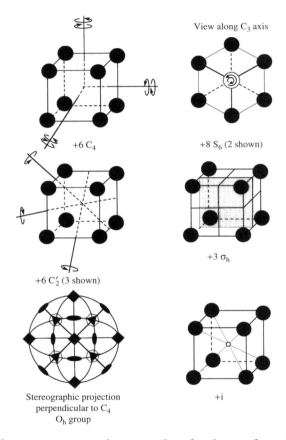

Figure 2.19 New symmetry operations appearing after the transformation of the unit

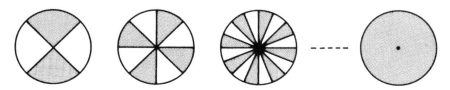

Figure 2.20 Representation of an infinite rotation axis (C_∞)

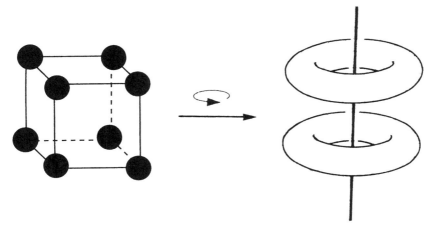

Figure 2.21 Unit containing a C_∞ symmetry axis

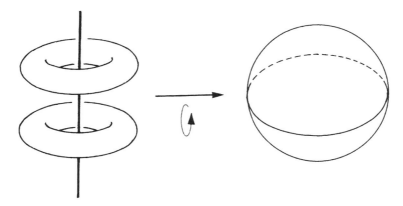

Figure 2.22 The presence of two noncolinear C_∞ axes leads to a spherical unit

The unit can then be modified in order to generate a C_∞ axis (Figure 2.21). The addition of another C_∞ axis leads to a sphere (Figure 2.22). The sphere contains all possible symmetry operations: all proper rotations C_n ($n = 0, 1, 2, \ldots, \infty$), all inverse reflexion $\bar{n}$ ($n = 0, 1, 2, \ldots, \infty$) and consequently all symmetry elements already described in $\bar{n}$ but expressed in a different way: i, σ, S_n.

It can therefore be seen that sets of compatible symmetry operations can be defined. Group theory demonstrates that only a limited number of classes of sets is possible: they have been classified according to the number (and type) of symmetry elements present and are called *point symmetry groups*. Two sets of symbols (Hermann-Mauguin and Schoenflies) can be employed to describe them. The *order* of a group is the total number of symmetry operations of this

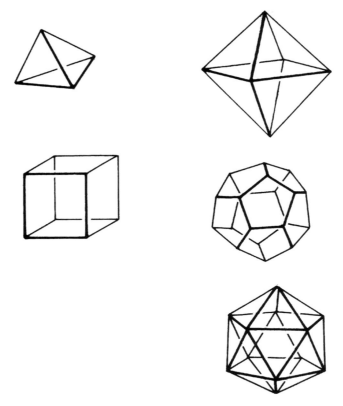

Figure 2.23 The five platonic solids. (Modified from Ref. [7])

group. The main groups as well as the corresponding notations can be found in Appendix 1.

Platonic solids are regular polyhedra made of 4, 8 or 20 equilateral triangles (tetrahedron, octahedron, icosahedron), 6 squares (cube) or 12 pentagons (dodecahedron). These are examples of highly symmetrical objects which possess several high-order axes (Figure 2.23). The icosahedron, for example, possesses $12C_5$, $20C_3$, $20S_6$ and 15σ among other symmetry elements. It is not possible to build a more symmetrical figure constituted of regular and equivalent polygons also possessing equivalent vertices and edges [7].

2.5 DETERMINATION OF THE SYMMETRY OF A MOLECULAR UNIT

The following procedure must be followed to determine the point symmetry group of a given molecular unit (Figure 2.24). The first step consists of determining whether the molecular unit possesses:

A FEW NOTIONS OF SYMMETRY

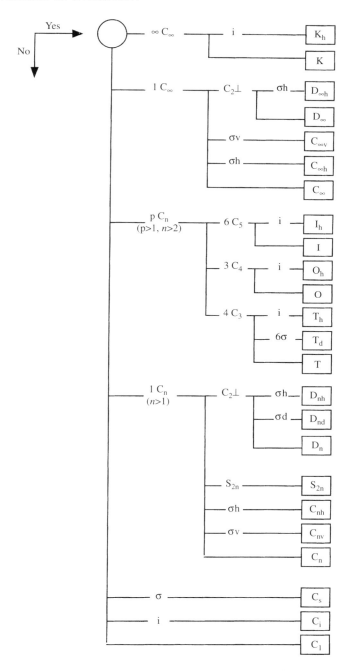

Figure 2.24 Procedure for finding the point symmetry group of a molecular unit (for notation see Appendix 1)

(a) Several C_∞:
   K_h (if i), K
(b) One C_∞:
   together with a $C_2 \perp$ to C_∞:
      $D_{\infty h}$ (if σ_h), D_∞
   without $C_2 \perp$ to C_∞:
      $C_{\infty v}$ (if σ_v), $C_{\infty h}$ (if σ_h), C_∞
(c) Several C_n ($n > 2$):
   together with $6C_5$:
      I_h (if i), I
   together with $3C_4$:
      O_h (if i), O
   together with $4C_3$:
      T_h (if i), T_d (if 6σ), T
(d) One C_n ($n > 1$):
   together with $C_2 \perp$ to C_n:
      D_{nh} (if σ_h), D_{nd} (if σ_d), D_n
   without $C_2 \perp$ to C_n:
      S_{2n} (if S_{2n}), C_{nh} (if σ_h), C_{nv} (if σ_v), C_n
(e) Without any rotation axis
   C_s (if σ), C_i (if i), C_1

One can consider for examples the molecules of triptycene and ferrocene (Figure 2.25). The triptycene molecular unit possesses only one C_3; it also has $3C_2$ perpendicular to the C_3 and $1\sigma_h$. The symmetry group is therefore D_{3h}. The ferrocene (staggered conformation) has a C_5 axis with $5C_2$ perpendicular; it has no σ_h but five dihedral planes of symmetry. It is therefore D_{5d}. In this symmetry group one can note the presence of one S_{10} axis.

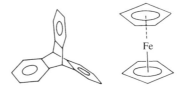

Figure 2.25 The molecules of triptycene and ferrocene (staggered conformation)

2.6 MATRIX NOTATION FOR GEOMETRIC TRANSFORMATIONS

Let us consider a point with coordinates x, y, z. The application of the identity E does not change its position. Therefore:

$$x \xrightarrow{E} x \qquad x = 1x + 0y + 0z$$

A FEW NOTIONS OF SYMMETRY

$$y \xrightarrow{E} y \qquad y = 0x + 1y + 0z$$

$$z \xrightarrow{E} z \qquad z = 0x + 0y + 1z$$

A condensed notation can be used

$$\begin{pmatrix} x \\ y \\ z \end{pmatrix} = \begin{pmatrix} 1 & 0 & 0 \\ 0 & 1 & 0 \\ 0 & 0 & 1 \end{pmatrix} \begin{pmatrix} x \\ y \\ z \end{pmatrix}$$

The notation associated with the symmetry operation identity is then

$$E = \begin{pmatrix} 1 & 0 & 0 \\ 0 & 1 & 0 \\ 0 & 0 & 1 \end{pmatrix}$$

Points, vectors or, more generally, functions are expressed in a given frame of reference defined by orthonormal basis vectors $\mathbf{e}_x$, $\mathbf{e}_y$ and $\mathbf{e}_z$:

$$\mathbf{v} = x\mathbf{e}_x + y\mathbf{e}_y + z\mathbf{e}_z \tag{2.3}$$

The application of the identity E does not change the position; therefore $\mathbf{v} = E\mathbf{v}$. In previous formula the vector $\mathbf{v}$ has the components $\begin{pmatrix} x \\ y \\ z \end{pmatrix}$ in the reference frame $\mathbf{e}_x$, $\mathbf{e}_y$, $\mathbf{e}_z$ and the notation of the symmetry operation is a *matrix*.

The matrix corresponding to a rotation of angle θ around the z axis (Figure 2.26) may be established:

$$\mathbf{v}' = C(\theta)\mathbf{v} \tag{2.4}$$

$$x' = x\cos\theta - y\sin\theta$$

$$y' = x\sin\theta + y\cos\theta$$

$$z' = z$$

$$\begin{pmatrix} x' \\ y' \\ z' \end{pmatrix} = \begin{pmatrix} \cos\theta & -\sin\theta & 0 \\ \sin\theta & \cos\theta & 0 \\ 0 & 0 & 1 \end{pmatrix} \begin{pmatrix} x \\ y \\ z \end{pmatrix}$$

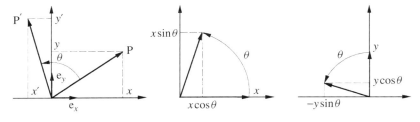

Figure 2.26 Effect of a rotation θ around z of a point P or coordinates x and y transformed into P'

By similar procedures, the matrices corresponding to any symmetry operation may be found, in particular:

$$S(\theta) = \begin{pmatrix} \cos\theta & -\sin\theta & 0 \\ \sin\theta & \cos\theta & 0 \\ 0 & 0 & -1 \end{pmatrix} \quad i = \begin{pmatrix} -1 & 0 & 0 \\ 0 & -1 & 0 \\ 0 & 0 & -1 \end{pmatrix} \quad (2.5)$$

The matrices previously described are applied on components x, y, z of a vector **v**.

In a general sense, the basis can be described not only by vectors but also by functions. The *trace* of the matrix is the sum of the diagonal elements. If f is a function of x, y or z, then for any element of symmetry operation R, the trace can be determined by applying:

$$Rf_1 = f_1 \Rightarrow +1$$
$$Rf_1 = -f_1 \Rightarrow -1 \quad (2.6)$$
$$Rf_1 = f_2 \Rightarrow 0$$

The trace of the matrix is called the *character*.

2.7 NOTION OF IRREDUCIBLE REPRESENTATION

In the course of using group theory, it is important to distinguish between variable and functions. In most publications, for example, the notation x, y, z, $x^2 - y^2$, ... recovers in fact functions.

Let us consider the symmetry group D_2 and three *basis functions* represented by positive (black) and negative (white) spheres (Figures 2.27 and 2.28). The set (x, y, z) can be transformed by to the symmetry operations of the symmetry group D_2 (Figure 2.28).

The *dimension* associated with the matrix representing the symmetry operation is given by the number of basis functions employed to describe it. The ensemble of matrices for all symmetry operations of a given group is called a *representation* (Γ). The dimension of the representation is given by the character associated with the identity.

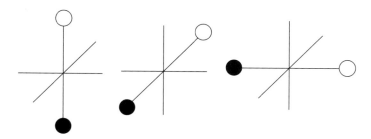

Figure 2.27 Functions of symmetries x, y or z

A FEW NOTIONS OF SYMMETRY

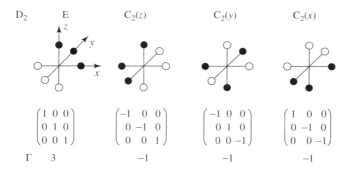

Figure 2.28 The various symmetry operations of D_2 and the corresponding matrices and their characters

The representation Γ shown in Figure 2.28 is said to be *reducible* since it is possible to consider x, y, z separately. In this case there is no symmetry operation of D_2 that permutes these three functions: three representations of dimension 1 can be determined. (Table 2.2). The characters of Γ are the sum of the characters of Γ_1, Γ_2, Γ_3: $\Gamma = \Gamma_1 + \Gamma_2 + \Gamma_3$.

The representations Γ_i are said to be *irreducible*; their basis functions are not interconnected by any of the symmetry operations of the group. They behave independently in the D_2 symmetry group.

Let us now consider, as another example, four s-type orbitals considered to be in a C_{4v} symmetry (Figure 2.29). It can be seen that the system of four orbitals with the same sign is unchanged by all the symmetry operations. It is a basis of the representation Γ_1. One can also see that various arrangements of positive (black) and negative (white) spheres are basis functions of the irreducible representations Γ_2 and Γ_3 (Table 2.3). A *table of characters* is thus constructed. The corresponding symmetry operations are indicated on a row:

$$2C_4(C_4^1, C_4^3), C_4^2 = C_2, 2\sigma_v(\sigma_{v1}, \sigma_{v2}), 2\sigma_d(\sigma_{d1}, \sigma_{d2})$$

In the case of Γ_3 (Table 2.3) only E leaves the two basis functions unchanged whereas C_2 transforms them into their opposites. The other symmetry operations interconvert the basis functions.

Table 2.2 The reducible representation Γ of Figure 2.28 is decomposed into three irreducible representations $\Gamma_i (i = 1–3)$

D_2	E	$C_2(z)$	$C_2(y)$	$C_2(x)$	
Γ_1	1	1	−1	−1	z
Γ_2	1	−1	1	−1	y
Γ_3	1	−1	−1	1	x
Γ	3	−1	−1	−1	(x, y, z)

DESIGN OF MOLECULAR MATERIALS

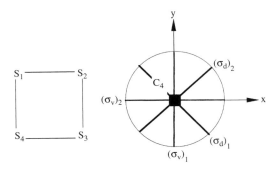

Figure 2.29 Symmetry elements of C_{4v}

Table 2.3 Irreducible representations of C_{4v}

C_{4v}	E	$2C_4$	C_2	$2\sigma_v$	$2\sigma_d$		
Γ_1	1	1	1	1	1		$x^2 + y^2$
Γ_2	1	$\bar{1}$	1	$\bar{1}$	1		xy
Γ_3	2	0	$\bar{2}$	0	0		(x, y)

Table 2.4 One example of a representation of dimension 4

C_{4v}	E	$2C_4$	C_2	$2\sigma_v$	$2\sigma_d$	
Γ	4	0	0	0	2	

Any given arrangement of positive and negative spheres gives a reducible representation which can be decomposed into a sum of irreducible representations (Table 2.4). The characters of Γ correspond to the sum of the characters of Γ_1, Γ_2 and Γ_3: $\Gamma = \Gamma_1 + \Gamma_2 + \Gamma_3$. More generally, the number of times (a_i) that the ith irreducible representation occurs in a reducible representation is given by

$$\Gamma = \sum_i a_i \Gamma_i \text{ with}$$

$$a_i = \frac{1}{h} \sum_R \chi(R)\chi_i(R) \tag{2.7}$$

where

h = order of the group
$\chi(R)$ = character of Γ for the operation of symmetry R

A FEW NOTIONS OF SYMMETRY

Each basis function φ of Γ can be expressed as a linear combination of the basis functions of $\Gamma_1, \Gamma_2, \Gamma_3$. An operator P_{Γ_i}, called a *projection operator*, may be used to find the decomposition into functions associated with the irreducible representations Γ_i:

$$P_{\Gamma_i}(\varphi) = \frac{l_i}{h} \sum_R [X_i(R)R(\varphi)] \qquad (2.8)$$

where

l_i = dimension of Γ_i (character of E)
$X_i(R)$ = character of the symmetry operation R
$R(\varphi)$ = application of R on the function φ

(When X_i is an imaginary term, however, see Chapter 6). P_{Γ_i} has the following properties:

$$P_{\Gamma_i}(\varphi_i) = \varphi_i \quad \text{and} \quad P_{\Gamma_i}(\varphi_j) = 0$$

when φ_i and φ_j are respectively the basis functions of Γ_i and Γ_j. An example of the decomposition of a basis function of Γ into its irreducible components is given in Figure 2.30.

The same process could be followed using the projection operator for any functions such as

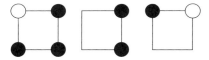

p-type orbitals may also be considered, in this case two new representations of dimension 1 are generated (Table 2.5).

By using all other types of orbitals — or even any object belonging to any point symmetry group — one would not find other representations than $\Gamma_1-\Gamma_5$. They

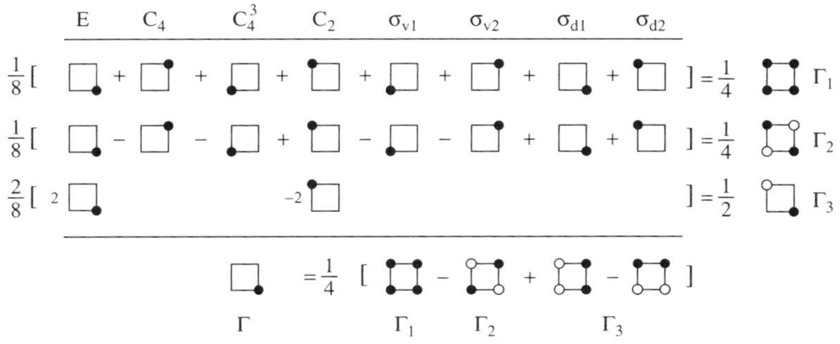

Figure 2.30 Decomposition of a basis function of Γ into its irreducible components in the symmetry group C_{4v}

Table 2.5 Representations associated with p-type orbitals

C_{4v}	E	$2C_4$	C_2	$2\sigma_v$	$2\sigma_d$		
Γ_4	1	1	1	$\bar{1}$	$\bar{1}$		R_z
Γ_5	1	$\bar{1}$	1	1	$\bar{1}$		$x^2 - y^2$

completely describe the symmetry C_{4v}. A list of the usual tables of characters is given in Appendix 2 [9]. The nomenclature used for the irreducible representations as first proposed by R. S. Mulliken has been employed. Mathematical demonstrations of the previous results can be found in reference books [5, 7].

2.8 GROUP–SUBGROUP RELATIONSHIPS

From objects with no symmetry up to the sphere, stepwise addition of symmetry elements generates *supergroups*. By following the path in the reverse direction, various families may be distinguished each member of which is a subgroup of its parent group: all the symmetry elements of a subgroup are contained in the corresponding group. A fairly detailed scheme is given in Appendix 3 [10].

A distinction can be made between groups derived from K_h, which includes symmetry planes and/or improper rotations and K, which does not. The two families include achiral and chiral objects, respectively (Figure 2.31).

Other important families are those allowing the presence of a dipole moment in the ground state of the molecular unit. The symmetry of a polar vector is

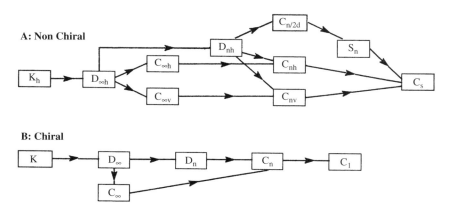

Figure 2.31 The subgroups of K and K_h which include one main rotation axis of order superior to 2

A FEW NOTIONS OF SYMMETRY

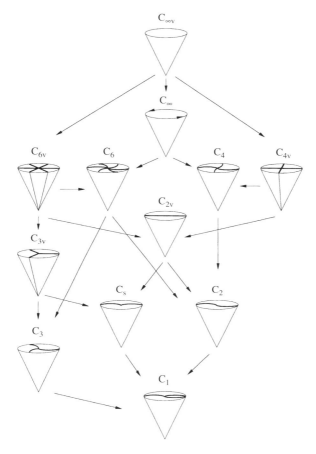

Figure 2.32 Group–subgroup relationships derived from $C_{\infty v}$ (polar groups). Only the crystallographic groups are shown (except infinite groups). Derived from Ref. [11])

$C_{\infty v}$, which is the most symmetrical polar object that can be found. All other polar molecular units must belong to this symmetry or one of its subgroups (Figure 2.32).

2.9 REFERENCES AND NOTES

1. Derived from Edward Burne Jones.
2. Periwinkle, *Vinca minor* L.
3. M. Field and M. Golubitsky, *La symétrie du chaos*, Interéditions, Paris (1993).
4. Diatom: unicellular algae (more than 10 000 are known).
5. D. S. Schonland, *La symétrie moléculaire*, Gauthiers-Villars, Paris (1971).
6. Model of a crystal vase, Emile Gallé (1846–1904).

7. F. A. Cotton, *Chemical Applications of Group Theory*, Wiley-Interscience, New York (1963).
8. K. F. Purcell and J. C. Kotz, *Inorganic Chemistry*, W. B. Saunders, Philadelphia (1997).
9. O. Kahn and M. F. Koenig, *Données fondamentales pour la chimie*, Hermann, Paris (1972).
10. Y. Sirotine and M. Chaskolskaïa, *Fondements de la physique des cristaux*, Ed. Mir, Moscow (1984).
11. G. Burns and A. M. Glazer, *Space Groups for Solid State Scientists*, Academic Press, New York (1978).

3 Supramolecular Engineering: Symmetry Aspects

3.1 Introduction	70
3.2 One-Dimensional Space Groups	71
3.2.1 Monocolor Stripes (One-Sided Bands)	71
3.2.2 Bicolor Stripes (Two-Sided Bands)	78
3.3 Two-Dimensional Space Groups	79
3.3.1 Introduction	79
3.3.2 Relationship between One-Dimensional and Two-Dimensional Space Groups	82
3.3.3 Bicolor Two-Dimensional Space Groups	84
3.3.4 Kitaigorodskii's Approach	84
3.3.5 Tiles and Tilings	90
3.4 Notion of Two-Dimensional Molecular Shape	93
3.4.1 Tiling with Polygons	93
3.4.2 Generalized Two-Dimensional Shape	96
3.4.3 Relationship Molecular Unit/Tile	103
3.4.4 Examples of Mesophases	104
3.4.4.1 15-Crown-5 Substituted Phthalocyanine	104
3.4.4.2 Paraffinic Chain Substituted Triptycene	106
3.4.4.3 Cupric Complex of Amphiphilic Tetraazaundecane	108
3.5 Three-Dimensional Case	111
3.5.1 Introduction	111
3.5.2 Plastic Crystals	111
3.5.3 Three-Dimensional Crystals	119
3.5.3.1 Space Group Occurrence	119
3.5.3.2 Examples	121
3.6 References and Notes	128

> Это правило следует рассматривать как обобщение всех существующих опытных данных и с полным основанием можно называть основным правилом органической кристаллохимии.
> (Reference: А.И.Китайгородский. Органическая кристаллохимия, Москва, 1955, стр. 83).

> *This rule should be considered as a generalization from all existing experimental data; it can with complete justice be taken as the fundamental law of organic chemical crystallography.*
>
> A. I. Kitaigorodskii
> *Organic Crystallography* (1961)

3.1 INTRODUCTION

It is becoming traditional to start any chapter concerning the prediction of crystal structures with the statement of J. Maddox [1]: 'One of the continuing scandals in the physical sciences is that it remains in general impossible to predict the structure of even the simplest crystalline solids from a knowledge of their chemical compositions'.

It is not our ambition to stop this scandal in the next sections. However, the fundamental nature of the difficulties encountered will be outlined. Only symmetry arguments will be developed; the influence of the chemical nature of the molecular unit is treated in a next section. The presentation of the space groups at one, two and three dimensions allows us to introduce the notation and symbolism used in further sections.

A precise definition of the notion of a molecular shape will then be given. It will be demonstrated that, at least in two dimensions, guidelines can be given for some supramolecular engineering principles [2].

A few hints will be given to help the design of molecular units in view of their organization in condensed phases. At first, only periodic structure (^{1}D, ^{2}D or ^{3}D crystals) will be considered. A final section is devoted to mesomorphic materials.

Nature has a dilection for ordered and symmetrical structures. Many minerals and organic molecules can be found in the form of single crystals, sometimes of enormous size (Figure 3.1).

Nicely shaped organic single crystals are difficult to grow compared to mineral ones. There are at least two reasons for this: first, weak van der Waals interactions are dominant in molecular crystals whereas strong covalent or ionic interactions are encountered in the mineral world; second, the molecules are almost always unsymmetrical and this does not favour the formation of highly symmetrical structures.

The growth of single crystals from a solution involves two essential processes: (a) self-recognition of the molecular units (it is well known that

SUPRAMOLECULAR ENGINEERING: SYMMETRY ASPECTS

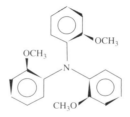

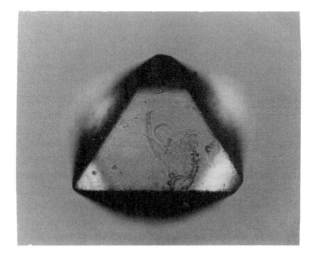

Figure 3.1 Single crystal of 2,2′,2″-trimethoxytriphenylamine. (After Ref. [3])

crystallization is one of the best means of purification), (b) among all structures that are possible — some of them having comparable or sometimes higher stabilities — one of them will grow preferentially. There is therefore both a molecular and a supramolecular selection. These two parameters are very probably interconnected.

In the next sections the notion of one-dimensional, two-dimensional and three-dimensional space groups is recalled for subsequent uses in supramolecular engineering.

3.2 ONE-DIMENSIONAL SPACE GROUPS

3.2.1 Monocolor Stripes (One-Sided Bands)

(Note that in Refs. [2] and [4], the term 'strips' is used.)

The truly one-dimensional case has not been examined since only two space groups are possible possessing binary axes. We will therefore first consider the symmetries associated with one-sided bands.

72 DESIGN OF MOLECULAR MATERIALS

The unit cell constitutes the repeating unit of the structure. In the simplest case no operation other than translation is involved. The cell may contain one or several identical or different molecular units (Figure 3.2). The symmetry elements of the molecular unit cannot be retained in the structure for this space group.

In stripes, only mirror planes, parallel (m_1) or perpendicular (m_2) to the direction of translation and binary axes are compatible with the lattice (Figure 3.3). Symmetry elements between and within the unit cells cannot be dissociated. This is exemplified in Figure 3.4 for a mirror plane. The points 1 and 1' (Figure 3.4) are equivalent because of the postulated mirror planes (m) within the unit cell. However, 1 and 2 and 1' and 2' are also equivalent because of the translation; therefore 1 and 2' are also equivalent and related by a mirror plane (m') relating two unit cells.

An infinite number of molecular units can be found. However, not *all* molecular units can be used. Some have not the necessary mirror plane, while others have additional symmetry elements (diamonds in Figure 3.5).

The *site symmetry* (or *induced symmetry*) is the symmetry around a particular point of the structure. (No symmetry elements involving translation are involved, and glide plane and screw axes cannot be taken into account.) The

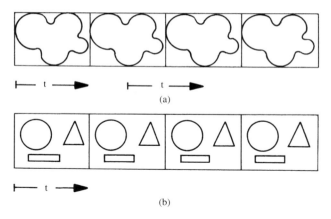

Figure 3.2 One-dimensional structures: (a) with a single molecular unit of arbitrary shape or (b) with several molecules per unit cell. The space group is denoted t

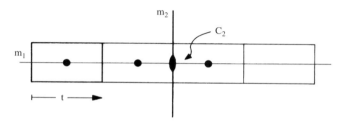

Figure 3.3 Symmetry elements compatible with a one-sided band

SUPRAMOLECULAR ENGINEERING: SYMMETRY ASPECTS

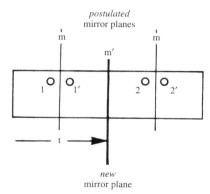

Figure 3.4 The presence of a mirror plane in the unit cell leads to another mirror plane between the unit cells ($t/2$)

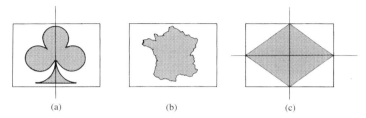

Figure 3.5 Molecular units which are, or which are not, appropriate to form the lattice shown in Figure 3.4: (a) one mirror plane (appropriate); (b) no mirror plane (not appropriate); (c) one mirror plane as expected (m_2), and one additional mirror plane (m_1) (not appropriate)

highest symmetry one can have in one dimension comprises altogether a twofold axis (C_2) and mirror planes (m). They constitute the elements of symmetry of the corresponding crystal class (or the symmetry of orientation): 2 mm (C_{2v}).

Several molecules per unit cell may be considered: the relative molecular arrangements are such that a desirable site symmetry is obtained. The site symmetry of the molecular unit is taken at the center of the point symmetry group of the molecule (Figure 3.6).

The order of the crystal class is given by the number of symmetry operations of the group. In the same way, one can define the order of the site symmetry group. There is a relationship between the number of molecules (Z) per primitive unit cell and the ratio = order of the crystal class/order of the site symmetry group. In the present case, the order of the group of the crystal class is four: E, $1C_2$, 2 m. The site symmetry of the molecular unit is 2 mm, m or 1 depending on the number of molecules per unit cell (1, 2 or 4).

By combining translation and symmetry elements, seven one-sided band space groups are found. These are illustrated in Figure 3.7 [4, 5] by Hungarian Folk needlework [4].

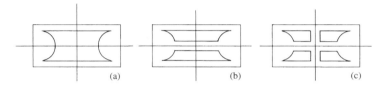

Figure 3.6 Unit cell with a C_2 axis and two mirror planes. One, two or four molecules per unit cell are considered. The shape of the constitutive molecular units is drastically different depending on the number of molecules per unit cell. The respective site symmetries of the molecular units are: (a) 2 mm, (b) m, (c) 1

Only a limited number of translations or pseudotranslations are possible in one-sided bands:

- $\vec{t}$ = simple translation
- $\tilde{t}$ = glide plane perpendicular to the plane of the band including the translation axis
- $(2)_t$ = repetition using C_2 axes perpendicular to the band plane
- $(m2)_t$ = a mirror is applied on the molecular unit (or the molecular unit is symmetrical relatively to this plane) and a C_2 axis perpendicular to the band plane is applied. The resulting figure is then translated. The final arrangement can be alternatively described with a C_2 axis — equivalent to i in this case — and a glide plane (see below, Figure 3.14) [6]:

$$\boxed{\vec{t} \ (only)}$$

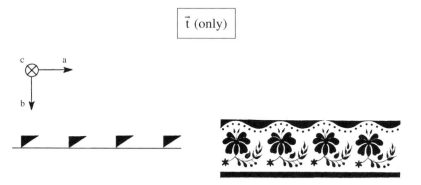

Figure 3.7 Symbol: $\vec{t}$ (111) (for notation see Appendix 1). (Reproduced with permission from [4]; copyright © 1984 Division of Chemical Education, Inc.)

$$\boxed{\vec{t} + m_\parallel \vec{t}/2}$$

A glide reflection plane is present in this group (see Figure 3.8). For one-sided band, one cannot distinguish between a glide plane and a 2_1 screw axis. The

SUPRAMOLECULAR ENGINEERING: SYMMETRY ASPECTS

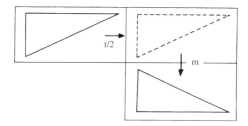

Figure 3.8 A glide plane for a one-sided chain

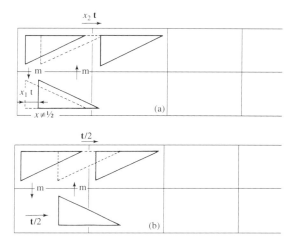

Figure 3.9 Case (a): translations $x_1\vec{t}$ and $x_2\vec{t}$ ($x_1 \neq x_2$); case (b): $x_1 = x_2$

figure comes into coincidence after translation through half of the lattice period and by applying a reflection. In the case where the translation associated with the glide plane is different from $\vec{t}/2$, two different translations $x_1\vec{t}$ and $x_2\vec{t}$ must be considered (Figure 3.9).

Figure 3.10 Symbol: $\tilde{t}$ (1a1). (Reproduced with permission from [4]; copyright © 1984 Division of Chemical Education, Inc.)

$$\boxed{\vec{t} + C_2}$$

The presence of a C_2 axis between the unit cells affords a C_2 axis at $\vec{t}/2$ within the cell. For one-sided bands, the elements $\bar{1}$ and 2 are identical (Figure 3.11).

Figure 3.11 Symbol: $(2)_t$ (112). (Reproduced with permission from [4]; copyright © 1984 Division of Chemical Education, Inc.)

$$\boxed{\vec{t} + m_\perp}$$

The translation is accompanied of tranverse symmetry planes within and between the unit cells (Figure 3.12).

Figure 3.12 Symbol: $(m)_t$ (m11). (Reproduced with permission from [4]; copyright © 1984 Division of Chemical Education, Inc.)

$$\boxed{\vec{t} + m_\parallel}$$

Figure 3.13 Symbol: t · m (1m1). (Reproduced with permission from [4]; copyright © 1984 Division of Chemical Education, Inc.)

$$\boxed{\vec{t} + m_\parallel(\vec{t}/2) + m_\perp + C_2}$$

Figure 3.14 Symbol: $(m2)_t$ (ma2). (Reproduced with permission from [4]; copyright © 1984 Division of Chemical Education, Inc.)

$$\boxed{\vec{t} + m_\parallel + m_\perp + C_2}$$

SUPRAMOLECULAR ENGINEERING: SYMMETRY ASPECTS

All the possible symmetry elements for stripes are found in this space group.

Figure 3.15 Symbol: $(m2)_t \cdot m$ (mm2). (Reproduced with permission from [4]; copyright © 1984 Division of Chemical Education, Inc.)

The *symmetry of orientation* (or *crystal class*) of structures is a widely used notion. Only the directions of the symmetry elements are taken into account independently of their position: the symmetry elements are considered to have a common point. The corresponding point symmetry group is then determined in a conventional way. The glide planes and screw axes are transformed into the corresponding point symmetry elements, which do not contain a translation.

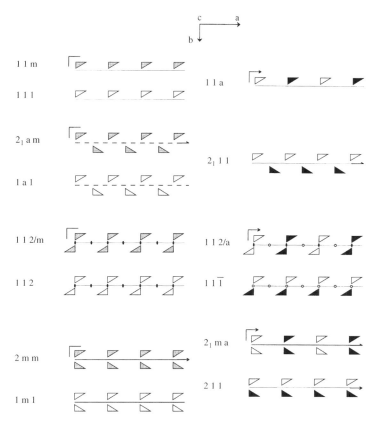

Figure 3.16 The 31 colored stripes space groups (for the notation used see Appendices 1 and 4)

78   *DESIGN OF MOLECULAR MATERIALS*

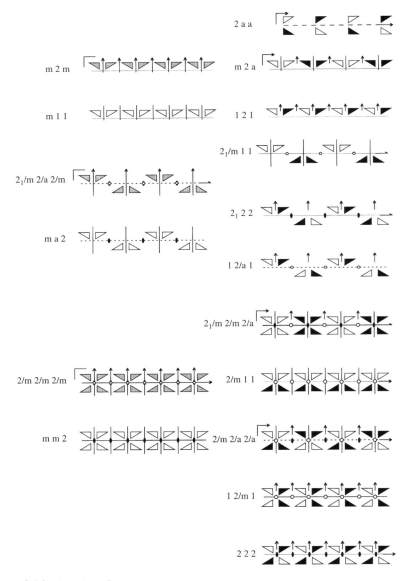

Figure 3.16 (*continued*)

3.2.2 Bicolor Stripes (Two-Sided Bands)

In the previous section, no differenciation has been made between the two sides of the plane that contains the band (one-sided bands). In the general case, three-dimensional molecular units should be considered. This consequently leads to new types of arrangement and the symmetry elements, that were equivalent for one-sided bands are no longer ($C_2 \neq \bar{1}$ or i; screw axis $\neq$ glide plane). In order to take into account their three-dimensional character, the molecular units can be colored in black or in white depending on which side of the molecule is above the plane. This will be examined in a subsequent paragraph. The previous seven one-dimensional space groups will be conserved with all the molecular units possessing the same color (and therefore the same orientation of the molecular units relative to the plane of the band).

In-plane mirror and C_2 axes contained in the stripe plane will turn the molecular units upside down and therefore transform the white figure into a black one. In the case where the two sides of the molecular units are identical and related by a mirror plane contained in the band plane, the molecular units will be colored in grey. For straightforward reasons, there are also seven grey space groups.

The bicolor groups arise from the new possible arrangements which can be found when the band plane can change (arbitrarily from black to white) the two different sides of the molecular unit. The following one-dimensional space groups are thus found [2, 7]:

(a) 7 monocolor groups,
(b) 7 grey groups,
(c) 17 bicolor groups.

An illustration of the corresponding 31 space groups is given in Figure 3.16. The notation of the space groups as well as the main definitions used in crystallography are given in Appendices 1 and 4.

3.3 TWO-DIMENSIONAL SPACE GROUPS

3.3.1 Introduction

In one dimension, only a single translation vector was considered, yielding a unique lattice (system). In two dimensions, depending on the relative orientations of the two translation vectors and their length, four different systems can be distinguished (Figure 3.17).

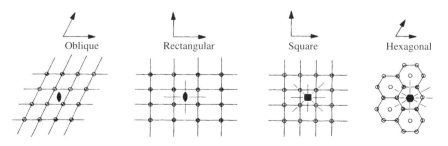

Figure 3.17 Two-dimensional systems and the corresponding lattices

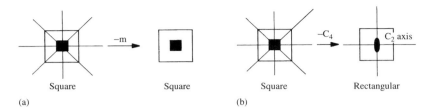

Figure 3.18 (a) The removal of mirror planes does not change the system. (b) The transformation of the C_4 axis into C_2 axis changes the system

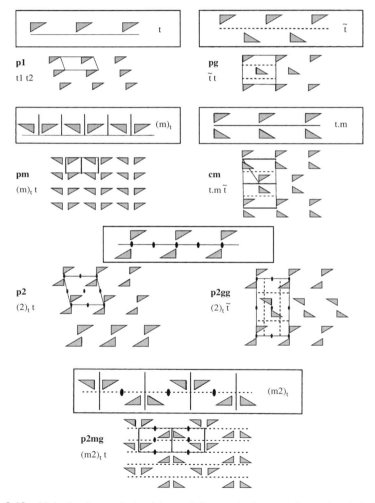

Figure 3.19 Unicolor layers derived (or not) from unicolor one-dimensional chains. The methods of translation are indicated following Kitaigorodskii [6]. ($t \cdot m$ = mirror parallel to the translation; $(m)_t$ = mirror perpendicular to the translation)

SUPRAMOLECULAR ENGINEERING: SYMMETRY ASPECTS

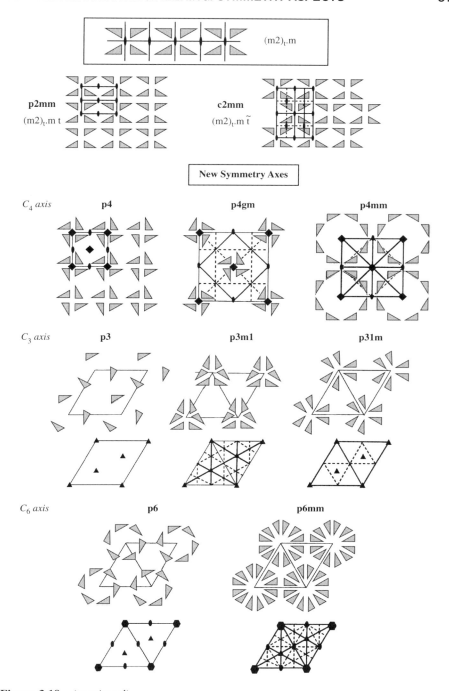

Figure 3.19 (*continued*)

In each system, different crystal classes are found which are subgroups of the lattice symmetry. All subgroups of the lattice symmetry group are possible under the conditions that the removal of the considered symmetry element does not transform a given system to another one (Figure 3.18). A limited number of subgroups may thus be found for each system.

3.3.2 Relationship between One-Dimensional and Two-Dimensional Space Groups

It has been seen that one-sided stripes can be built by using a limited number of translation or pseudotranslation compatible with the one-dimensional character of the arrangement. In the simplest case, these chains may be duplicated by a simple translation to give monocolor layers. In the other cases, the method of translation $\tilde{t}$ is sufficient to generate the layers from the corresponding chains. However, C_3, C_4 and C_6 axes perpendicular to the plane cannot be present in a chain whereas they can be used in layers (Figure 3.19). A list of the various space groups is given in Table 3.1.

The two-dimensional space groups (monocolor) associated with the square lattice will be detailed as an example. A complete description of all space groups is given in Appendix 4.

The common origin of the two translation vectors defining the lattice is chosen arbitrarily. It is possible to define a unit cell in such a way that the elementary molecular unit is entirely or partly inscribed within it. The area of the unit cell is larger than the cross-section of the molecular unit. The ratio of these two numbers gives the compactness of the structure:

$$\text{Compactness} = \frac{\text{area of molecular unit}}{\text{area of unit cell}} \text{(1 molecule per unit cell)}$$

The unit cell may also contain several identical or different molecules. In this case, the total cross-section of the molecules must be considered.

The point symmetry group of the molecule is equal to the group or is a supergroup of the site symmetry. Consequently, it is only through the site symmetry that one can find some relationship between the space group of the lattice and the shape and symmetry characteristics of the constituting molecular units.

In oblique and rectangular space groups only the symmetry elements $2(C_2)$ or $m(\sigma)$ are allowed. The square lattice permits the use of a higher order symmetry axis: $4(C_4)$. In the p4 space group, molecules or groups of molecules must give rise to a C_4 axis. However, the shape of the molecular unit is not directly connected to the space group, especially when the number of molecules per unit cell is varied (see Figure 3.20).

This approach therefore seems difficult to use for supramolecular engineering since differently shaped molecular units can lead to the same space group.

SUPRAMOLECULAR ENGINEERING: SYMMETRY ASPECTS

Table 3.1 Systems, crystal classes and space groups in two dimensions (monocolor) (see also Appendix 4)

System	Crystal classes	Space groups
Oblique	2	p2
	1	p1
Lattice symmetry: 2		
Rectangular	2mm	p2mm, p2gg, p2mg, c2mm
	m	pm, pg, cm
Lattice symmetry: 2mm		
Square	4mm	p4mm, p4gm
	4	p4
Lattice symmetry: 4mm		
Hexagonal	6mm	p6mm
	6	p6
	3 m	p31 m
	3 m	p3 m1
Lattice symmetry: 6mm	3	p3

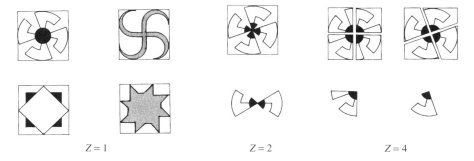

Figure 3.20 Relationship between the molecular unit characteristics and the number of molecules per unit cell

3.3.3 Bicolor Two-Dimensional Space Groups

In two-dimensional space groups, three categories may be distinguished:

(a) The layer plane is not a symmetry element of the space group. One then finds 17 (monocolor) groups, as previously described.
(b) The two sides of the layer are identical; the mirror plane of the layer is a symmetry element of the space group. One finds again 17 (grey) groups with the same characteristics as the previous ones.
(c) The application of three symmetry elements: C_2 (parallel to the plane), the inversion centre and glide plane (in the plane) interconvert black and white sides of a given molecular unit yielding 46 space groups.

In total 80 space groups are found [7]. The process for finding the bicolor space groups is identical to the one previously described for chains. A list of the two-dimensional space groups is given in Table 3.2.

In the hexagonal system, besides the monocolor and grey space groups directly deduced from the previous monocolor hexagonal system, six other arrangements can be found (Figure 3.21).

3.3.4 Kitaigorodskii's Approach

Following Kitaigorodskii [6], a three-dimensional structure may be constructed in three steps:

Table 3.2 Two-dimensional space groups and the corresponding colored groups. The three-dimensional space group symbols are given for the translations **a** and **b** in the layer plane and **c** perpendicular to the layer. (Modified from Ref. [8])

Two dimensions	One color	Grey	Two colors
p1	P1	P11m	P11b
p2	P112	P112/m	$P\bar{1}$, P112/b
pm	Pm11	Pm2m	Pb2b, $Pm2_1b$, P121, Pm2a, Cm2a
pg	Pb11	$Pb2_1m$	$P12_11$, $Pb2_1a$
cm	Cm11	Cm2m	C121, Pb2n, $Pm2_1n$
p2mm	Pmm2	Pmmm	Pmmb, P2/m11, Cmma, Pbmb, P222
p2mg	Pma2	Pmam	Pmab, P12/al, Pbab, $P2_1/m11$, $P2_122$
p2gg	Pba2	Pbam	$P2_1/b11$, $P2_12_12$
c2mm	Cmm2	Cmmm	Pban, C2/m11, Pman, C222, Pmmn
p3	P3	P3/m	
p31m	P31m	$P\bar{6}2m$	P321
p3m1	P3m1	$P\bar{6}m2$	P312
p4	P4	P4/m	P4/n, $P\bar{4}$
p4mm	P4mm	P4/mmm	P4/nbm, P422, $P\bar{4}2m$, $P\bar{4}m2$, P4/nmm
p4gm	P4bm	P4/mbm	$P42_12$, $P\bar{4}2_1m$, $P\bar{4}b2$
p6	P6	P6/m	$P\bar{3}$
p6mm	P6mm	P6/mmm	$P\bar{3}12/m$, $P\bar{3}2/ml$, P622

SUPRAMOLECULAR ENGINEERING: SYMMETRY ASPECTS

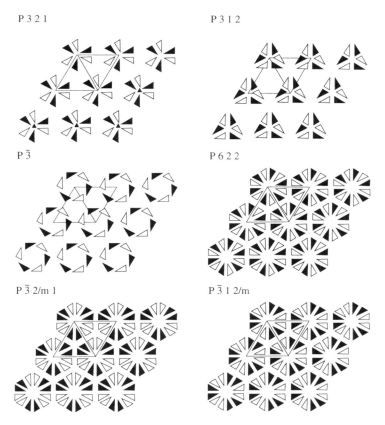

Figure 3.21 The bicolor space groups derived from the hexagonal system. (Modified from Ref. [2])

(a) Molecular units are used to form chain structures. These chains differ from the one-sided or two-sided stripes examined so far in that the molecular units are considered to be truly three-dimensional.

(b) Layers are formed from the previous chains. As already mentioned, new symmetry elements perpendicular to the plane (threefold, fourfold and sixfold axes) are allowed for layers that were not for chains.

(c) The stacking of layers yields crystals (three-dimensional periodic assemblies) and ultimately one finds the 230 different space groups.

This approach has been partially substantiated by electron microscope observations: the growth of thin films of zinc phthalocyanine deposited under vacuum shows that molecular columns form at the early stages, before full crystallization occurs [9]. On the other hand, Monte Carlo calculations have been carried out that seem to confirm the generality of this observation [10].

In 1929, B. P. Orelkin pointed out that the molecules in crystals are stacked in such a way that the bumps of one molecular unit is inserted in the hollows of the neighbours [6].

In 1945, A. I. Kitaigorodskii demonstrated 'quantitatively that the close-packing concept was correct for organic crystals' in which directional interactions (hydrogen bonds, dipole–dipole interactions, etc.) are not the main driving force of the stacking. Further analyses led to the proposal that 'the number of contacts between a molecule and all its neighbours is of prime importance for close-packing.... A coordination number of 12 is assumed to provide adequately close packing' [6].

It has also been stated [6] that it is always possible to localize layers within three-dimensional crystalline structures. A coordination number in the layer of 6 ensures *closely packed layers*. It will be seen further on that this assumption on the coordination number is correct for most structures made of arbitrarily shaped molecular units. It is, however, not a general principle and the compactness criteria depend on the shape of the molecular unit. Moreover, no general definition of an arbitrary-shaped molecular unit can be given.

One can first examine a chain made by a simple translation of arbitrary-shaped molecular units. An arrangement can be found in which the bump of one unit is inserted in the hollow of another one (Figure 3.22). The translation parameter (t) is unknown. However, a bump/hollow fit ensures the highest compactness. The geometrical arrangement thus found leads to the determination of t. The bump/hollow fit is difficult or impossible when the molecules of the chain are related by a symmetry element. This is exemplified in Figure 3.23 for a mirror plane or a C_2 axis.

Any method of translation may be considered for constructing a chain. However, since the simple translation offers a wider choice of relative arrangements of the molecular units, only this case will be discussed in detail in the next paragraphs.

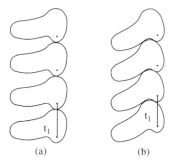

Figure 3.22 Formation of a chain from a translation of an arbitrary-shaped molecular unit. In case (b) the arrangement is found from the bump/hollow requirement. (Modified from Ref. [6])

Figure 3.23 Molecular units related via a mirror plane or a C_2 axis

A layer can be built by using a second simple translation ($t_1 t_2$ layer). The second parameter t_2 is chosen in such a way that two contacts with two different molecular units belonging to the original chain are established (Figure 3.24). This mode of construction ensures a sixfold coordination (six contacts) in the layer: two contacts within a given chain and two others (multiplied by 2 because of the translation), with the molecular units belonging to neighbouring chains.

Layers can be derived from any combination of the methods of translation. In the case of a simple translation combined with a binary axis, sixfold coordination may also be found (Figure 3.25).

Binary axes relate the molecular unit of one chain to the next one. There are therefore two molecular units per unit cell. As previously, if there are two points of contact of a given molecular unit with two different molecular units of another chain, this ensures a sixfold coordination in the layer. However, the use of mirror

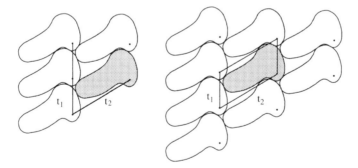

Figure 3.24 Close-packed $t_1 t_2$ (p1) layer with an oblique cell. (After Ref. [6])

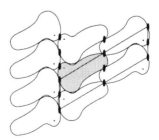

Figure 3.25 Close-packed $(2)_t t$ layer with an oblique cell, p2. (After Ref. [6])

planes (for example the layer $(m)_t t$) prevents sixfold coordination being obtained (Figure 3.26) [6].

There are four systems possible at two dimensions: oblique, rectangular, square and hexagonal. In the case of an arbitrary shape, the two last systems cannot be considered (Figure 3.27). In the oblique and rectangular systems, all the corresponding space groups, except pm and p2mm, can yield sixfold coordination. However, the site symmetries must also be taken into account (Table 3.3).

The presence of a mirror plane common to the molecule and the space group allows a sixfold coordination to be obtained only if the space group contains a glide plane in addition to this mirror plane (Figure 3.28). At least nine types of two-dimensional layers, characterized by their space group and the site symmetry of the molecule, can afford sixfold coordination (Table 3.3).

Kitaigorodskii's classification distinguishes four categories for two-sided layers [6]:

(a) *Coordination close-packed layers*: those in which figures (molecules) of arbitrary shape and symmetry can be packed with sixfold coordination.
(b) *Closest-packed layers*: those in which one can, by selecting the orientation for figures of given shape (and symmetry) and the repeat periods, produce a cell of minimal dimensions.

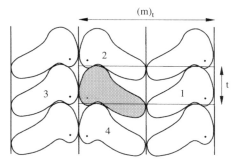

Figure 3.26 An $(m)_t t$ layer, pm, with a fourfold coordination number (the molecules in contact are numbered from 1 to 4). (After Ref. [6])

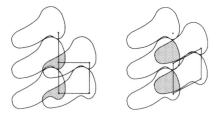

Figure 3.27 Illustration of the impossibility of packing an arbitrary-shaped molecular unit with square and hexagonal cells

SUPRAMOLECULAR ENGINEERING: SYMMETRY ASPECTS

Table 3.3 Two-dimensional space groups and methods of translation which allow sixfold coordination for a molecular unit (Modified from Ref. [6])

Two-dimensional space groups	Methods of translation	Site symmetry			
		1	2	m	2mm
p1	$t_1 t_2$	+	−	−	−
p2	$(2)_t\, t$	+	+	−	−
pm	$(m)_t\, t$		−	−	−
pg	$\tilde{t}\,t$	+	−	−	−
cm	$t \cdot m\tilde{t}$		−	+	−
p2mm	$(m2)_t \cdot mt$	−			
p2mg	$(m2)_t t$			+	−
p2gg	$(2)_t \tilde{t}$	+	+	−	−
c2mm	$(m2)_t \cdot m\,\tilde{t}$				+

−, site symmetry not compatible with the space groups.
+, sixfold coordination is possible.
blank, sixfold coordination is impossible.

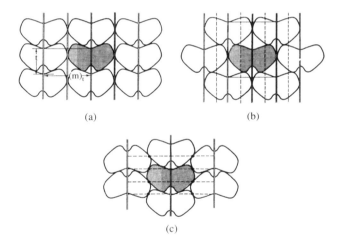

Figure 3.28 Layers formed from a molecular unit possessing a mirror plane (site symmetry m) constructed via the following methods of translation: (a) $(m)_t\, t$ (pm), (b) $t \cdot m\,\tilde{t}$ (cm), (c) $(m2)_t\, t$ (p2mg)

(c) *Limitingly close-packed layers*: (for a given symmetry) are closest-packed ones in which the figure occupies a special position.

(d) *Permissible layers*: those which, while falling in the first class, do not fall in the second or third.

For each two-dimensional space group allowing close-packed layers, one-color, grey and bicolor space groups may be determined (Table 3.4).

Table 3.4 Two-dimensional colored space groups of close-packed layers of an arbitrary shape. The site symmetry is given in parentheses. The superscript indicates closest-packed layers (1), permissible layers (2), limitingly close-packed layers (3)

Two-dimensional space group	One color		Grey		Two colors	
p1	P1	$(1)^2$	P11m	$(m)^3$	P11b	$(1)^1$
p2	P112	$(1)^2$	P112/m	$(m)^3(2/m)^3$	P$\bar{1}$	$(1)^1(\bar{1})^1$
					P112/b	$(1)^2(\bar{1})^2(2)^3$
pg	Pb11	$(1)^2$	Pb2_1m	$(m)^3$	P$12_1$1	$(1)^1$
cm					C121	$(2)^2$
					Pm2_1n	$(m)^2$
p2mg					P12/a1	$(2)^2$
					P2_1/m11	$(m)^2$
p2gg	Pba2	$(1)^2$	Pbam	$(2/m)^3$	P2_1/b11	$(1)^2(\bar{1})^1$
					P2_12_12	$(1)^2(2)^3$
c2mm	Cmm2	$(mm2)^2$	Cmmm	$(mmm)^3$	Pban	$(222)^3$
					C2/m11	$(2/m)^2$
					Pmmn	$(mm2)^3$

3.3.5 Tiles and Tilings

Polygons depending (a) on the shape of the molecular unit chosen and (b) on the type of packing may be found. They are defined by the points that are equidistant from three molecular units (Figure 3.29). In the case where there is one molecule per unit cell, the polygon described by the previous points tiles the plane. This is exemplified for the method of translation t_1t_2 shown in Figure 3.29. When there

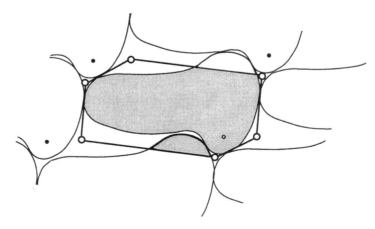

Figure 3.29 Polygons tiling the plane formed from a given arbitrary shape within a layer of type t_1t_2 (p1) and illustration of the Yin–Yang principle (see the text) [11]

SUPRAMOLECULAR ENGINEERING: SYMMETRY ASPECTS

are several molecules per unit cell, the associated polygon is defined taking into account all the molecules contained in the unit cell (Figure 3.30).

The molecular unit is not — in the general case — inscribed within the polygon. However, for all humps outside the polygonal tile there must correspond a hollow (see Figure 3.29). The outer part of the molecular unit must be associated to empty space in the polygonal tile: this can be called the *Yin–Yang principle* [11].

An approach has been developed to define the notion of *generalized shape* using principles found in the theory of tilings [2, 12]. Tilings are made of flat units which can cover a plane without gap or overlap. When a tiling is periodic its symmetry properties are described by one of the 17 two-dimensional space groups. We will consider a special type of periodic tiling named *isohedral tiling* [2]. In this case the tiling is built from one elementary tile and all the other tiles are related by a symmetry operation of the corresponding two-dimensional space group. Topological arguments permit 11 different elementary tiles to be defined (Laves tiles) (Figure 3.31). Each tile is defined by a number of *singular points*, which are the vertices common to at least three different tiles.

A topological class is named by the abbreviation X^y where X is the number of tiles sharing one singular point and y is the number of singular points of this type

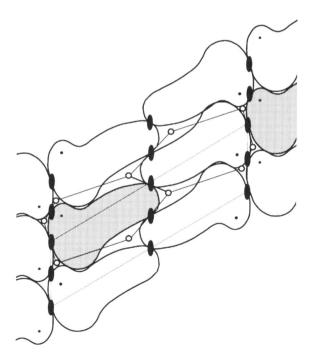

Figure 3.30 Method of translation $(2)_t t$ (p2) leading to two molecules per unit cell: definition of the corresponding polygon tiling the plane

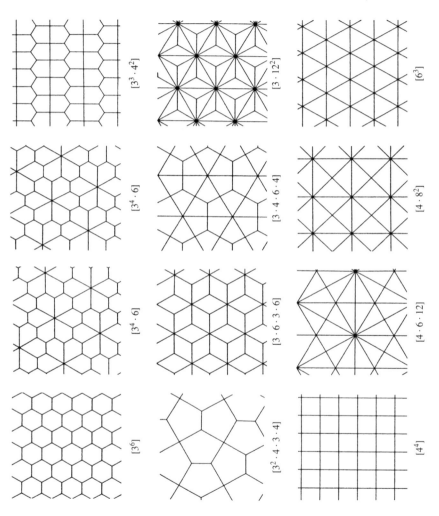

Figure 3.31 The 11 different topological classes (Laves tile). The tiling $[3^4 \cdot 6]$ occurs in two enantiomorphic forms. For the notation, see the text. (From: *Tilings and Patterns* by Grunbaum and Shephard © 1987 by W. H. Freeman and Company. Used with permission [2])

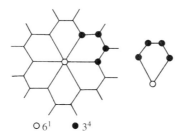

Figure 3.32 An example of a Laves tile. The class $[6^1 \cdot 3^4]$ has two kinds of singular points: ○ 6^1, 6 = number of tiles sharing the singular point, 1 = number of singular points of this type in a tile; ● 3^4, four singular points with three adjacent tiles. (Modified from Ref. [2])

in a tile. These classes have a topological significance since they are unchanged by any distortion applied to the plane of the tiling. An example of a Laves tile is given in Figure 3.32.

Among all the topological classes, only the $[3^6]$ tile demonstrates a sixfold coordination number (Figure 3.31) by assuming that coordination is ensured by a common edge. This topological class will therefore be more thoroughly detailed in the following sections. The other topological classes must, however, be found. This approach is a starting point of the notion of *generalized shape*.

3.4 NOTION OF TWO-DIMENSIONAL MOLECULAR SHAPE

3.4.1 Tiling with Polygons

A first approach of the notion of *generalized molecular shape* can be obtained by considering the various polygons that can cover the plane without gaps or overlaps [2, 12]. The ancient Greeks already knew that only three regular polygons can tile the plane: the equilateral triangle, the square and the hexagon. The hexagons yield a unique way of packing whereas triangles and squares lead, on the contrary, to several structures (Figure 3.33).

The three regular polygons tile the plane and the corresponding compactness is consequently 100%. At that stage, a molecular unit must be associated with one of them. The difference in area between the molecule and the tile gives the empty spaces within the structure. The compactness is simply given by the ratio of the cross-section of the molecule over the area of the tile.

In fact, any triangle and any quadrilateral can tile the plane [13]. This can be shown for triangles: it is merely necessary to fit together two identical triangles in such a way that a parallelogram is generated. A stripe can then be made by putting the previous parallelograms side by side. Layers are finally formed from the stripes (Figure 3.34).

94 DESIGN OF MOLECULAR MATERIALS

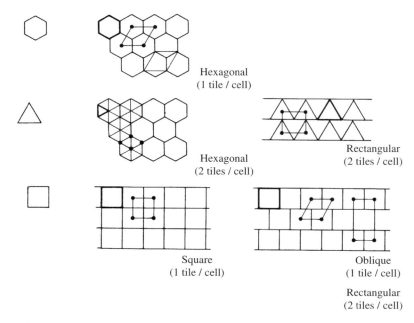

Figure 3.33 A few structures derived from hexagons, equilateral triangles and squares. For the square, an oblique lattice is obtained in the general case. If the shift is equal to t/2, a rectangular lattice can be found (see Figure 3.37)

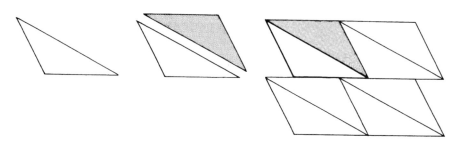

Figure 3.34 Any triangle can tile the plane (see Ref. [13])

The same process may be repeated with quadrilaterals: two of them are associated via a binary axis to obtain a hexagon in which each edge is equal and parallel to the opposite edge. Simple translations then yield a two-dimensional structure. Concave quadrilaterals may also be used (Figure 3.35).

The preceding process cannot be made with the mirror image of the quadrilateral, (shown in Figure 3.36).

The case of a hexagon is more complicated [13]. It has been shown that a convex hexagon that tiles the plane must belong to one of these three classes:

SUPRAMOLECULAR ENGINEERING: SYMMETRY ASPECTS

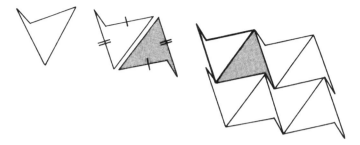

Figure 3.35 Any concave or convex quadrilateral can tile the plane

Figure 3.36 The two mirror images of a given quadrilateral cannot tile the plane

(a) $A + B + C = 360°$
   $a = d$
(b) $A + B + D = 360°$
   $a = d, c = e$
(c) $A = C = E = 120°$
   $a = b, c = d, e = f$

On tessellating the plane with convex pentagons, eight different classes are possible [13].

In summary, any triangle, any quadrilateral, three different classes of hexagons and eight different classes of pentagons may be used to tile the plane. No convex polygon of more than six sides can be used and the series is limited to the polygons previously mentioned.

The two-dimensional layers used by Kitaigorodskii must be *dense* and the coordination number of the molecular unit cannot be less than six. The packing of hexagons permits this coordination number to be reached.

Quadrilateral and pentagons lead, most often, to the coordination numbers 4 or 5, respectively. Exceptions are, however, possible, as in the case of the oblique system of squares.

The previous approach presents several severe limitations. First, the same tile may yield different structures with the same compactness (Figure 3.37). Second, the flexibility of the tile is not sufficient to be compared with actual cross-sections of molecular compounds. Third, the number of molecules (or tiles) per unit cell is not predictible from the characteristics of the molecular unit.

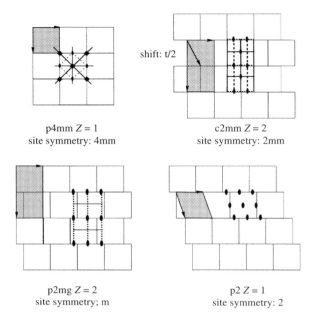

Figure 3.37 Various possibilities of tiling with the same square tile. The space group, the site symmetry and the number of tiles per unit cell are indicated

In consequence, another method must be followed. However, the polygon approach is an easy way to introduce the topological approach, which is the object of the next section.

3.4.2 Generalized Two-Dimensional Shape

The polygon approach largely suffers from the fact that the elementary tiles cannot follow the convolutions of the molecular cross-section. A topological approach allows this problem to be solved.

Topological arguments [2] allow a finite number of *elementary tiles* to be defined which cover the plane without gap or overlap. The elementary tiles can be deduced from the 11 Laves classes shown in Figure 3.31. Contrarily to the previous case, one is not limited to the use of polygons: defined deformations of the tile are, this time, allowed. A difference must be made between monohedral and isohedral tilings. In the first case, the tiles are of the same size and shape but are not compulsorily equivalent by symmetry, whereas the isohedral tilings are. In further discussions only the isohedral case is considered.

The *shape* of a molecule is a concept widely used to interpret phenomena in many fields: biology (molecular recognition), physicochemistry (cocrystallizations, formation of molecular assemblies), chemistry (enzyme-like catalysis), etc. Surprisingly, a precise definition of the shape is never given. In a more or less

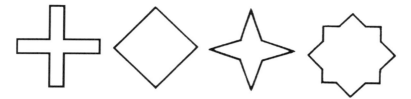

Figure 3.38 Various 'shapes' derived from the D_{4h} (4/mmm) point symmetry group (notation in three dimensions)

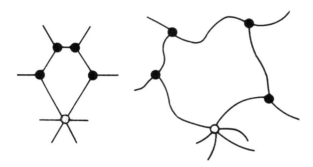

Figure 3.39 A Laves tile can be distorted without changing the topological class

diffuse way, the following notions are mixed: symmetry, molecular volume, bulkiness, etc. It is easy to see that a same point symmetry group can lead to very different shapes. This is exemplified in Figure 3.38. More or less consciously, the notion of shape is rarely considered for a single molecular unit but rather for assemblies of molecules or associations.

In a given topological class (Laves tiling) the elementary tile can be distorted without changing the class (Figure 3.39). Another constraint is added if we consider only periodic arrangements of tiles. In this case, one of the 17 monocolor two-dimensional space groups must be taken into account. The tilings may be classified by their topological class, their space group and the site symmetry of the tile. The tile can be distorted but the site symmetry and the space group must be preserved. This is exemplified in Figure 3.40.

In the example shown in Figure 3.40, the tile can be distorted under the conditions that:

(a) The singular points must be arranged following a $2(C_2)$ symmetry.
(b) The edge distortions of opposite sides (noted a, b and c) are related via the C_2 axis.

Moreover, the space group p2 leads to C_2 symmetry axes at the middle of all sides of the tile. The edge distortions must therefore respect this symmetry requirement (Figure 3.41).

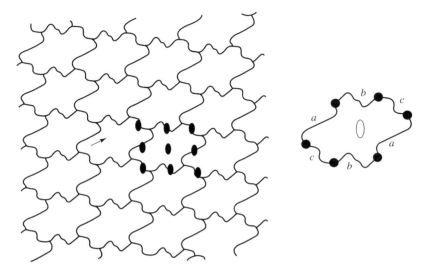

Figure 3.40 Tiling and corresponding tile. Topological class, 3^6; space group, p2; site symmetry, $2(C_2)$ (From: *Tilings and Patterns* by Grunbaum and Shephard © 1987 by W. H. Freeman and Company, Used with permission [2])

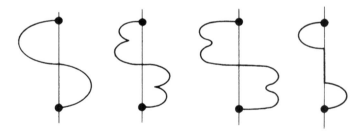

Figure 3.41 Edge distortions in the space group p2 must have a $2(C_2)$ symmetry (S-curve distortion)

Other types of edge distortions are possible depending on the space groups and the related symmetry elements. In the case where the side is parallel to a mirror plane, no distortion is possible and the side is represented by a straight line. When the edge is divided by a perpendicular mirror plane, a C-curve type is obtained (Figure 3.42).

However, another limitation must be considered concerning the edge distortions: any concave edge deformation must correspond to a similar convex distortion in order to avoid gaps or overlaps between the tiles. It is another example of the Yin–Yang principle. The correlated deformations may be for adjacent or opposite sides depending on the type of tiling considered. A line connecting the two edges is generally drawn to indicate this correlation. A +

SUPRAMOLECULAR ENGINEERING: SYMMETRY ASPECTS

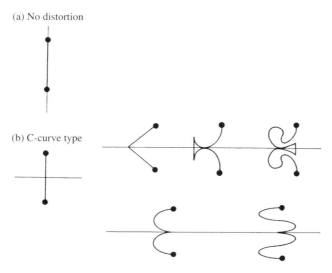

Figure 3.42 Two types of edge distortion: (a) mirror plane parallel to the edge, (b) mirror plane perpendicular to the edge (C-curve type)

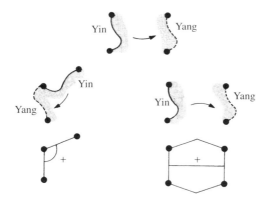

Figure 3.43 Yin–Yang deformations (+ type) for adjacent or opposite edges

sign is added when the two deformations correspond by a mere translation or rotation (Figure 3.43). A − sign is used to indicate a reflection after the rotation or translation (Figure 3.44). The relationships shown in Figures 3.43 and 3.44 indicate two deformations among others that are possible (for a complete study see Ref. [2]).

By combining the 17 two-dimensional space groups, the 11 topological classes, the arrangement of the singular points following the site symmetry and the possibilities of edge distortions made possible by the symmetry and the Yin–Yang principle, 93 different elementary tiles can be defined: 20 in the [3^6] topology,

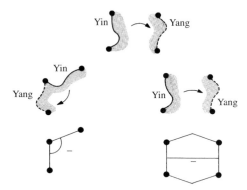

Figure 3.44 Yin–Yang deformations (− type): the rotation or translation is followed by the application of a mirror plane

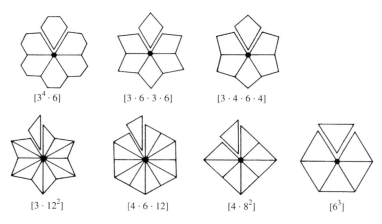

Figure 3.45 The topological classes in which a singular point is shared by more than four tiles

1 in the $[3^4 \cdot 6]$, 5 in the $[3^3 \cdot 4^2]$, 3 in the $[3^2 \cdot 4 \cdot 3 \cdot 4]$, 3 in the $[3 \cdot 4 \cdot 6 \cdot 4]$, 5 in the $[3 \cdot 6 \cdot 3 \cdot 6]$, 3 in the $[3 \cdot 12^2]$, 36 in the $[4^4]$, 1 in the $[4 \cdot 6 \cdot 12]$, 5 in the $[4 \cdot 8^2]$ and 11 in the $[6^3]$ (see Appendix 5). The coordination number remains unchanged within a given topological class.

One can associate a molecular unit to an elementary tile. It is then considered that the elementary tile is the *generalized shape* of the corresponding molecular unit. The generalized shape can be distorted via 'allowed' deformations to fit as well as possible the molecular unit and as to increase the compactness in relation.

Seven topological classes are highly improbable: those in which more than four tiles have a singular point in common (Figure 3.45). In these cases severe structural limitations are imposed on the molecular unit.

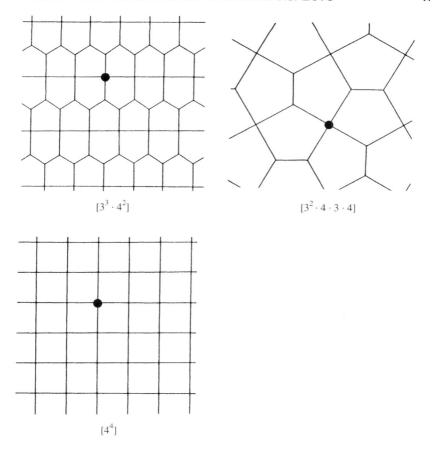

Figure 3.46 The topological classes in which a singular point is shared by four tiles (From: *Tilings and Patterns* by Grunbaum and Shephard © 1987 by W. H. Freeman and Company. Used with permission [2])

Only four topological classes are left: $[3^6]$, $[3^34^2]$, $[3^2 \cdot 4 \cdot 3 \cdot 4]$ and $[4^4]$. Among them three tilings show an arrangement in which a singular point is shared by four tiles (Figure 3.46).

The requirements on the molecular unit needed to obtain a good adequacy with the elementary tile are less stringent than in the previous cases, but these types of tilings are still improbable for molecular compounds. The topological class $[3^6]$ leads to two decisive advantages in this respect: (a) the coordination number of the elementary tile is equal to 6, as required from the extrapolated Kitaigorodskii's approach and (b) the limitations imposed by the presence of singular points shared by more than three tiles is avoided. A square, considered as a tile, can be arranged in a $[4^4]$ tiling but also in the topological class $[3^6]$ by varying the relative positions of the elementary tiles (Figure 3.47). In conse-

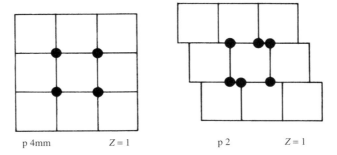

Figure 3.47 A square can afford a [4^4] or [3^6] topological class depending on the relative arrangement

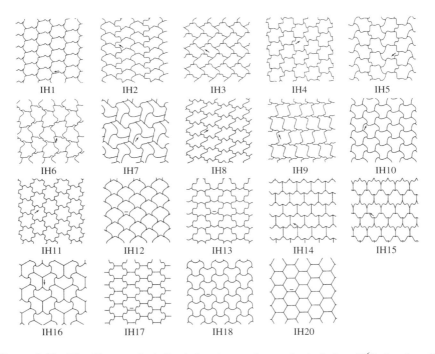

Figure 3.48 The 19 unmarked tiles belonging to the topological class [3^6]. Another tile (marked) is possible (see Appendixes 5 and 6) (From: *Tilings and Patterns* by Grunbaum and Shephard © 1987 by W. H. Freeman and Company. Used with permission [2])

quence, the [3^6] topological class will mainly be further examined in the next paragraphs (Figure 3.48).

It is worth mentioning that the number of tiles per unit cell is not constant in a topological class. It varies from 1 to 4 in the class [3^6]. The elementary tile, considered to be at the origin of a family of shapes, contains information both on the tile characteristics and on the way they are arranged in two dimensions. In Appendix 6, the space group, the site symmetry, the distortions allowed, the crystal class and the symmetry elements of the unit cell for the 20 different tiles of the topological class [3^6] have been given. It has been stated [6] that the structures of highest site symmetry are preferred at constant compactness. However, the use of a highly symmetrical generalized shape imposes severe geometrical constraints and compatibility between the tile and the molecular unit is more difficult to achieve. The two-dimensional structure will therefore be given by the balance between two factors: (a) the best fit in area between the elementary tile (generalized shape) and the actual molecular unit and (b) the highest site symmetry possible.

The molecular unit/elementary tile association does not depend on the number of tiles per unit cell. In all cases, it will be considered that a single molecule (or its cross-section) is associated with a single elementary tile. The degree of adequacy can be estimated by comparing the derived compactness from the compactness of closely packed circles (compactness $= \pi R^2/2\sqrt{3}R^2$ where $R =$ radius of the circle), which is 90.7%.

When inscribing a molecular unit within an elementary tile, the site symmetry must be respected. In other words, the molecular unit symmetry point group is a *supergroup* or is the *same group* as the site symmetry. Since all possible distortions are already taken into account in a given elementary tile, the two-dimensional molecular unit must be entirely inscribed within it. All molecular units that give a good compactness with an elementary tile will be considered to be members of the same family. In that respect, the elementary tile can be considered as a generalized shape.

Kitaigorodskii's [6] approach which uses the criterion of sixfold contacts to ensure close packing led to the conclusions shown in Table 3.3. The correspondence with the elementary tiles of the topological class [3^6] is shown in Table 3.5.

It can be seen that every time sixfold coordination (contacts) criteria are fulfilled, one (or several) [3^6] tile with proper site symmetry is allowed. The space groups pm (t (m)$_t$) and p2mm ((m)$_t$(m)$_t$) which cannot fulfil Kitaigorodskii's criterion of close packing do not allow isohedral tiling. There is therefore a perfect relationship between the two approaches.

3.4.3 Relationship Molecular Unit/Tile

Molecules are three-dimensional objects whereas the previous formalism has been given for two-dimensional tilings. In subsequent considerations a projection of the molecular unit (or cross-section see page 117) on an arbitrary plane will be taken unless the structure affords special planes, as in layered condensed phases.

Only the symmetry of the projected molecular unit can be used *a priori* to postulate a given arrangement. As seen in the previous chapters, the only strict requirement is that the site symmetry is equal to or is a subgroup of the symmetry

Table 3.5 Possibility of having a [3^6] isohedral tiling with specified site symmetries and two-dimensional space groups: —, site symmetry not allowed; blank, no [3^6] isohedral tiling possible

Two-dimensional space group	Site symmetry			
	1	2	m	2mm
p1	IH 1	—	—	—
p2	IH 4	IH 8	—	—
pm				—
pg	IH 2			
	IH 3	—	—	—
cm			IH 12	
			IH 14	—
p2mm		—		
p2mg			IH 13	
			IH 15	—
p2gg	IH 5	IH 9		
	IH 6	—	—	
c2mm				IH 17

of the projection of the molecular unit. A first hypothesis can therefore be that the site symmetry is equal to the projection of the molecular symmetry: this defines a certain number of isohedral tilings. However, this does not ensure that the compactness is optimized. This requirement necessitates site symmetry of lower order to be considered also. The topological class [3^6] is often encountered although the other classes are not impossible for special shapes of molecular units. The site symmetry is lowered until a *satisfactory* compactness is obtained. For a same compactness, the higher site symmetry will be favored [6].

No other conclusion can be given at that stage without additional experimental evidence. X-ray diffraction can yield the three-dimensional structure in the case of single crystals and, in many cases, the lattice parameters (and unit cell) for polycrystalline materials. These two cases will be treated in the following sections. With liquid crystalline phases, the definition of two-dimensional lattices is often possible and examples will be taken from this category. The knowledge of the two-dimensional unit cell permits, in general, together with the site symmetry, an easy selection of a few isohedral tilings.

3.4.4 Examples of Mesophases

3.4.4.1 15-Crown-5 Substituted Phthalocyanine

The 15-crown-5 substituted phthalocyanine molecule, (15-crown-5)$_4$PcM, will be used as a first example. Metastable mesophases have been characterized by X-ray diffraction, indicating a tetragonal structure with $a = b = 20.8$ Å [14]. The molecular symmetry is D_{4h} and the site symmetry will be considered to be 4(C_4), a subgroup of D_{4h}. Only one tile, IH62 (see Appendix 7) preserves this site symmetry. It is constituted of four singular points forming a square. The first problem to be faced is to find the correct position of the singular points relative to the molecular unit (Figure 3.49).

The position of the singular points that yield the higher molecular unit/tile area ratio is not necessarily the one that will give the highest compactness after the allowed edge distortions have been taken into account. The edge distortions follow the Yin–Yang principle as previously described (Figure 3.50). It can be seen that the edge deformation is an S-curve type with a C_2 axis at the middle of the side.

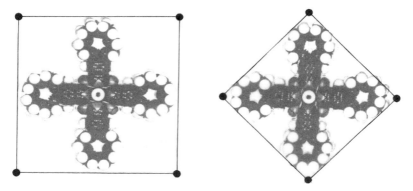

Figure 3.49 The fitting of the $(15\text{-crown-}5)_4$Pc molecular unit with the elementary tile IH62 (undistorted)

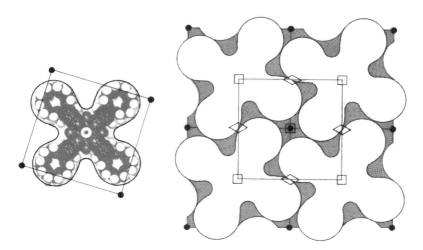

Figure 3.50 Application of the Yin–Yang principle to fit the tile IH62 to the molecular unit $(15\text{-crown-}5)_4$PcM. The lattice and the symmetry elements of the tiling have been figured

This example could have been treated in the topological class [3⁶] but without preserving the C_4 site symmetry.

3.4.4.2 Paraffinic Chain Substituted Triptycene

Triptycene was synthesized in 1942 by P. D. Bartlett [15]. This molecule has a D_{3h} symmetry which is rarely encountered in organic chemistry (Figure 3.51). Triptycene can be used as the rigid core of mesogens by substituting with paraffinic side chains susceptible to inducing the formation of smectic liquid crystals by segregation of the rigid and flexible moieties (Figure 3.52).

A penta-substituted triptycene derivative has been synthesized via a nine-step procedure starting from diaminoanthrarufin [16–19]. Differential scanning calorimetry and X-ray diffraction experiments indicate that four different

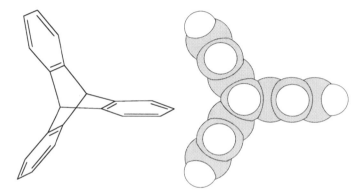

Figure 3.51 Chemical formula of triptycene and corresponding van der Waals representation (view along the C_3 axis). (After Ref. [16])

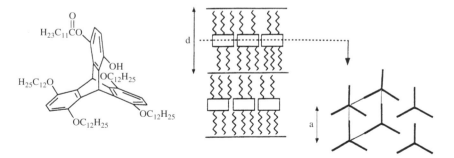

Figure 3.52 Substituted triptycene derivative and the type of smectic liquid crystals it can form by segregation of the rigid and flexible moieties, where d = interlamellar spacing and a = lattice parameter within the layer and corresponding unit cell (space group p31m). (After Ref. [16])

SUPRAMOLECULAR ENGINEERING: SYMMETRY ASPECTS

mesophases with lamellar order are formed [18]. The transition temperatures and the interlayer spacing are indicated below:

$$M_4 \xrightarrow{153\,°C} M_3 \xrightarrow{156\,°C} M_2 \xrightarrow{163\,°C} M_1 \xrightarrow{169\,°C} I$$
$$d = 30\,\text{Å} \qquad 30.8\,\text{Å} \qquad 31.5\,\text{Å} \qquad 35.3\,\text{Å}$$

The observation of the mesophase M_1 by optical microscopy with crossed polarizers indicates a typical conic focal texture (Figure 3.53). In the phase M_1, only rays indicating a lamellar order can be observed by X-ray diffraction; a halo around 4.5 Å shows a quasi-molten state of the paraffinic chains. The agreement between theoretical and experimental results on the interlamellar spacing is very satisfactory (see Chapter 5).

By cooling down the sample from the isotropic phase, one observes below 159 °C (beside the previous smectic order) a hexagonal organization within the molecular layer. Three narrow rays in a distance ratio $1:1/\sqrt{3}:1/\sqrt{4}$ are seen, which correspond to an hexagonal parameter $a = 8.04$ Å. The closest packing of triptycene cores as deduced from van der Waals models gives approximately the same value (8 Å).

In the topological class [3^6] only two tiles, IH16 and IH18, correspond to the space group p31m, which is the most probable to be derived from X-ray data (see Figure 3.52). However, only IH18 conserves a site symmetry C_{3v} (3 mm) [16–19] (Figure 3.54).

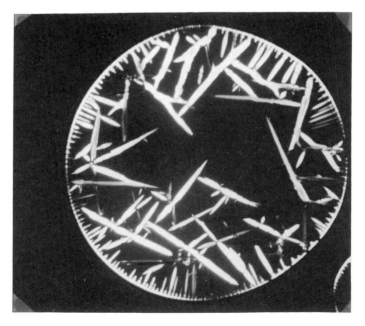

Figure 3.53 Texture of the substituted triptycene derivative shown in Figure 3.52 ($T = 166.7\,°C$)

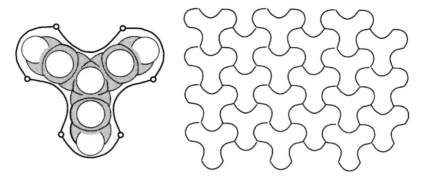

Figure 3.54 Isohedral tiling IH18 and association of the corresponding tile with the triptycene subunit

The packing of the triptycene cores leads to an area of 55.4 Å^2. For a pentasubstituted derivative, this allows, on average, an area of 22.2 Å^2 per paraffinic chain. This value is sufficiently large to allow a quasi-molten state of the chains. On the contrary, the hexa-substituted derivative leads to 18.5 Å^2 per paraffinic chain, which is a value found in solid crystals. In consequence, this last compound preferentially forms highly ordered mesophases or even crystals [16–19].

It has been found by X-ray diffraction that the space group for the triptycene subunits is p31m, which is an subgroup of p6mm usually found for an hexagonal stacking of paraffinic chains. There is therefore a perfect compatibility in symmetry between the rigid core and the flexible chain domains. However, steric requirements of the two subsystems must also be compatible. This is indeed the case for substituted triptycenes (Figure 3.55).

In order to have a geometrical fit between the chains and the triptycene cores, one must have

$$a_c = a'_c = \frac{a_T}{\sqrt{3}} \qquad (3.1)$$

X-ray structures of various triptycene structures indicate that the intramolecular parameter a_c is of the order of 4.5 Å [16]. X-ray diffraction studies carried out on the liquid crystalline phases of the substituted triptycene show a diffuse halo around 4.6 Å typical of intermolecular quasi-molten paraffin chain distances ($a'_c = 4.6$ Å). As a consequence, $a_c \sim a'_c$. Moreover, it has been previously seen that $a_T = 8.04$ Å and then $a_T/\sqrt{3} = 4.64$ Å. Equation (3.1) is therefore almost perfectly fulfilled.

3.4.4.3 Cupric Complex of Amphiphilic Tetraazaundecane

The second example will be furnished by the cupric complex of a 1,4,8,11-tetraazaundecane derivative substituted with polyethyleneoxide and

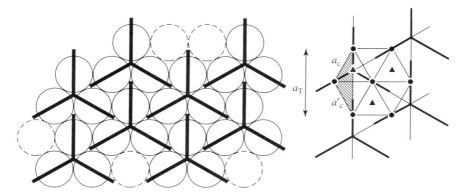

Figure 3.55 Schematic representation of the packing of the triptycene rigid core and the paraffinic side chains and geometrical parameters of the two types of packing, with a_T = hexagonal unit cell parameter of the triptycene core, a_c = intramolecular paraffinic distance and a'_c = intermolecular paraffinic distance

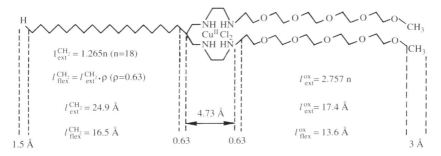

Figure 3.56 Chemical formula of the substituted cupric complex of 1,4,8,11-tetraaza undecane [20], where n = number of methylene units, l_{ext} = extended chain length, l_{flex} = length in a quasi-molten state (see Section 8.2), (CH_2 = paraffinic chain and OX = polyoxyethylene chain). (After Ref. [20])

paraffinic chains [20] (Figure 3.56). X-ray diffraction studies on a lyotropic phase of partially orientated copper complex (3% H_2O) permitted two sets of peaks to be assigned, the first indicating a lamellar order with an interlayer periodicity of 59 Å. The other four rays correspond to a periodicity axis perpendicular to the one associated with the interlamellar order. They may be assigned to an oblique two-dimensional unit cell with $a = 8.3$ Å, $b = 6.8$ Å, $\beta = 75°$ (with the two-dimensional space group p1 or p2) [20] (Figure 3.57).

Knowing the unit cell, fitting to an elementary tile may be attempted by postulating that the higher site symmetry possible (2 or C_2) is retained (with the space group p2). The isohedral tiles IH8 (class [3^6]) and IH57 (class [4^4]) both lead to the two-dimensional space group p2 with a site symmetry 2. For reasons of simplicity, it has been chosen to treat the problem with only four

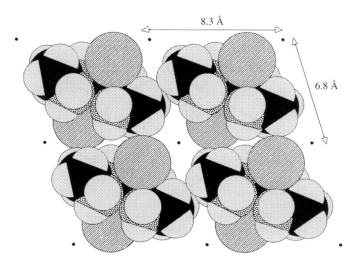

Figure 3.57 The substituted 2,3,2-tet, $CuCl_2$ subunits arranged in a two-dimensional oblique lattice ($a = 8.3$ Å, $b = 6.8$ Å, $\beta = 75°$). The atomic van der Waals radii have been shown. View of the intralamellar order

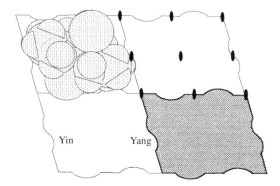

Figure 3.58 Molecular models of the substituted 2,3,2-tet, $CuCl_2$ subunit and the corresponding generalized shape (IH57 with the topological class $[4^4]$)

singular points (class $[4^4]$). The distorted IH57 tile which better corresponds to the present molecular unit is shown in Figure 3.58.

The four edges are distorted following an S-curve type deformation (IH57) with a C_2 axis at the edge midpoint. Opposite sides are obtained via a simple translation. The area of the monoclinic cell is 54.5 Å^2, compared with the molecular unit surface of 47.4 Å^2. This affords a two-dimensional compactness of about 87%. Since polyethyleneoxide and paraffinic chains are linked on the upper and lower sides of the metallo-organic subunit, these side chains must occupy the same area, i.e. approximately 27.7 Å^2 per lipophilic or hydrophilic chains (see Figure 1.43).

3.5 THREE-DIMENSIONAL CASE

3.5.1 Introduction

The complete description of all the polyhedra susceptible to fill periodically three-dimensional space has not yet been given. In 1885, E. S. Fedorov discovered that only five convex polyhedra can fill the space via simple translations [21] (Figure 3.59).

The limitation to simple translations implies that the opposite sides of the polyhedra must be parallel. The analogy with the two-dimensional case leads to a very partial view of the overall number of solutions possible. The Kitaigorodskii's approach may, however, be used: the knowledge of the structure may be employed to determine the corresponding layers and chains constituting this structure. In the following sections two types of three-dimensional periodic arrangements will be studied: plastic crystals and three-dimensional crystals.

3.5.2 Plastic Crystals

High symmetry molecules have been demonstrated to form plastic crystals. These materials are three-dimensional ordered phases in which there is a more or less rapid reorientation about their geometrical centre. $C(SMe)_4$, SiF_4, Cl_3C-CCl_3, $Me_3Si-SiMe_3$ or some derivatives of terpenes may form, for example, plastic crystalline phases. In all cases, the maximum diameter of the molecular unit is longer than the average size of the effective sphere in which the rotational reorientation occurs; the rotation is thus not free [22]. A few examples of these mesophases will be given below (see also Chapter 1).

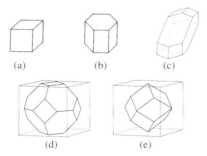

Figure 3.59 The convex polyhedra that can fill the space via simple translations: (a) cube, (b) hexagonal prism, (c) Fedorov's dodecahedron, (d) truncated octahedron, (e) rhombic dodecahedron

A criterion has been established by M. Postel and J. G. Riess [23] to determine whether a molecular unit is globular enough to yield plastic crystalline phases. By rotation, any molecular unit may afford a sphere of diameter $D_{\max}$. The globularity criterion is then defined by the ratio

$$R = \frac{d_m}{D_{\max}}$$

where d_m is the minimum distance between two molecular units in the crystal. It has been experimentally shown that a plastic crystal is obtained when $R \geq 0.81$ [23]. A few examples of globular or quasi-globular molecules yielding or not yielding plastic crystals are shown in Figure 3.60.

A few compounds have been listed in Table 3.6. All the molecular units with $R \geq 0.81$ do indeed show a platic crystalline phase and the others lead to more ordered solids.

However, the Postel and Riess criterion does not allow the formation of plastic crystals to be predicted for a given molecular unit because d_m must be determined from crystallographic data. A new globularity criterion (G) has been defined [24, 25] as the ratio between the volume of the molecular unit over the volume of the sphere defined by $D_{\max}$ (free rotation of the molecular unit). With only one

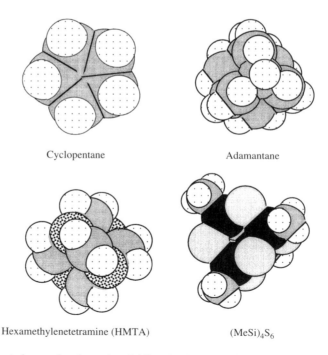

Cyclopentane   Adamantane

Hexamethylenetetramine (HMTA) $(MeSi)_4S_6$

Figure 3.60 A few molecular units yielding (top) or not yielding (bottom) plastic crystalline phases

SUPRAMOLECULAR ENGINEERING: SYMMETRY ASPECTS

Table 3.6 Various molecular units classified according to the Postel and Riess parameter [23]. The double line indicates the frontier between the compounds that afford plastic crystals and the others that do not

Compound	R (Postel–Riess)	$V_{molecule}$ (Å^3)	D_{max} (Å)	V_{max} (Å)	$G = V_{molecule}/V_{max}$
Cyclohexane	0.94	100.50	6.57	148.49	0.68
Cyclohexanol	0.89	107.44	7.01	180.37	0.60
Cyclopentane	0.88	83.75	6.61	151.22	0.55
Adamantane	0.88	144.54	7.55	225.34	0.64
Quinuclidine	0.88	110.31	7.22	197.07	0.56
$(CH_3)_3CCl$	0.85	93.12	7.00	179.59	0.52
Cyclobutane	0.84	74.06	6.22	126.00	0.59
Camphor	0.84	179.14	8.45	315.91	0.57
P_4S_{10}	0.82	268.39	9.34	426.62	0.63
TED	0.82	115.44	7.10	187.40	0.62
HMTA	0.77	132.37	7.29	202.85	0.65
UF_6	0.75	113.51	7.82	259.39	0.45
$(CCH_2Cl)_4S_6$	0.74	287.33	11.77	853.74	0.34
P_4O_{10}	0.65	142.68	9.31	422.52	0.34
$(CH_3Si)_4S_6$	0.65	261.92	10.28	568.82	0.46
$Te(OH)_6$	0.62	124.19	9.19	406.39	0.31

TED = triethylenediamine, HMTA = hexamethylenetetramine.

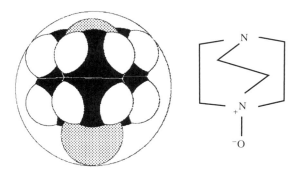

Figure 3.61 Molecular representation of the N-oxide of triethylenediamine (NOTED) and the corresponding spherical envelope

exception (HMTA), the new globularity criterion G distinguishes between the compounds generating plastic crystals ($G \geq 0.50$) and the others (Table 3.6).

X-ray diffraction studies have been carried out on the N-oxide of triethylene diamine (NOTED) as a good example of a plastic crystal with correlated positions of the molecular units [26–28] (Figure 3.61). X-ray diagrams of powders of NOTED could be indexed with a face-centred cubic lattice ($a_c = 17.35$ Å at 20 °C). At temperatures higher than 145 °C a primitive cubic lattice was observed with a lattice parameter ($a_c = 8.85$ Å at 152 °C) approximately half the previous value. The low-temperature phase is birefringent, which is incompatible with a cubic structure. The relative orientations of the molecule must therefore lead to a lower symmetry. On the contrary, the high-temperature phase is not birefringent and therefore can really belong to the cubic system.

In order to gain information on the low-temperature phase structure, octahedral single crystals have been obtained by sublimation under vacuum and X-ray diffraction studies have been carried out [26, 27]. The space group R3c ($Z = 8$) with the parameters $a_r = 12.26$ Å, $\alpha = 60.07°$ could be determined. As a consequence, only one of the four C_3 axes present in a face-centred cubic lattice is really a symmetry axis of the structure. A schematic representation of the low-temperature phase is given in Figure 3.62.

Atomic coordinates show that the NOTED molecule is not distorted in the solid state. The molecular envelope has a diameter of 6.8 Å. The closest distance

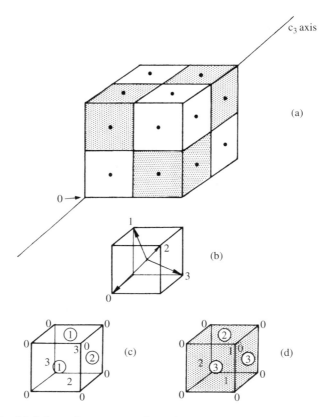

Figure 3.62 (a) Schematic representation of the low-temperature phase of NOTED. (b) The orientation of the dipole moment of the molecules is shown by the numbers 0, 1, 2, 3. (c,d) The cubic cell is composed of two kinds of elementary cubes (white and black) which differ by the relative orientations of the molecules

PLATE 1

Figure 1.49 Crossed polarizer optical microscopy photograph of $(C_{12}OCH_2)_8PcH_2$, showing the flower-like texture [65] (C_{12} stands for $C_{12}H_{25}$)

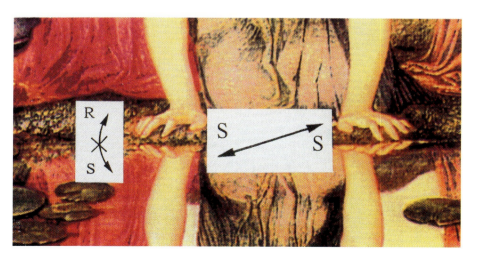

Figure 2.2 A mirror plane [1]. The arrows indicate the conservation (or not) of the chirality of the two corresponding hands

PLATE 2

Figure 3.69 Chemical formula of the first generation of polyamidoamine and the corresponding molecular model. (Reproduced by permission of Wiley-VCH from [35])

PLATE 3

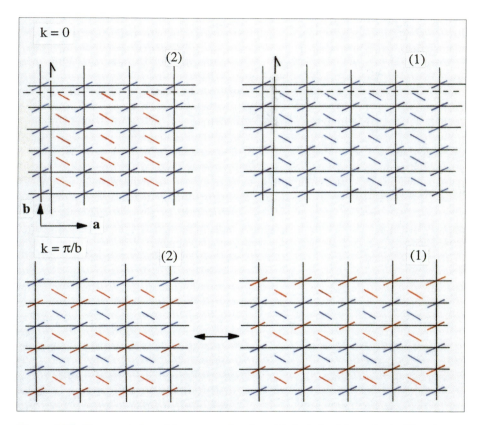

Figure 6.25 Representation of the supramolecular orbitals derived from the A_u (1) and B_u (2) orbitals at $k = 0$ and $k = \pi/|b|$ in the (a, b) plane (along the b^{-1} direction). Convention used: A_u application of 2_1: blue → blue, B_u application of 2_1: blue → red

PLATE 4

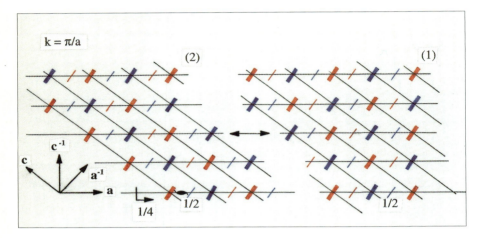

Figure 6.26 Representation of the supramolecular orbitals derived from A_u (1) and B_u (2) at $k = \pi/|a|$ in the (a, c) plane (along the a^{-1} direction)

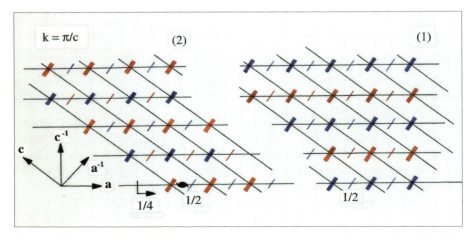

Figure 6.27 Representation of the supramolecular orbitals derived from A_u (1) and B_u (2) at $k = \pi/|c|$ in the (a, c) plane (along the c^{-1} direction)

PLATE 5

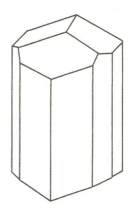

Figure 7.15 Schematic representation of a tourmaline crystal and photograph of rubellite (red tourmaline): $Na(Li, Al)_3Al_6B_3Si_6O_{27}(O, OH, F)_4$ ($9 \times 8 \times 6 cm^3$). Coll. Minéraux Jussieu. Photo: G. Mourguet

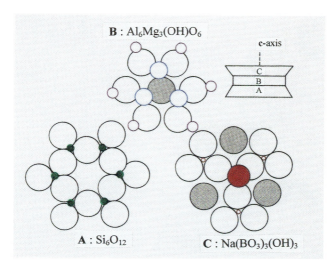

Figure 7.16 The three rings which constitute the structure of tourmaline: green, silicon; violet, aluminum; blue, magnesium; red, sodium; grey, (OH). (Modified from Ref. [26])

PLATE 6

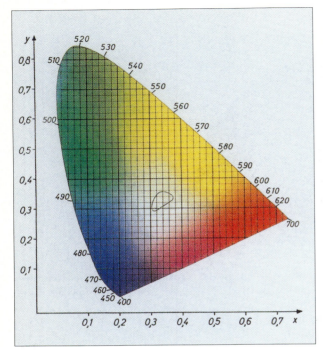

Figure 8.23 Chromaticity diagram (x, y) according to CIE 1931. (Reproduced by permission of Wiley-VCH from Ref. [13])

Figure 8.24 Fragment of a cave painting (Salle des Taureaux-Lascaux, 17000 BC; (adapted from Ref. [14])

PLATE 7

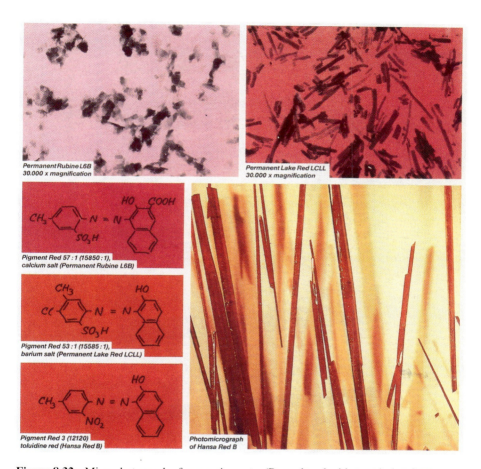

Figure 8.32 Microphotograph of some pigments. (Reproduced with permission from Ref. [23])

PLATE 8

Figure 8.83 Canon ferroelectric liquid crystal display (1998): display area, 29.5 x 23.5 cm^2; screen size 38 cm (14.8 inches, diagonal); number of pixels, 1280 x 1024; pixel pitch, 0.23 mm; pixel structure, 4 dots per pixel (red, green, blue, white); contrast ratio, 60:1; operable temperature range, 10–35°C; power consumption, 60W. (Reproduced by permission of Mr Junichiro Kanbe, Canon Inc R & D) [99]

SUPRAMOLECULAR ENGINEERING: SYMMETRY ASPECTS

between two neighbours is 6.13 Å, leading to a globular factor $R = 0.83$. Calculation of G leads to a value of 0.55. Both parameters therefore predict the appearance of a plastic crystalline phase before the melting point.

The angle of the rhombohedral cell is close to 60°; the structure can therefore be described by a face-centred cubic lattice with 32 molecules per unit cell (Figure 3.62). This cubic cell may be subdivided into eight elementary cubes having molecules at all vertices and at the centre of all faces leading to a cubic close packing. Two types of elementary cubes are present differing by the relative orientations of the molecules. The space group R3c is then obtained.

By considering the structure along the C_3 axis, we can gain a more detailed insight into the orientations of the molecules (Figure 3.63). Six different (111) layers, ABCA'B'C', may be distinguished. One-fourth of the molecules are on the ternary axis (site symmetry 3) whereas three-fourths are in a general position (site symmetry 1). In the ideal position, the dipole moment of the molecule at

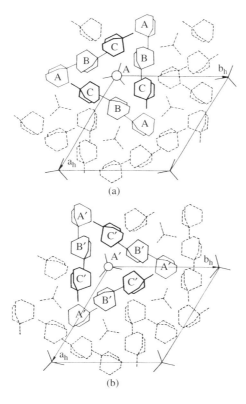

Figure 3.63 Representation of the structure of the low-temperature phase of NOTED: views along the C_3 axis. There are six different layers that can be distinguished: (a) A(0), B(1/6), C(2/6); (b) A'(3/6), B'(4/6), C'(5/6) (the heights are given as multiples of the C_h hexagonal axis). (After Ref. [27])

site symmetry 3 is exactly compensated by the sum of the three others. In the real system, a residual dipole moment of $0.39D$ per molecule is calculated from the atomic positions.

The relative contributions (van der Waals and dipolar) to the crystalline energy have been calculated. The dipolar term contributes for -19.6 kJ/mol (22%) to the total crystalline energy (-90.0 kJ/mol) and stabilizes the structure.

The transition at 145 °C to the high-temperature phase is not destructive and single-crystal studies can be obtained. However, exposure times were not long enough to determine precisely the intensities of the reflections because of the flowing of the crystal under gravity. At 152 °C, the crystalline lattice is cubic with $a_c = 8.850 \pm 0.006$ Å and the space group is $P2_13$. By comparison with the low-temperature phase, a plausible model for the structure may be proposed. First, the lattice parameter found corresponds to one of the eight elementary cubes previously defined in Figure 3.62. Some reorientational disorder must therefore have taken place. However, the symmetry $P2_13$ implies that the orientations of the four molecules within one unit cell are correlated (Figure 3.64). The association of a tile with the NOTED molecular unit may be attempted, but it is difficult to find a suitable two-dimensional representation of a three-dimensional structure and no easily distinguishable planes (as for smectic liquid crystals) can be found.

It has been observed [29] that the frequency of occurrence of a crystal face is higher when the corresponding interreticular distance d_{hkl} is larger. Thus, a larger d_{hkl} value corresponds to a higher density of lattice nodes. This is true for all types of lattices, but a simple geometrical illustration is given in Figure 3.65.

It therefore seems logical to associate close-packed layers to planes corresponding to large interreticular distances. However, this approach is valid only for primitive lattices and not for a face-centred cubic lattice, as in the present case. For a face-centred cubic lattice, as observed for NOTED, the (111) layer is the densest one and corresponds to a two-dimensional hexagonal lattice. Knowing its orientation, the plane selected in the structure is chosen in order to contain the elements of symmetry of the layer. There is therefore not too much ambiguity concerning the definition of the plane into which a correlation molecular unit/tile may be attempted. At that stage, two possibilities may be considered for this correlation:

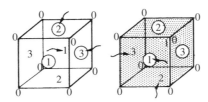

Figure 3.64 Molecules which must be reorientated (arrows) during the phase transition from the crystal to the plastic crystalline phase in order to obtain the space group $P2_13$ (same notation as in Figure 3.62)

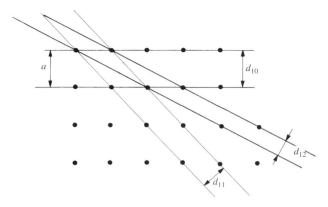

Figure 3.65 Illustration of a two-dimensional square lattice where the density of the lattice nodes is higher when the interreticular distance (d_{hk}) is larger. In this example $d_{hk} = a/\sqrt{h^2 + k^2}$

(a) The cross-section of the molecular units. This way does not ensure the closest-packed layer and therefore it *minimizes the compactness*.

(b) The projection of the molecular units on to the previously defined plane. This procedure can yield unreal molecular overlaps whenever the molecular unit is inclined relative to the plane. This method therefore *maximizes the compactness*.

The real compactness in two dimensions cannot be determined with certainty. However, the use of molecular cross-sections or projections allows upper and lower limits of compactness to be estimated.

The (111) plane of NOTED structure contains two types of molecular units in site symmetries C_{3v} and C_1, respectively. The projection of the NOTED molecules on to the (111) plane is shown in Figure 3.66.

The site symmetry (1) and space group (p3) lead to the tile (IH33) belonging to the topological class [3.6.3.6]. The two adjacent edges of IH33 are related to each other via a simple rotation. In this process the NOTED molecular units in site symmetry C_3 have been ignored and are considered to fill the empty space of the layer arrangement of the previous type of NOTED molecules.

Two other illustrative examples of plastic crystals will be given in the next paragraphs.

By vaporizing graphite with a laser beam, H. Kroto, R. Smalley and their coworkers [30] obtained in 1985 new compounds (C_{60}, C_{70}) named *fullerenes*, in analogy with R. Buckminster Fuller's constructions (Architect: 1895–1983) (Figure 3.67).

The C_{60} molecule has an icosahedral symmetry (I_h) and its shape is close to a sphere. At 249 K a phase transition is observed from a primitive cubic ($a = 14.04$ Å) to a face-centred cubic arrangement ($a = 14.17$ Å, $Z = 4$) [31, 32]. The

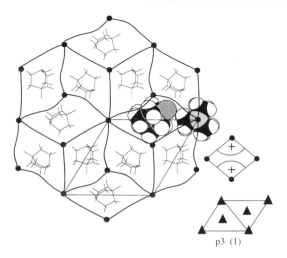

Figure 3.66 Projection of the NOTED molecular units on to the (111) plane of the structure and corresponding tile (IH33) in the topological class [3.6.3.6] (the NOTED molecules in site symmetry C_{3v} are not considered)

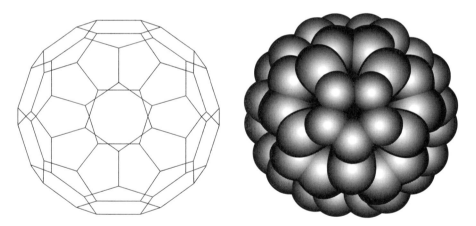

Figure 3.67 Molecular representation and molecular models (Hyperchem) of the C_{60} molecule: view along the C_5 axis

lattice parameter is practically unchanged during the transition (Figure 3.68). The van der Waals diameter of the C_{60} molecule is calculated to be 10 Å and the lattice parameter of the cubic primitive cell is 14.04 Å. The value $14 \times \sqrt{2}/2 = 9.93$ Å (half the diagonal of a face) represents the closest approach between two C_{60} molecular units.

The molecular volume is, however, not always precisely defined and voids and internal cavities may render the globularity criterion previously defined unviable.

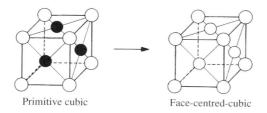

Figure 3.68 Transformation of a primitive cubic to a face-centred cubic lattice. The units which are orientationally different in the former case become equivalent

This is the case for dendrimers. This class of molecules was first synthesized by F. Vögtle and coworkers in 1978 [33]. It is based upon a repetitive synthesis providing a branch multiplicity with a simple mathematical progression (see, for example, Refs. [34] and [35]). As an example, the case of starbust polyamidoamine will be mentioned [35] (Figure 3.69 Plate 2).

For such molecular units, the definition of a molecular volume is not straightforward. A method has been proposed in which a sphere of radius r_p (probe radius) is rolled around the van der Waals surface of the molecular unit [35]. A measure of the internal cavities is achieved by analysing the volume (V_p) defined by the rolling surface as a function of the probe radius. With an ideal sphere, a plot of $\sqrt[3]{V_p}$ versus r_p is linear with a slope of $\sqrt[3]{4/3\pi}$. This relationship is satisfied at large r_p but becomes unrealistic for small r_p values. The discrepancy between the curve determined for the ideal sphere and the one observed with the molecular unit considered gives a measure of the possibility of intermolecular gear effects in the corresponding condensed media.

3.5.3 Three-Dimensional Crystals

3.5.3.1 Space Group Occurrence

It has been seen before (see Section 3.3.4) that the presence of certain symmetry elements — in particular mirror planes — within the structure decreases the probability of obtaining a close-packed arrangement or is even incompatible with a sixfold coordination in two dimensions. In Kitaigorodkii's approach, three-dimensional structures are obtained by close packing of layers; therefore, the same limitations concerning symmetry are obtained. Close packing of layers is generally not possible if the layers are related by a mirror plane or a translation perpendicular to their plane. Moreover, site symmetries mmm (D_{2h}) or 222 (D_2) completely determine the position and the orientation of the molecular unit. They therefore necessitate fairly rare molecular shapes to be compatible with a close-packing requirement. Following this methodology, A. I. Kitaigoroskii [6] has proposed a list of the most probable three-dimensional space groups (Table 3.7).

Table 3.7 The most probable three-dimensional space groups allowing close-packed structures as a function of the molecular site symmetry. In each case, the number of molecules per unit cell Z is given. (For a more detailed description see Ref. [6])

Molecular site symmetry							
1 (C_1)	Z	$\bar{1}$ (C_i)	Z	2 (C_2)	Z	m (C_s)	Z
P$\bar{1}$	2,4	P$\bar{1}$	1,2	C2/c	4	Pmc2_1	4
P2_1	2,4	P2_1/c	2,4	P$2_1 2_1 2$	2,4	Cmca	4
P2_1/c	4	C2/c	4	Pbcn	4	Pnma	4
Pca2_1	4	Pbca	4				
Pna2_1	4						
P$2_1 2_1 2_1$	4						

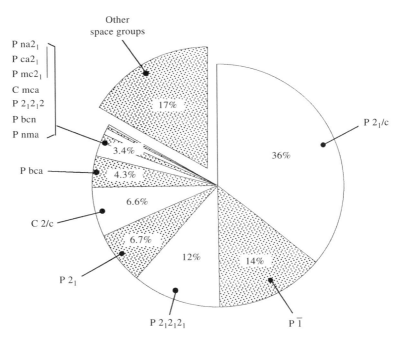

Figure 3.70 Three-dimensional space groups observed statistically by examining 30 000 organic crystals [36]

A statistical study of the occurrence of space groups for molecular crystals has been carried out on 30 000 crystalline structures [36] (Figure 3.70). Six space groups (P2_1/c, P$\bar{1}$, P$2_1 2_1 2_1$, P2_1, C2/c, Pbca) represent 80% of the total number of structures found experimentally. None of these space groups admit the mirror plane as symmetry element. On the contrary, glide planes are commonly encountered (P2_1/c, C2/c, Pbca).

SUPRAMOLECULAR ENGINEERING: SYMMETRY ASPECTS

The space group P1, which contains only translations with no other symmetry elements, is found in 1.05% of the cases. There is a satisfactory agreement between the list of the most probable space groups proposed by Kitaigorodskii (Table 3.7) and experimental findings (see Figure 3.70).

3.5.3.2 Examples

Metallophthalocyanines

Metallophthalocyanines (PcM) — and many other organic and metallo-organic compounds — crystallize in the commonly found space group $P2_1/c$. In the case of PcCu, the following cell parameters have been found: $a = 14.628$ Å, $b = 4.790$ Å, $c = 19.507$ Å, $\beta = 120°56$ [37]. The symmetry of the molecular unit PcCu is D_{4h}; only the inversion centres are retained in the structure (site symmetry $\bar{1}$). Two views of the structure with the corresponding symmetry elements are shown in Figure 3.71.

At that stage, it is possible to give a scheme of analysis:

$$\text{Structure} \longleftrightarrow \text{Layer} \longleftrightarrow \text{Chain}$$
$$\updownarrow$$
$$\text{Isohedral tiling}$$
$$\updownarrow$$
$$\text{Tile}$$

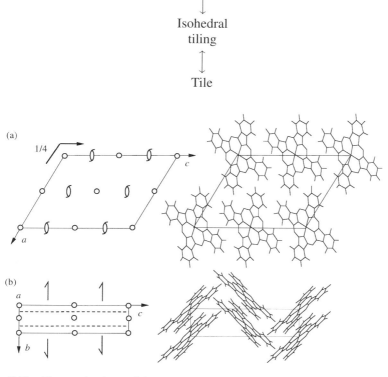

Figure 3.71 Two projections of the structure of PcCu in the (a) *ac* and (b) *bc* planes (β modification, space group $P2_1/c$, $Z = 2$)

Figure 3.72 Chain of molecules along the b axis. The corresponding color stripe group is $11\bar{1}$ (see Figure 3.16)

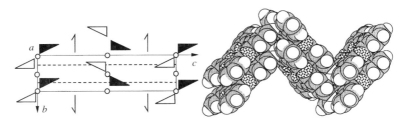

Figure 3.73 Layer in the bc plane obtained by applying the 2_1 symmetry operation on the chain shown in Figure 3.72. The bicolor layer symmetry group is $P12_1/c1$

The chain considered can be chosen along the b axis where the intermolecular distance is the shortest (Figure 3.72). A layer can be constructed from the previous chain in the bc plane by applying a 2_1 symmetry operation (Figure 3.73).

Finally, the three-dimensional structure is obtained by simply translating the layers found along the a axis. The phthalocyanine stacking in the bc layers is an example of the herringbone arrangement very commonly encountered in molecular crystals.

In order to relate bc layers thus found to an isohedral tiling, it is necessary to make the projection of the constituent atoms in the layer. In such a process, the bicolor layer must be transformed into a monocolor one; the inversion centres and the 2_1 axes have no precise definition in two dimensions and must be transformed into binary axes and glide planes, respectively (Figure 3.74). The two-dimensional monocolor space group is p2gg and the molecular site symmetry is 2 (or C_2). In the [3^6] topological class, only the isohedral tiling IH9 presents the necessary requirements. IH9 has the characteristics shown in Figure 3.75.

It is now possible to fit IH9 with the projected molecular unit by applying the Yin–Yang principle and by transforming the tile using allowed distortions (Figure 3.76).

Thiourea

Thiourea is another representative example of molecular stacking. Thiourea crystallizes in the three-dimensional space group Pnma and the molecule

SUPRAMOLECULAR ENGINEERING: SYMMETRY ASPECTS

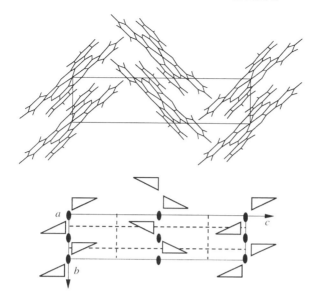

Figure 3.74 Projection on the *bc* plane. The corresponding two-dimensional monocolor space group is p2gg

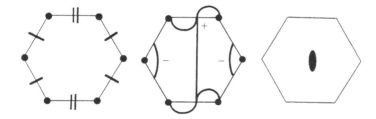

Figure 3.75 General characteristics of the IH9 (see Appendix 6)

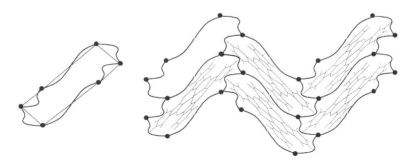

Figure 3.76 Fit of the projected molecular unit with the tile IH9 and corresponding tiling (two tiles per unit cell)

(symmetry 2mm (C_{2v})) retains only a mirror plane in the structure (site symmetry m) [38] (Figure 3.77).

The crystal is made of layers parallel to the bc plane, the mirror being perpendicular to this plane. In the bc plane, the corresponding two-dimensional bicolor space group is P12$_1$/m1 (Figure 3.78). The three-dimensional structure contains successive layers related by a 2$_1$ axis parallel to the layer plane. The corresponding two-dimensional monocolor space group of the layers is p2mg (Figure 3.79).

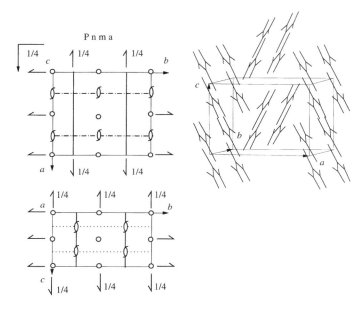

Figure 3.77 Three-dimensional structure of thiourea (space group is Pnma ($Z = 4$), $a = 7.66$ Å, $b = 8.59$ Å, $c = 5.48$ Å). (After Ref. [38])

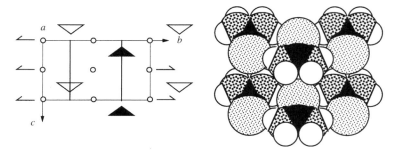

Figure 3.78 Two-dimensional bicolor space group and representation of the projection of thiourea on the bc plane (space group P12$_1$/m1)

SUPRAMOLECULAR ENGINEERING: SYMMETRY ASPECTS 125

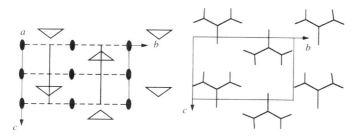

Figure 3.79 The projection of the structure on the *bc* plane. The corresponding two-dimensional monocolor space group is p2mg

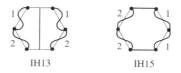

Figure 3.80 The two possible tiles for urea (see Appendix 6)

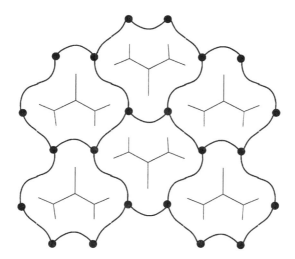

Figure 3.81 Tiling IH13 associated with the structure of thiourea projected on the *bc* plane

In the topological class [3^6], two tiles are possible, IH13 and IH15, with the space group p2mg and site symmetry m. IH15 may be eliminated since two opposite sides contain a mirror plane and must remain undistorted (flat), significantly decreasing the possibility of finding a satisfactory fit to the molecular shape (Figure 3.80). In IH13, two opposite edges of the tile are perpendicular to the mirror plane and related by a translation. The other edges belong to the S-type. Each unit cell contains two tiles (or molecules) associated by a binary axis or a glide plane (Figure 3.81).

2,2′,2″-Trimethoxytriphenylamine

In the crystalline state [3], 2,2′,2″-trimethoxytriphenylamine is chiral (symmetry C_3). The methoxyphenyl groups linked to the nitrogen atom are arranged like the blades of a screw propeller (Figure 3.82).

Octahedral crystals (see Figure 3.1) have been obtained. This crystalline shape is characteristic of cubic structures with dense (111) planes perpendicular to the ternary axes. X-ray determinations [3] indicate a space group $Pa\bar{3}$ and a cubic cell with $a = 15.98$ Å (eight molecules per unit cell). The molecular units

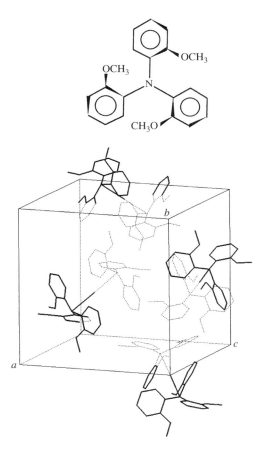

Figure 3.82 Crystalline structure of 2,2′,2″-trimethoxyphenylamine (space group $Pa\bar{3}$, $a = 15.39$ Å). (After Ref. [3])

SUPRAMOLECULAR ENGINEERING: SYMMETRY ASPECTS 127

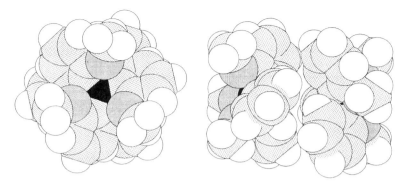

Figure 3.83 The van der Waals volume of a single molecule of 2,2′,2″-trimethoxy-triphenylamine (viewed along the C_3 axis) and corresponding packing of couples of molecular units

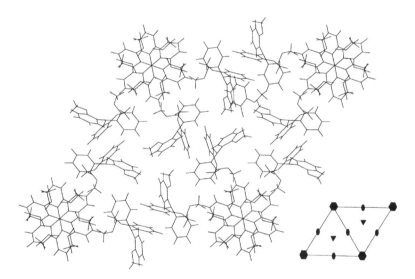

Figure 3.84 Molecular arrangement in the (111) plane of projected trimethoxytriphenylamine (two-dimensional space group p6)

are associated by pairs of enantiomers related by a centre of symmetry with a common C_3 axis (site symmetry 3). The van der Waals volume of a pair of molecules has a quasi-spherical shape (Figure 3.83). The pairs of molecular units afford dense (111) layers in a compact cubic structure with a sequence ABCABC.... The molecular arrangement in the (111) planes is given in Figure 3.84. The two-dimensional symmetry group of this layer is p6. Molecules with ternary axes perpendicular to the layer plane are centered on the C_6 axes. These molecules are not related to the other pairs of molecules by the symmetry

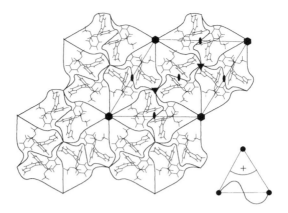

Figure 3.85 Isohedral tiling IH88 (see Appendix 5), topological class [6^3], six tiles per unit cell fitting the molecular units of trimethoxytriphenylamine in the (111) plane. The nonoccupied space occurs at the C_6 axes

operations of the layer group. In this case a single tile *cannot* be associated to each molecule. Nevertheless, it is possible to use an isohedral tiling to describe the arrangement of the molecules whose ternary axes are not perpendicular to the layer (Figure 3.84). This tiling of the (111) plane leads to a gap in each tile which can be occupied by the remaining nonequivalent molecules (Figure 3.85). Structures of cubic symmetry are scarcely observed when there is no molecular orientational disorder, as in the case of plastic crystals. This example can be considered as an exception in which pairs of molecules are formed yielding an overall quasi-spherical unit, which in turn affords a cubic symmetry.

3.6 REFERENCES AND NOTES

1. J. Maddox, *Nature*, **335**, 201 (1988).
2. B. Grünbaum and G. C. Shepard, *Tilings and Patterns*, W. H. Freeman, New York (1986).
3. C. Soulié, P. Bassoul and J. Simon, *New J. Chem.*, **17**, 787 (1993).
4. I. Hargittai and G. Lengyel, *J. Chem. Educ.*, **61**, 1033 (1984).
5. *International Tables of Crystallography*, Vol. A, *Space Group Symmetry*, Kluwer Academic, Dordrecht (1989).
6. A. I. Kitaigorodskii, *Organic Chemical Crystallography*, Consultant Bureau, New York (1961).
7. Y. Le Corre, *Bull. Soc. Franç. Minér. Crist.*, **81**, 120 (1958).
8. S. Goshen, D. Mukamel and S. Shtrikhman, *Molec. Cryst. Liquid Cryst.*, **31**, 171 (1975).
9. T. Kobayashi, Y. Fujiyoshi and N. Uyeda, *Acta Cryst.*, **A38**, 356 (1982).
10. J. Perlstein, *J. Am. Chem. Soc.*, **114**, 1955 (1992); **116**, 455 (1994).
11. In Chinese the term Yin–Yang expresses the principle of complementarity.
12. J. Simon and P. Bassoul, *Phthalocyanines*, Vol. II, (eds. C. C. Leznoff and A. B. P. Lever) p. 223, VCH, Weinheim (1993).

13. M. Gardner, *Scientific American*, July, 112 (1975); August, 112 (1975).
14. C. Sirlin, L. Bosio, J. Simon, V. Ahsen, E. Yilmazer and Ö. Bekâroglu, *Chem. Phys. Lett.*, **139**, 362 (1987).
15. P. D. Bartlett, M. J. Ryan and S. G. Cohen, *J. Am. Chem. Soc.*, **64**, 2649 (1942).
16. S. Norvez, Thèse de Doctorat, ESPCI-Paris (1991).
17. S. Norvez and J. Simon, *J. Chem. Soc. Chem. Commun.*, 1398 (1990).
18. S. Norvez, *J. Org. Chem.*, **58**, 2414 (1993).
19. S. Norvez and J. Simon, *Liquid Cryst.*, **14**, 1389 (1993).
20. P. Bassoul and J. Simon, *New J. Chem.*, **20**, 1131 (1996).
21. J. Sivardière, *La Symétrie en mathématiques, physique et chimie*, Presses Universitaires de Grenoble (1995).
22. J. N. Sherwood, *The Plastically Crystalline State*, John Wiley & Sons, Chichester (1979).
23. M. Postel and J. G. Riess, *J. Phys. Chem.*, **81**, 2634 (1977).
24. C. Soulié, Thèse de Doctorat d'état, ESPCI-Paris (1993).
25. P. Bassoul, J. Simon and C. Soulié, *J. Phys. Chem.*, **100**, 3131 (1996).
26. P. Mihailovic, P. Bassoul and J. Simon, *Chem. Phys. Lett.*, **141**, 462 (1987).
27. P. Mihailovic, P. Bassoul and J. Simon, *J. Phys. Chem.*, **94**, 2815 (1990).
28. P. Mihailovic, Thèse de Doctorat, ESPCI-Paris (1988).
29. G. Friedel, *Bull. Soc. Fr Mineral*, **30**, 326 (1907); cited in Z. Berkovitch-Yellin, *J. Am. Chem. Soc.*, **107**, 8239 (1985).
30. H. W. Kroto, J. R. Heath, S. C. O'Brien, R. F. Curl and R. E. Smalley, *Nature*, **318**, 162 (1985).
31. P. A. Heiney, J. E. Fischer, A. R. McGhie, W. J. Romanow, A. M. Denenstein, J. P. McCauley Jr and A. B. Smith III, *Phys. Rev. Lett.*, **66**, 2911 (1991).
32. G. S. Hammond and V. J. Kuck, *Fullerenes*, ACS Symp. Series 481, American Chemistry Society, Washington DC (1992).
33. F. Vögtle, E. Buhleier and W. Wehner, *Synthesis*, 155 (1978).
34. G. R. Newkome and G. R. Baker, *Molecular Engineering for Advanced Materials* (eds. J. Becher and K. Schaumburg) Kluwer Academic Dordrecht (1995).
35. D. A. Tomalia, A. M. Naylor and W. A. Goddard III, *Angew. Chem. Int. ed. Engl.*, **29**, 138 (1990).
36. A. D. Mighell, V. L. Himes and J. R. Rodgers, *Acta Cryst.*, **A39**, 737 (1983).
37. C. J. Brown, *J. Chem. Soc. A*, 2488 (1968).
38. D. Mullen, G. Heger and W. Treutmann, *Z. Kristallogr.*, **148** 95 (1978).

4 Symmetry and Physicochemical Properties: The Curie Principle

4.1 Electrical Potentials — 131
   4.1.1 Potential Due to One Charge — 131
   4.1.2 Potential Due to Two Point Charges — 133
   4.1.3 Potential Due to a Distribution of Charges — 134
   4.1.4 Legendre's Polynomials and Spherical Harmonics — 134
4.2 Electrical Multipoles in Chemistry — 139
4.3 Cause/Effect Symmetry Relationships — 140
   4.3.1 The Curie Principle — 140
   4.3.2 Applications of the Curie Principle — 141
   4.3.3 The Curie Principle Using Irreducible Representations — 146
4.4 Generalization of the Curie Principle — 153
4.5 References — 155

> *Il ne peut y avoir davantage dans l'effet qu'il n'y a dans la cause.*
> R. Descartes (1596–1650)

When fabricating new materials the chemist needs to know the physical bases of various phenomena, at least on qualitative grounds. In this chapter an attempt is made to transcribe often complicated physical and mathematics notions into simple terms.

SYMMETRY AND PHYSICOCHEMICAL PROPERTIES

4.1 ELECTRICAL POTENTIALS

The potential in vacuum generated by a single point charge at a distance r is given by:

$$\phi = \frac{1}{4\pi\varepsilon_0} \frac{e}{r}$$

Where

e = charge (in coulombs)
r = distance (in meters)
$1/(4\pi\varepsilon_0) = 9 \times 10^9$ V m/C
ϕ = potential (in volts)

In the following discussion, the constant $1/(4\pi\varepsilon_0)$ will not be written explicitly.

The potential created by two charges is simply the sum of two of the previous potentials:

$$\phi = \frac{e_1}{r_1} + \frac{e_2}{r_2}$$

In the same way, a distribution of i charges gives the potential

$$\phi = \sum_i \frac{e_i}{r_i} \tag{4.1}$$

where

r_i = distance from the charge i considered

The expression of the potential of one charge is more complicated whenever the origin is not chosen at the centre of the charge.

4.1.1 Potential Due to One Charge (see Refs. [1] and [2])

The potential ϕ generated by the charge e at a point $\mathcal{P}$ (Figure 4.1) is given by

$$\phi(\mathcal{P}) = \frac{e}{r_e} = e(s^2 - 2rs\cos\alpha + r^2)^{-1/2}$$

$$= \frac{e}{r}\left(1 - 2\frac{s}{r}\cos\alpha + \frac{s^2}{r^2}\right)^{-1/2} \tag{4.2}$$

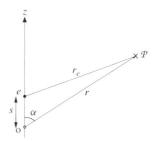

Figure 4.1 Notation used to describe the potential of a charge

and
$$\phi(P) = \frac{e}{r}(1+x)^{-1/2} \qquad (4.3)$$

with
$$x = \left(\frac{s^2}{r^2} - 2\frac{s}{r}\cos\alpha\right)$$

The term $(1+x)^n$ may be developed in series:
$$(1+x)^n = 1 + nx + \frac{n(n-1)}{2!}x^2 + \cdots$$
$$+ \frac{n(n-1)\cdots(n-p+1)}{p!}x^p \cdots \qquad x^2 < 1$$

$$\phi(P) = \frac{e}{r}\left[1 - \frac{1}{2}\left(\frac{s^2}{r^2} - 2\frac{s}{r}\cos\alpha\right) + \frac{3}{8}\left(\frac{s^2}{r^2} - 2\frac{s}{r}\cos\alpha\right)^2\right.$$
$$\left. - \frac{15}{48}\left(\frac{s^2}{r^2} - 2\frac{s}{r}\cos\alpha\right)^3 + \cdots\right]$$
$$= \frac{e}{r}\left[1 + \cos\alpha\left(\frac{s}{r}\right) + \frac{3\cos^2\alpha - 1}{2}\left(\frac{s}{r}\right)^2\right.$$
$$\left. + \frac{5\cos^3\alpha - 3\cos\alpha}{2}\left(\frac{s}{r}\right)^3 + \cdots\right]$$

This can be written in a general form as

$$\phi(P) = \frac{e}{r}\sum_{l=0}^{\infty}\left(\frac{s}{r}\right)^l P_l(\cos\alpha) \qquad (4.4)$$

where $P_l(x)$ is the Legendre's polynomial with $x = \cos\alpha$:

$$P_l(x) = \frac{1}{2^l l!}\frac{d^l}{dx^l}(x^2 - 1)^l \qquad (4.5)$$

and
$$P_0(x) = 1$$
$$P_1(x) = x$$
$$P_2(x) = \tfrac{1}{2}(3x^2 - 1)$$
$$P_3(x) = \tfrac{1}{2}(5x^3 - 3x)$$

Each electrical moment is associated with angular and radial parts that depend on the parameter l of the Legendre's polynomial (Table 4.1).

SYMMETRY AND PHYSICOCHEMICAL PROPERTIES

Table 4.1 Moment, angular and radial parts as a function of the nature of the multipole

l		Moment	Angular part	Radial part
0	Charge	e	1	$1/r$
1	Dipole	es	$\cos \alpha$	$1/r^2$
2	Quadrupole	es^2	$\dfrac{3\cos^2 \alpha - 1}{2}$	$1/r^3$
3	Octopole	es^3	$\dfrac{5\cos^3 \alpha - 3\cos \alpha}{2}$	$1/r^4$

The function $\phi(\mathcal{P})$ depends on the position of the origin. When the origin is situated at the centre of the charge ($s = 0$), it is simply given by

$$\phi(\mathcal{P}) = \frac{e}{r}$$

The shift of the origin allows the multipolar components to be introduced [2].

4.1.2 Potential Due to Two Point Charges

The potential ϕ generated by the charges e_1 and e_2 at a point $\mathcal{P}$ (Figure 4.2) is given by

$$\phi(\mathcal{P}) = \frac{e_1}{r_1} + \frac{e_2}{r_2} = \frac{e_1}{r}\sum_{l=0}^{\infty}\left(\frac{s_1}{r}\right)^l P_l(\cos \alpha_1) + \frac{e_2}{r}\sum_{l=0}^{\infty}\left(\frac{s_2}{r}\right)^l P_l(\cos \alpha_2)$$

In the case of a dipole ($e_1 = -e_2 = e$) with the origin taken at the midpoint between the two charges ($s_1 = s_2 = s$) (Figure 4.3),

$$\phi(\mathcal{P}) = \frac{e}{r}\sum_{l=0}^{\infty}\left(\frac{s}{r}\right)^l [P_l(\cos \alpha) - P_l(-\cos \alpha)]$$

where

$P_1(\cos \alpha) = P_1(-\cos \alpha)$ for l even
$P_1(\cos \alpha) = -P_1(-\cos \alpha)$ for l odd (see Table 4.1)

and

$$\phi(\mathcal{P}) = \frac{2es}{r^2}\cos \alpha + \frac{2es^3}{r^4}\frac{5\cos^3 \alpha - 3\cos \alpha}{2} + \cdots$$

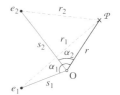

Figure 4.2 Potential at $\mathcal{P}$ due to two charges e_1 and e_2

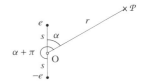

Figure 4.3 The case of a dipole

Only odd moments remain in the polynomial. The first nonzero moment is the dipole moment ($2es$).

4.1.3 Potential Due to a Distribution of Charges

A distribution of N charges e_i at point (x_i, y_i, z_i) for a given origin O is described by the vectors $\mathbf{s}_i$ (Figure 4.4). The potential at a point $\mathcal{P}$ described by the vector $\mathbf{r}$, when $|\mathbf{r}| \gg |\mathbf{s}_i|$, is the sum of the potential of each charge:

$$\phi(\mathcal{P}) = \sum_{i=1}^{i=N} \frac{e_i}{|\mathbf{r} - \mathbf{s}_i|} = \sum_{i=1}^{i=N} \frac{e_i}{|\mathbf{r}|} \sum_{l=0}^{l=\infty} \left(\frac{|\mathbf{s}_i|}{|\mathbf{r}|} \right)^l P_l(\cos \alpha_i) \qquad (4.6)$$

For a given value of l, the multipolar moment of the distribution of charges is the sum of the contribution of each charge to the multipole considered.

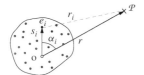

Figure 4.4 Notations used for a distribution of charges

4.1.4 Legendre's Polynomials and Spherical Harmonics

Spherical harmonics (Y) are functions which, under the application of any proper rotation operation, are transformed into a function also belonging to spherical harmonics (see Appendix 9). A relationship may be found between Legendre's polynomials and spherical harmonics [3]:

$$P_l(\cos \alpha_i) = \frac{4\pi}{2l+1} \sum_{m=-l}^{m=+l} (-1)^m Y_l^{-m}(\theta_i, \varphi_i) Y_l^m(\Theta, \Phi) \qquad (4.7)$$

The spherical coordinates are defined in Figure 4.5.

$$\mathbf{r} \Rightarrow r, \Theta, \Phi$$

$$\mathbf{s}_i \Rightarrow s_i, \theta_i, \varphi_i$$

$$(\mathbf{r}, \mathbf{s}_i) \text{ angle} \Rightarrow \alpha_i$$

SYMMETRY AND PHYSICOCHEMICAL PROPERTIES

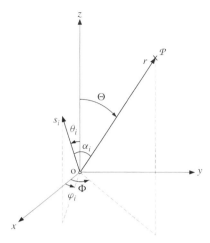

Figure 4.5 Notations used for spherical harmonics

In equation (4.7), $Y_l^{-m}(\theta_i, \varphi_i)$ is related to the charge i considered, whereas $Y_l^m(\Theta, \Phi)$ depends on the point $\mathcal{P}$ where the potential is calculated. For a distribution of charges, using equations (4.6) and (4.7), it can be shown that

$$\phi(\mathcal{P}) = \sum_{i=1}^{N} \frac{e_i}{r} \sum_{l=0}^{l=\infty} \left(\frac{s_i}{r}\right)^l \frac{4\pi}{2l+1} \sum_{m=-l}^{m=+l} (-1)^m Y_l^{-m}(\theta_i, \varphi_i) Y_l^m(\Theta, \Phi) \qquad (4.8)$$

In equation (4.8), the two integer numbers m and l appear; l has already been encountered (Table 4.1) and is associated with the type of electrical moment and angular part for the various terms found in $\phi(\mathcal{P})$. The point symmetry group of space to be considered is K, whose table of characters is shown in Table 4.2.

In Table 4.1 only some of the angular parts of the functions have been found since only the potential due to a single charge was considered. In the case where a three dimensional distribution of electrical charges is studied new angular contributions appear, which depend on the two angles Θ and Φ. Their number — for a given value of l — is equal to the order of the corresponding irreducible representation and numerically equal to $(2l + 1)$. These $(2l + 1)$ different values

Table 4.2 Table of characters of the point symmetry group K and association with the electrical moments

	E	∞C_Φ	...		l
S	1	1	...	Charge	0
P	3	$1 + 2\cos\Phi$	...	Dipole	1
D	5	$1 + 2\cos\Phi + 2\cos 2\Phi$	...	Quadrupole	2
F	7	$1 + 2\cos\Phi + 2\cos 2\Phi + 2\cos 3\Phi$	...	Octopole	3
...	...	...	...	...	

constitute the various m which must be taken into account for each l. In other words, the $(2l + 1)$ functions differing by m constitute the basis of a given irreducible representation of the group K. Any proper rotation in three dimensions transforms a function $Y_l^j(\Theta, \Phi)$ into a linear combination of the $(2l + 1)$ functions with the same l value [3–6].

In Table 4.3 the angular dependences previously found by developing in Taylor series the potential generated by a single charge (Table 4.1) are shown in bold. The functions can also be expressed in cartesian coordinates (Table 4.4). It is relatively easy to make a geometrical representation of a multipole since there is a direct correlation with the corresponding angular part of the single electron atomic orbitals (Table 4.3).

A polar molecule must have a symmetry group compatible with the symmetry of a vector. This means, expressed in terms of group theory, that at least one

Table 4.3 Expressions of the first spherical harmonics (polar coordinates). A geometrical representation of some associated atomic orbitals is given

		l	m	$Y_l^m(\Theta, \Phi)$
S		0	0	$Y_0^0 = \dfrac{1}{2\sqrt{\pi}}$
P		1	0	$Y_1^0 = \dfrac{\sqrt{3}}{2\sqrt{\pi}}\cos\Theta$
			± 1	$Y_1^{\pm 1} = \dfrac{\sqrt{3}}{2\sqrt{2\pi}}\sin\Theta e^{\pm i\Phi}$
D		2	0	$Y_2^0 = \dfrac{\sqrt{5}}{4\sqrt{\pi}}(3\cos^2\Theta - 1)$
			± 1	$Y_2^{\pm 1} = \dfrac{\sqrt{15}}{2\sqrt{2\pi}}\sin\Theta\cos\Theta e^{\pm i\Phi}$
			± 2	$Y_2^{\pm 2} = \dfrac{\sqrt{15}}{4\sqrt{2\pi}}\sin^2\Theta e^{\pm 2i\Phi}$
F		3	0	$Y_3^0 = \dfrac{\sqrt{7}}{4\sqrt{\pi}}(5\cos^3\Theta - 3\cos\Theta)$
			± 1	$Y_3^{\pm 1} = \dfrac{\sqrt{21}}{8\sqrt{\pi}}\sin\Theta(5\cos^2\Theta - 1)e^{\pm i\Phi}$
			± 2	$Y_3^{\pm 2} = \dfrac{\sqrt{105}}{4\sqrt{2\pi}}\sin^2\Theta\cos\Theta e^{\pm 2i\Phi}$
			± 3	$Y_3^{\pm 3} = \dfrac{\sqrt{35}}{8\sqrt{\pi}}\sin^3\Theta e^{\pm 3i\Phi}$

SYMMETRY AND PHYSICOCHEMICAL PROPERTIES

Table 4.4 Expressions in cartesian coordinates of the first spherical harmonics

Y_0^0	s	$\dfrac{1}{\sqrt{4\pi}}$	s	$\dfrac{1}{\sqrt{4\pi}}$	
Y_1^0	p_0	$\sqrt{\dfrac{3}{4\pi}}\left(\dfrac{z}{r}\right)$	p_z	$\sqrt{\dfrac{3}{4\pi}}\left(\dfrac{z}{r}\right)$	
$Y_1^{\pm 1}$	$p_{\pm 1}$	$\sqrt{\dfrac{3}{8\pi}}\left(\dfrac{x\pm iy}{r}\right)$	p_x	$\sqrt{\dfrac{3}{4\pi}}\left(\dfrac{x}{r}\right)$	
			p_y	$\sqrt{\dfrac{3}{4\pi}}\left(\dfrac{y}{r}\right)$	
Y_2^0	d_0	$\sqrt{\dfrac{5}{16\pi}}\left[\dfrac{2z^2-(x^2+y^2)}{r^2}\right]$	d_{z^2}	$\sqrt{\dfrac{5}{16\pi}}\left[\dfrac{2z^2-(x^2+y^2)}{r^2}\right]$	
$Y_2^{\pm 1}$	$d_{\pm 1}$	$\sqrt{\dfrac{30}{16\pi}}\left[\dfrac{xz\pm iyz}{r^2}\right]$	d_{xz}	$\sqrt{\dfrac{60}{16\pi}}\left(\dfrac{xz}{r^2}\right)$	
			d_{yz}	$\sqrt{\dfrac{60}{16\pi}}\left(\dfrac{yz}{r^2}\right)$	
$Y_2^{\pm 2}$	$d_{\pm 2}$	$\dfrac{1}{2}\sqrt{\dfrac{30}{16\pi}}\left[\dfrac{(x\pm iy)^2}{r^2}\right]$	$d_{x^2-y^2}$	$\sqrt{\dfrac{15}{16\pi}}\left(\dfrac{x^2-y^2}{r^2}\right)$	
			d_{xy}	$\sqrt{\dfrac{60}{16\pi}}\left(\dfrac{xy}{r^2}\right)$	

of the functions x, y or z is a basis of the totally symmetric representation of the group. The three components of the dipole have the symmetry of p orbitals. Quadrupole components correspond to polynomials of second order (d orbitals, see Table 4.3) and the octopole ones (f orbitals) are associated to third-order polynomials.

All multipoles are compatible with a molecule which has no elements of symmetry (C_1 group), but only the multipole of zero order corresponding to a point charge remains in the case of isotropic groups (K and K_h). Between these two limits, multipole components may be cancelled by the presence of some symmetry operation. Multipole components of first and second order are described in conventional character tables (see Appendix 2). C_n and C_{nv} groups could have a dipolar moment with orientation along the z axis, d_{z^2} and $d_{x^2-y^2}$ quadrupolar moments for $n=2$ and only d_{z^2} if $n>2$.

The first nonzero moment of groups, C_i, D_n, D_{nh}, D_{nd} and S_n is quadrupolar. In the case of symmetry groups containing several axes of order greater than 2, dipolar and quadrupolar terms are cancelled. The first moment of the group T is octopolar (order 3, seven components) and hexadecapolar (order 4, nine components) for the group O.

Real functions can be calculated from the two complex conjugate functions Y_l^m and Y_l^{-m}:

$$Y_l^{\text{real}} = \dfrac{1}{\sqrt{2}}\left[Y_l^{|m|} + Y_l^{-|m|}\right] \quad (4.9)$$

Table 4.5 Free energy of interaction of a solute molecule (as indicated) dissolved in cyclohexane ($\varepsilon = 2.015$) or acetone ($\varepsilon = 20.70$). The dipolar (G_{dip}) and quadrupolar (G_{quad}) contributions are distinguished. G_{dip} and G_{quad} are in kcal/mole. (After Ref. [7])

| Symmetry | HCl | HBr | CH$_3$X X= | | | H$_2$O | CH$_3$CN | (pyridine) | (benzene) | (C$_6$F$_6$) | O=C=O | S=C=S |
			Cl	Br	I							
Symmetry	$C_{\infty v}$	$C_{\infty v}$	$C_{\infty v}$	$C_{\infty v}$	$C_{\infty v}$	C_{2v}	C_{3v} ($C_{\infty v}$)	C_{2v}	D_{6h}	D_{6h}	$D_{\infty h}$	$D_{\infty h}$
Dipolar Mt 10^{18} esu cm	1.08	0.79	1.90	1.84	1.62	1.85	3.91	2.21				
Quadrupolar Mt 10^{26} esu cm^2	3.8	4.0	−1.2	−3.6	−5.4	2.6	−1.8	9.7	−8.7	9.5	−4.49	3.60
	−1.9	−2.0	0.6	1.8	2.7	−0.1	0.9	−6.2	4.35	−4.75	2.25	−1.80
	−1.9	−2.0	0.6	1.8	2.7	−0.5	0.9	−3.5	4.35	−4.75	2.25	−1.80
G_{dip} cyclohexane	−0.27	−0.16	−0.48	−0.44	−0.31	−1.39	−2.11	−0.44				
G_{quad} cyclohexane	−0.62	−0.74	−0.02	−0.20	−0.38	−0.89	−0.06	−0.82	−0.54	−0.42	−0.49	−0.18
G_{quad}/G_{dip} (%) cyclohexane	230	463	4	45	123	64	3	186				
G_{dip} acetone	−0.64	−0.36	−1.10	−1.01	−0.71	−3.21	−4.86	−1.02				
G_{quad} acetone	−1.50	−1.79	−0.06	−0.48	−0.92	−2.18	−0.14	−1.99	−1.31	−1.02	−1.19	−0.43
G_{quad}/G_{dip} (%) acetone	234	497	5	48	130	68	3	195				

$$Y_2^{\text{real}} = \frac{1}{i\sqrt{2}} \left[Y_l^{|m|} - Y_l^{-|m|} \right] \tag{4.10}$$

For instance:

$$p_z = Y_1^0 = \frac{\sqrt{3}}{2\sqrt{\pi}} \cos \Theta$$

$$p_x = \frac{1}{\sqrt{2}} \left[Y_1^1 + Y_1^{-1} \right] = \frac{\sqrt{3}}{\sqrt{4\pi}} \sin \Theta \cos \Phi$$

$$p_y = \frac{1}{i\sqrt{2}} \left[Y_1^1 - Y_1^{-1} \right] = \frac{\sqrt{3}}{\sqrt{4\pi}} \sin \Theta \sin \Phi$$

The various energy levels of a monoelectronic state are represented conventionally as (1s) (2s 2p) (3s 3p 3d) (4s 4p 4d 4f)$\cdots$.

4.2 ELECTRICAL MULTIPOLES IN CHEMISTRY

The multipolar character of a molecule is very important for determining its chemical properties. Calculations have been carried out in which the energy of interaction of a solute with different surrounding solvent molecules (cyclohexane dielectric constants, $\varepsilon = 2.015$; acetone, $\varepsilon = 20.70$) has been determined [7] (Table 4.5).

CH_3CN has a larger dipole moment $(3.91D)$ than HCl $(1.08D)$. In the latter the quadrupolar contribution dominates the interaction with the solvent both in cyclohexane and in acetone, whereas it is negligible for CH_3CN. In the series CH_3X (X = Cl, Br, I), the quadrupolar contribution increases with the van der Waals radius of the heteroatom and becomes predominent for the iodo derivative.

In the case where the molecular unit has no dipole moment for symmetry reasons (D_{6h}, $D_{\infty h}$), the first nonzero contribution is quadrupolar (for the compounds shown) and is of the same order of magnitude as the one previously found for dipolar contributions. It is therefore clear that, even for purely chemical properties, the multipolar contributions of the molecular units must be taken into account.

Table 4.6 Free energy of interaction of CH_4 dissolved in cyclohexane ($\varepsilon = 2.015$) or acetone ($\varepsilon = 20.70$) (after Ref. [7]). The octopolar moments are determined in references [8] and [9], respectively

Solute	Octopolar Mt Ω_{xyz} (10^{34} esu cm^3)		G_{oct} (kcal/mole)	Ref.
CH_4	1.94	Cyclohexane	−0.05	8
		Acetone	−0.14	
CH_4	4.5	Cyclohexane	−0.29	9
		Acetone	−0.72	

In tetrahedral symmetry, the first electrical moment different from zero is the octopolar one. As previously, the energy of interaction of CH_4 dissolved in acetone or cyclohexane can be calculated (Table 4.6). The uncertainty associated with the octopolar moment is fairly high. The numerical value of the octopolar free energies of interaction are, however, of the same order of magnitude as the ones found for lower-order multipoles.

4.3 CAUSE/EFFECT SYMMETRY RELATIONSHIPS

4.3.1 The Curie Principle

There are many ways to express the relationship existing between the symmetry of a cause (perturbation of a system) and the symmetry of the effect. Neumann's principle states [10, 11]: 'The symmetry elements of any physical property of a crystal must include the symmetry elements of the point symmetry group of the crystal'. It must be noticed that the physical properties may have additional symmetry elements to the crystal.

Another way to express such a relationship is to consider the couple cause/effect. The 'Principe de Curie' [12] states: 'Lorsque certaines causes produisent certains effets, les éléments de symétrie des causes doivent se retrouver dans les effets produits'. In its original statement, P. Curie indicated that the 'cause' is in fact the material submitted to a perturbation or a field. It is consequently the symmetry of the couple perturbation/material which must be considered as the symmetry of the 'cause'. The set of symmetry elements of the 'cause' must not contain elements incompatible with the expected effect.

One can consider, for example, three basic perturbations: magnetic field ($C_{\infty h}$), electrical field ($C_{\infty v}$) and torsion (D_{∞}) (Figure 4.6). One can, for instance, apply an electrical field ($C_{\infty v}$) on a molecular unit of symmetry D_{3h} along the C_3 axis. The 'cause' must be considered to be the intersection $C_{\infty v} \cap D_{3h} = C_{3v}$, this latter symmetry group possessing all the common symmetry operations (Figure 4.7). A polarization (symmetry $C_{\infty v}$) may therefore occur in the molecular unit since all the symmetry elements of the 'cause' (C_{3v}) are contained in the symmetry of

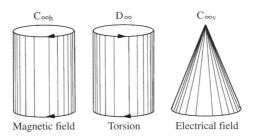

Figure 4.6 Schematic representation of three perturbations and the corresponding point symmetry groups

SYMMETRY AND PHYSICOCHEMICAL PROPERTIES

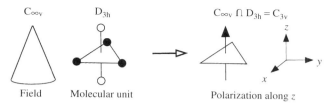

Figure 4.7 Application of an electrical field ($C_{\infty v}$) on a molecular unit (D_{3h}). The 'cause' is $C_{\infty v} \cap D_{3h} = C_{3v}$

the effect. In other words, the symmetry group of the 'cause' is a subgroup of the symmetry of the effect [10, 11, 12]:

$$\text{Symmetry (cause)} \subseteq \text{symmetry (effect)}$$

It is worth stressing that the symmetry group of the 'cause' depends on the relative orientations of the field and of the material on which it is applied.

4.3.2 Applications of the Curie Principle

In practical cases, the electrical field is applied on a molecular material rather than on a single molecular unit. For single crystals, the crystalline class — the symmetry of orientation — must be taken. For isotropic liquids, either the group K or K_h must be considered depending on whether the molecular unit is chiral or not. The various mesophases which have been encountered have been shown to possess the following symmetries: smectic A, $D_{\infty h}$; smectic C, C_{2h}; hexagonal columnar, D_{6h}; etc.

Another example can be considered in which a linearly polarized wave (in the xy plane) propagates in a liquid constituted of achiral or chiral molecules

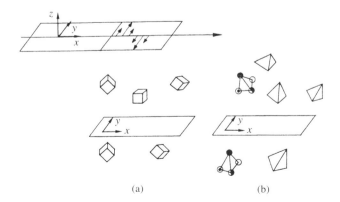

Figure 4.8 Propagation of a linearly polarized wave (in the xy plane) in (a) achiral and (b) chiral liquids. (From Ref. [13])

(Figure 4.8). The electrical field which is contained in the (x, y) plane polarizes the various molecular units of the liquid. The induced dipole moments in turn generate an electrical field. In the case where the plane (x, y) is a symmetry element of the medium (this is the case with an achiral molecular unit) the contributions arising from the molecular units situated above and below the plane must cancel. This is no longer true in chiral media where the (x, y) plane is not a symmetry element of the condensed phase: then, the sum of the various molecular induced dipole moments may be different from zero (at least from the symmetry point of view) along the z axis.

The symmetry groups associated with chiral molecular units are those that do not include improper rotations and, in particular, no centre of symmetry and no mirror plane (groups C_n, D_n, T, O, I). The same type of physical phenomenon may be examined using the 'Principe de Curie'. An electromagnetic wave, at any instant, consists of an electrical and a magnetic field perpendicular to each other. The respective symmetries of the external perturbations are therefore $C_{\infty v}$ and $C_{\infty h}$. However, in this notation the reference frame is not common; only the plane (x, z) is a common symmetry element of the electrical and magnetic fields (Figure 4.9).

The external perturbation (the electromagnetic wave) is consequently of symmetry $C_s(m)$ and an achiral medium is of symmetry K_h. The symmetry of the 'cause' is therefore

$$C_s \cap K_h = C_s$$

The plane (x, z) contains the axis along which the wave propagates (Ox) and it does not allow the appearance of a polarization along Oy. The electrical vector **E** must therefore remain in the (x, z) plane and no rotation of the plane of polarization is permitted.

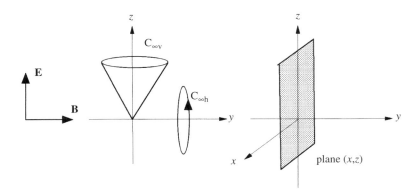

Figure 4.9 Determination of the common symmetry elements of an electrical field (**E**) and a magnetic field (**B**) (electromagnetic wave)

SYMMETRY AND PHYSICOCHEMICAL PROPERTIES

In the case of a chiral medium, the symmetry of the medium is K and the 'cause' is given by

$$C_s \cap K = C_1$$

The plane (x, z) is no longer preserved and a polarization *can* occur along Oy, giving the possibility of yielding a rotation of the polarization plane.

Piezoelectricity was discovered in 1880 by Jacques and Pierre Curie [14]. As an example, one can consider a mechanical constraint applied on a crystal of symmetry D_3 (Figure 4.10). The response of the material depends on the direction of the applied constraint. If it is applied along the z axis, which contains the C_3 axis, the symmetry of the constraint is z^2 ($D_{\infty h}$). The 'cause' is therefore

$$D_{\infty h} \cap D_3 = D_3$$

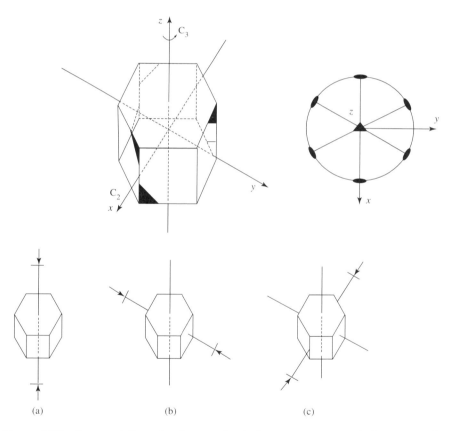

Figure 4.10 Representation of various mechanical stresses on a crystal of symmetry D_3: (a) along the z axis, D_3 is unchanged; (b) along the y axis, D_3 becomes C_2 (pol. along x); (c) along the x axis, D_3 becomes C_2 (pol. along x)

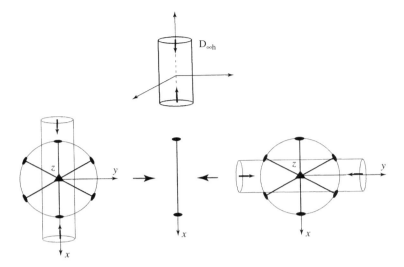

Figure 4.11 Representation of the $D_{\infty h}$ symmetry of a constraint applied along x (stress = x^2) and stereographical projection showing the symmetry elements conserved when the stress is applied along the x or the y axis

In order to find the common symmetry elements, the same reference axes must be used. D_3 is not a polar group (subgroups of $C_{\infty v}$):

$$\underset{\text{cause}}{D_3} \not\subseteq \underset{\text{effect}}{C_{\infty v}}$$

and therefore no polar vector can form. If the constraint is applied along the x or y axes, the cause has a symmetry element C_2 along the x axis since only this symmetry element is conserved (see Figure 4.11).

The symmetry group C_2 is compatible with an electrical vector ($C_{\infty v}$):

$$\underset{\text{cause}}{C_2} \subseteq \underset{\text{effect}}{C_{\infty v}}$$

Whatever the direction of the applied field some crystalline symmetry groups lead to a symmetry of the 'cause', which is never a subgroup of the expected effect. For instance, a piezoelectric effect cannot be obtained with condensed phases that possess a centre of symmetry. The centre is common to the stress and to the material and is not a symmetry element of the effect (electrical polarization).

A classification relating crystalline classes to some anisotropic effects is shown in Figure 4.12:

(a) Optical activity: rotation of the polarization plane of the light.
(b) Pyroelectricity: spontaneous polarization which can be induced by heating.
(c) Second harmonic generation: frequency doubling of a monochromatic light.
(d) Piezoelectricity: electric polarization induced by a stress.

SYMMETRY AND PHYSICOCHEMICAL PROPERTIES

SYSTEM	CRYSTALLINE CLASS		OPT.	PYRO.	PIEZO. S.H.G.
TRICLINIC 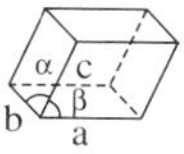 $a \neq b \neq c$, $\alpha \neq \beta \neq \gamma \neq 90°$	$\bar{1}$ (C_i)	1 (C_1)	+	+	+
MONOCLINIC 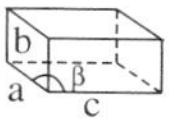 $a \neq b \neq c$, $\alpha = \gamma = 90° \neq \beta$	$\dfrac{2}{m}$ (C_{2h})	2 (C_2)	+	+	+
		m (C_s)	+	+	+
ORTHORHOMBIC 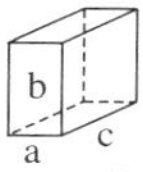 $a \neq b \neq c$, $\alpha = \beta = \gamma = 90°$	$\dfrac{2\,2\,2}{mmm}$ (D_{2h})	2 2 2 (D_2)	+	−	+
		mm2 (C_{2v})	+	+	+
TETRAGONAL 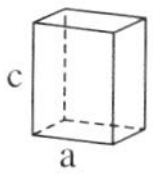 $a = b \neq c$, $\alpha = \beta = \gamma = 90°$	$\dfrac{4}{m}$ (C_{4h})	4 (C_4)	+	+	+
		$\bar{4}$ (S_4)	+	−	$(+)^a$
	$\dfrac{4\,2\,2}{mmm}$ (D_{4h})	4 2 2 (D_4)	+	−	$(+)^b$
		4 mm (C_{4v})	−	+	+
		$\bar{4}$ 2 m (D_{2d})	+	−	+
TRIGONAL  $a = b = c$, $\alpha = \beta = \gamma \neq 90°$ $a = b \neq c$, $\alpha = \beta = 90°$, $\gamma = 120°$	$\bar{3}$ (C_{3i})	3 (C_3)	+	+	+
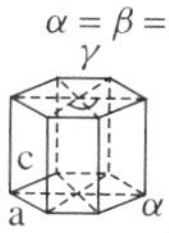	$\bar{3}\dfrac{2}{m}$ (D_{3d})	32 (D_3)	+	−	+
		3 m (C_{3v})	−	+	+
HEXAGONAL 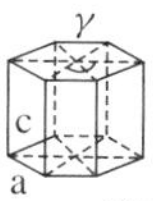 $a = b \neq c$, $\alpha = \beta = 90°$, $\gamma = 120°$	$\dfrac{6}{m}$ (C_{6h})	6 (C_6)	+	+	+
		$\bar{6}$ (C_{3h})	−	−	+
	$\dfrac{6\,2\,2}{mmm}$ (D_{6h})	6 2 2 (D_6)	+	−	$(+)^b$
		6 mm (C_{6v})	−	+	+
		$\bar{6}$ 2 m (D_{3h})	−	−	+
CUBIC 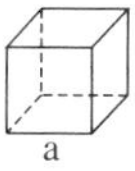 $a = b = c$, $\alpha = \beta = \gamma = 90°$	$\dfrac{2}{m}\bar{3}$ (T_h)	23 (T)	+	−	$(+)^a$
	$\dfrac{4}{m}\bar{3}\dfrac{2}{m}$ (O_h)	432 (O)	+	−	−
		$\bar{4}$ 3 m (T_d)	−	−	$(+)^a$

Figure 4.12 Relation between the symmetry of the crystalline classes and some physical properties. (After Refs. [13] and [15]). ^{a}phase matching cannot be realized in SHG. ^{b}negative for SHG (Kleinman relation)

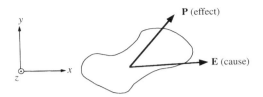

Figure 4.13 An electrical field generates an induced dipole (**P**) when applied to a given molecular unit

4.3.3 The Curie Principle Using Irreducible Representations

In the previous approach 'cause' and 'effect' were considered on an overall basis. A more detailed description may be given by using tensor notation.

For example, consider a molecular unit submitted to an electrical field. An induced dipole is generated and, in the general case, it is not colinear to the applied field (Figure 4.13). The polarizability may be described by a tensor of rank two:

$$\begin{bmatrix} P_x \\ P_y \\ P_z \end{bmatrix} = \begin{bmatrix} \alpha_{xx} & \alpha_{xy} & \alpha_{xz} \\ \alpha_{yx} & \alpha_{yy} & \alpha_{yz} \\ \alpha_{zx} & \alpha_{zy} & \alpha_{zz} \end{bmatrix} \begin{bmatrix} E_x \\ E_y \\ E_z \end{bmatrix} \quad (4.11)$$

In most cases, the molecular unit may be modelled by an ellipsoid and then $\alpha_{ij} = \alpha_{ji}$. The number of molecular unit polarizability tensor coefficients may be reduced when the axes chosen are colinear to the principal axes of the ellipsoid (Figure 4.14). In the other cases, crossed terms appear in the tensor.

The case where the reference axes follow the principal axes of the ellipsoid is shown in Figure 4.15. The polarization is parallel to the field only along the principal axes of the ellipsoid. In this case the tensor is in its diagonal form.

The applied electrical field (**E**) in the (x, z) plane generates an induced dipole (**P**) with components E_x and E_z:

$$P_x = \alpha_{xx} E_x$$
$$P_z = \alpha_{zz} E_z$$

In two dimensions, the tensor may therefore be written as

$$\begin{pmatrix} \alpha_{xx} & 0 \\ 0 & \alpha_{zz} \end{pmatrix}$$

The crossed terms α_{ij} $(i \neq j)$ are null.

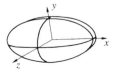

Figure 4.14 Representation of the principal axes of the ellipsoid of polarizability of a molecular unit

SYMMETRY AND PHYSICOCHEMICAL PROPERTIES

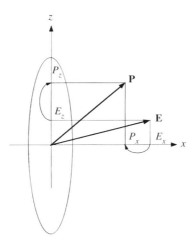

Figure 4.15 Induced polarization of a molecular unit with $\alpha_{xx} \neq \alpha_{zz}$. The reference frame follows the principal axes of the ellipsoid representing the molecular unit

When the reference frame is chosen in an arbitrary way, the external field generates induced dipoles along the main axes of the ellipsoid, which are not colinear with the reference frame. Consequently, the components E'_x and E'_z induce a polarization on both the x and z axes (Figure 4.16).

The polarization $P_{z'}$ is due to fields directed along both E_x and E_z. $P_{z'}$ in turn has z and x components. P_z is the sum of the projections on Oz of $P_{x'}$ and $P_{z'}$.

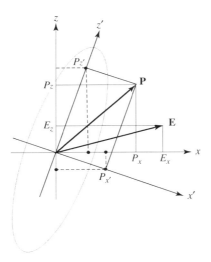

Figure 4.16 Case where the reference frame (xOz) does not follow the principal axes of the ellipsoid ($x'Oz'$) representing a molecular unit. In the case shown, P_z is the sum of the projections on Oz of the components $P_{z'}$ and $P_{x'}$

Consequently, P_x and P_z are due simultaneously to E_x and E_z and crossed terms appear:

$$P_x = \alpha_{xx}E_x + \alpha_{xz}E_z$$
$$P_z = \alpha_{zx}E_x + \alpha_{zz}E_z$$

In the general case, the tensor is then given by

$$\begin{pmatrix} \alpha_{xx} & \alpha_{xy} & \alpha_{xz} \\ \alpha_{yx} & \alpha_{yy} & \alpha_{yz} \\ \alpha_{zx} & \alpha_{zy} & \alpha_{zz} \end{pmatrix}$$

In the case where the axes are properly chosen, as shown before, the tensor of rank 2 is reduced to

$$\begin{pmatrix} \alpha_{xx} & 0 & 0 \\ 0 & \alpha_{yy} & 0 \\ 0 & 0 & \alpha_{zz} \end{pmatrix}$$

and the corresponding equation of the ellipsoid of polarizability is

$$\alpha_{xx}x^2 + \alpha_{yy}y^2 + \alpha_{zz}z^2 = +1 \tag{4.12}$$

The polarization can be simply written as

$$P_x = \alpha_{xx}E_x$$
$$P_y = \alpha_{yy}E_y$$
$$\underset{\text{effect}}{P_z} = \underset{\text{cause}}{\alpha_{zz}E_z}$$

If the components of the applied field and of the induced dipole are considered, there is a simple relationship between the cause and the effect. However, contrary to P. Curie's statement, the cause, this time, is merely the external field applied without consideration of the condensed phase on which it is applied. This way of expressing the Curie principle is different from the conventional one since the three components of the effect (projection of the polarization vector along x, y or z) and the three components of the cause (E_x, E_y, E_z) are distinguished and related to each others via individual tensor components α_{ij} ($i, j = x, y$ or z), which represent the physical properties of the medium.

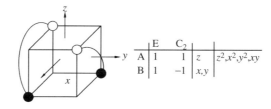

Figure 4.17 Symmetry operations and table of characters associated with the point symmetry group C_2

SYMMETRY AND PHYSICOCHEMICAL PROPERTIES

The α_{ij} coefficients different from zero can now be determined. Consider the example of a molecular unit of symmetry C_2. The corresponding symmetry operations are listed in Figure 4.17.

The application, for instance, of the C_2 symmetry operation on the corresponding molecular unit (while preserving the reference frame) leaves the molecule unchanged. The same symmetry operation applied on the induced dipole vector **P**—or any other vector—leads to the result shown in Figure 4.18. The application of the C_2 axis on the field **E** and on the induced polarization **P** gives

$$E_x \longrightarrow -E_x, \quad E_y \longrightarrow -E_y \quad \text{and} \quad E_z \longrightarrow E_z$$
$$P_x \longrightarrow -P_x, \quad P_y \longrightarrow -P_y \quad \text{and} \quad P_z \longrightarrow P_z$$

since $P_x = \alpha_{xx} E_x$, $-P_x = \alpha_{xx}(-E_x)$.

The coefficient α_{xx} remains unchanged after the rotation of the molecular unit around its symmetry axis. The same treatment can be applied to α_{yy}. Both E_z and P_z are invariant by the application of the C_2 axis, permitting the determination of α_{zz}. The tensor coefficients α_{ij} are unchanged only when the corresponding components of the field E_i (cause) and of the polarization P_i (effect) are associated with the same irreducible representation of the symmetry group of the molecular unit. This can be seen in the character table shown in Figure 4.17. The reference axes are taken with z along the C_2 symmetry axis. In this case,

$$\Gamma(E_z) = \Gamma(P_z) = \Gamma(z) = A$$
$$\Gamma(E_x) = \Gamma(P_x) = \Gamma(E_y) = \Gamma(P_y) = \Gamma(x) = \Gamma(y) = B$$

No polarization along z can be obtained by applying a field along x or y: Thus $\Gamma(x) = \Gamma(y) \neq \Gamma(z) \Rightarrow \alpha_{xz} = \alpha_{yz} = 0$. Since $\Gamma(x) = \Gamma(y)$, a nonzero coefficient α_{xy} is possible. This coefficient will be equal to zero only if the x and y axes are taken along the principal axes of the ellipsoid.

For a given symmetry group (and by considering the reference frame used to establish the corresponding table of character), a general condition relating cause

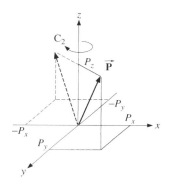

Figure 4.18 Application of $C_2(z)$ on the vector **P**

to effect may be written as

$$\Gamma(\text{cause}) = \Gamma(\text{effect}) \quad (4.13)$$

In the point symmetry group of a given molecular unit or condensed phase, the irreducible representation associated with the cause must be the same as the irreducible representation of the effect in order to have the corresponding tensor coefficient different from zero [16]. (See, however, the next section for a generalization.) It is important to note that the symmetry of the cause is not this time the intersection corresponding to the external perturbation and the medium, as in the 'Principe de Curie', but the symmetry of the perturbation itself.

In the case where two electrical fields are considered, the exciting radiation is no longer a vector but a tensor of rank 2. The product **EE** means that a novel type of constraint is applied on the molecular unit: the component E_x is effective along the x axis while E_y is simultaneously acting along the y axis (Figure 4.19). **EE** is a tensorial product and behaves as $E_i E_j$.

The hyperpolarizability is then given by

$$P_i = \sum_{j,k} \beta_{ijk} E_j E_k \quad (4.14)$$

where β_{ijk} is a third rank tensor with 27 components which can be reduced to 18 when $E_i E_j = E_j E_i$:

$$\begin{pmatrix} P_x \\ P_y \\ P_z \end{pmatrix} = \begin{pmatrix} \beta_{xxx} & \beta_{xyy} & \beta_{xzz} & \beta_{xyz} & \beta_{xxz} & \beta_{xxy} \\ \beta_{yxx} & \beta_{yyy} & \beta_{yzz} & \beta_{yyz} & \beta_{yxz} & \beta_{yxy} \\ \beta_{zxx} & \beta_{zyy} & \beta_{zzz} & \beta_{zyz} & \beta_{zxz} & \beta_{zxy} \end{pmatrix} \begin{pmatrix} E_x^2 \\ E_y^2 \\ E_z^2 \\ 2E_y E_z \\ 2E_x E_z \\ 2E_x E_y \end{pmatrix} \quad (4.15)$$

The non-zero β coefficients may be determined by the method previously used. In the C_2 symmetry group, the application of the $C_2(z)$ symmetry operation leads to

$$P_x \to -P_x$$
$$E_x \to -E_x$$

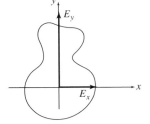

Figure 4.19 The $E_x E_y$ constraint appearing in the polarization of a molecular unit when two fields are considered (two-dimensional case)

SYMMETRY AND PHYSICOCHEMICAL PROPERTIES

$$E_y \to -E_y$$
$$E_z \to E_z$$

$$-(\beta_{xxx}E_x^2 + \beta_{xyy}E_y^2 + \beta_{xzz}E_z^2 + 2\beta_{xyz}E_yE_z + 2\beta_{xxz}E_xE_z + 2\beta_{xxy}E_xE_y)$$
$$= \beta_{xxx}(-E_x)^2 + \beta_{xyy}(-E_y)^2 + \beta_{xzz}E_z^2 + 2\beta_{xyz}(-E_y)(E_z) + 2\beta_{xxz}(-E_x)(E_z)$$
$$+ 2\beta_{xxy}(-E_x)(-E_y)$$

Therefore:

$$\beta_{xxx} = \beta_{xyy} = \beta_{xzz} = \beta_{xxy} = 0$$
$$\beta_{xyz} \neq 0; \quad \beta_{xxz} \neq 0$$

By repeating the same operation with the other components, P_y and P_z, one can find

$$\begin{bmatrix} 0 & 0 & 0 & \beta_{xyz} & \beta_{xxz} & 0 \\ 0 & 0 & 0 & \beta_{yyz} & \beta_{yxz} & 0 \\ \beta_{zxx} & \beta_{zyy} & \beta_{zzz} & 0 & 0 & \beta_{zxy} \end{bmatrix}$$

The new formulation of the 'Principe de Curie' can then be written as

$$\Gamma(P_i) = \Gamma(E_jE_k)$$
$$\text{cause} \quad \text{effect}$$

or, since P_i and E_jE_k behave as the corresponding vectors,

$$\boxed{\Gamma(i) = \Gamma(jk)} \tag{4.16}$$

Determination of the nonzero coefficients may be more readily carried out by using the table of characters. The irreducible representations associated with i, j and k are usually indicated in the tables, which allows determination of the nonzero hyperpolarizability coefficients β_{ijk}:

C_2	E	C_2	Effect	Cause	Coefficients
A	1	1	z	x^2, y^2, z^2, xy	zxx, zyy, zzz, zxy
B	1	-1	x, y	yz, xz	xyz, xxz, yyz, yxz

This apparent simplicity is broken whenever the corresponding irreducible representations are not of order 1. For example, in the group D_3:

D_3	E	$2C_3$	$3C_2$	Effect	Cause
E	2	-1	0	(x, y)	$(x^2 - y^2, xy), (xz, yz)$

It is then necessary to determine the matrices corresponding to the symmetry operation of the group D_3. Only two generating operations have to be considered:

a rotation of angle $2\pi/3$ around the z axis $C_3(z)$ and a rotation of the angle π around the x axis $C_2(x)$:

$$\begin{pmatrix} x' \\ y' \\ z' \end{pmatrix} = \begin{pmatrix} c & s & 0 \\ -s & c & 0 \\ 0 & 0 & 1 \end{pmatrix} \begin{pmatrix} x \\ y \\ z \end{pmatrix}$$

with

$c = \cos(2\pi/3) = -1/2$
$s = \sin(2\pi/3) = \sqrt{3}/2$

and

$$\begin{pmatrix} x' \\ y' \\ z' \end{pmatrix} = \begin{pmatrix} 1 & 0 & 0 \\ 0 & -1 & 0 \\ 0 & 0 & -1 \end{pmatrix} \begin{pmatrix} x \\ y \\ z \end{pmatrix}$$

Matrices based on the two-dimensional basis (x, y), $(x^2 - y^2, xy)$, (xz, yz) should have the same coefficients for each symmetry operation:

$$C_3(z) : \begin{pmatrix} c & s \\ -s & c \end{pmatrix} \begin{pmatrix} x \\ y \end{pmatrix} \Rightarrow \begin{pmatrix} c^2 - s^2 & -2cs \\ 2cs & c^2 - s^2 \end{pmatrix} \begin{pmatrix} x^2 - y^2 \\ -2xy \end{pmatrix}$$

$$\Rightarrow \begin{pmatrix} c & s \\ -s & c \end{pmatrix} \begin{pmatrix} yz \\ -xz \end{pmatrix}$$

The corresponding matrices are identical since $c^2 - s^2 = -\frac{1}{2} = c$ and $-2cs = \sqrt{3}/2 = s$:

$$C_2(x) : \begin{pmatrix} 1 & 0 \\ 0 & -1 \end{pmatrix} \begin{pmatrix} x \\ y \end{pmatrix} \Rightarrow \begin{pmatrix} 1 & 0 \\ 0 & -1 \end{pmatrix} \begin{pmatrix} x^2 - y^2 \\ -2xy \end{pmatrix} \Rightarrow \begin{pmatrix} 1 & 0 \\ 0 & -1 \end{pmatrix} \begin{pmatrix} yz \\ -xz \end{pmatrix}$$

The exact relationship between cause and effect may be written as

Effect	(x, y)	(x, y)
Cause	$(x^2 - y^2, -2xy)$	$(yz, -xz)$

It can be readily seen that the nonzero coefficients of the tensor are

$$\beta_{yxy} \quad \beta_{xyz} \quad \beta_{yxz}$$

The case of $(x^2 - y^2)$ is slightly more difficult to treat since for a P_x polarization both the causes $E_x E_x$ (equivalent to x^2) and $E_y E_y (y^2)$ intervene. It must therefore be written as

$$P_x = \beta_{xxx} E_x E_x - \beta_{xyy} E_y E_y$$

Moreover, in the group D_3, x and y cannot be distinguished and are associated with the same irreducible representation of dimension 2. In consequence,

$$|\beta_{xxx}| = |\beta_{xyy}|$$

SYMMETRY AND PHYSICOCHEMICAL PROPERTIES

Further symmetry considerations could permit the following relationships to be demonstrated:

$$\beta_{xxx} = -\beta_{xyy} = -\beta_{yxy}$$

$$\beta_{xyz} = -\beta_{yxz}$$

The piezoelectrical and nonlinear optical tensor coefficients corresponding to the most used groups of symmetry are shown in Appendix 8.

4.4 GENERALIZATION OF THE CURIE PRINCIPLE

The polarizability of a molecule is described by a tensor of rank 2 which relates two vectors: the field **E** (cause) and the polarization of the molecule **P** (effect). Vectors can be considered as tensors of rank 1 and scalars as tensors of rank 0. Tensors of rank n will have 3^n coefficients (see Ref. [17]). It is then possible to classify the cause/effect relationships for physicochemical properties according to the rank of the corresponding tensors. Such a classification has been given below for the main physicochemical properties of materials.

(a) Scalar/scalar
    Calorific capacity
(b) Scalar/vector: tensor of rank 1
    Pyroelectricity
    Heat of polarization
    Polarization due to a hydrostatic pressure
(c) Vector/vector: tensor of rank 2
    Dielectric permittivity
    Magnetic permeability
    Electrical conductivity
    Thermal conductivity
    Thermoelectricity
(d) Scalar/tensor of rank 2: tensor of rank 2
    Thermal dilatation
    Heat of deformation
    Peltier effect
(e) Vector/tensor of rank 2: tensor of rank 3
    Piezoelectric effect
    Electro-optical effect
(f) Tensor rank 2/tensor rank 2: tensor of rank 4
    Elasticity
    Piezo-optic coefficients

In some cases, the representations associated with the cause and/or the effect are a sum of irreducible representations and/or irreducible representations of dimension greater than one:

$$\Gamma(\text{cause}) = \sum_i \Gamma_i$$

$$\Gamma(\text{effect}) = \sum_j \Gamma_j$$

where Γ_i, Γ_j are irreducible representations. It this case, it is necessary and sufficient that one of the irreducible representations of the cause is identical to one of the irreducible representations of the effect. This can be considered as a generalization of the 'Principe de Curie'.

Any anisotropic physical property can be described by a tensor T:

$$\mathcal{E}_m = (T_{m+n})C_n$$

where

$\mathcal{E}$ = effect
C = cause
T = tensor

and n is the rank of the cause and m the rank of the effect.

As an example, we will consider the case of the elastic deformation of crystals [18]. The perturbation — the stress — is a tensor of rank 2. The various stresses are generally divided into uniaxial (tensile) and shear stresses (Figure 4.20).

The effect (the strain or deformation) is also a second rank tensor and the cause/effect relationship is of the type:

$$\varepsilon_{ij} = s_{ijkl}\sigma_{kl}$$

If the strain is applied on a crystal of a given crystalline class, the symmetry relationships must be sought in the corresponding symmetry group. In the group

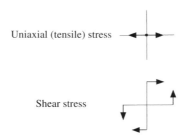

Figure 4.20 Two stresses which can be applied on a material of the types x^2 and xy

D_4 for example, the following can be seen:

D_4	Cause, effect
A_1	$x^2 + y^2, z^2$
B_1	$x^2 - y^2$
B_2	xy
E	(xz, yz)

In this case the cause and the effect always correspond to the same irreducible representations. Uniaxial stresses (x^2, y^2, z^2) cannot yield shear strains (xy, xz, yz).

4.5 REFERENCES

1. C. J. F. Böttcher, *Theory of Electric Polarization*, Elsevier, Amsterdam (1973).
2. A. D. Buckingham, *Quat. Rev.*, **13**, 183 (1959).
3. C. Cohen-Tannoudji, B. Diu and F. Laloë, *Mécanique quantique*, Hermann, Paris (1973).
4. A. Messiah, *Mécanique quantique*, Dunod, Paris (1962).
5. J. L. Rivail, *Eléments de chimie quantique à l'usage des chimistes*, Savoirs Actuels, Interéditions/éditions du CNRS, Paris (1989).
6. D. I. Blokhintsev, *Mécanique quantique*, Masson et c^{ie}, Paris (1967).
7. M. Claessens, L. Palombini, M.-L. Stien and J. Reisse, *Nouv. J. de Chimie*, **6**, 595 (1982).
8. R. D. Amos, *Mol. Phys.*, **38**, 33 (1979).
9. E. Stogryn and A. P. Stogryn, *Molec. Physics*, **11**, 371 (1966).
10. F. Neumann (1833) and Minnigerode (1884), see Ref. [19].
11. D. R. Lovett, *Tensor Properties of Crystals*, Adam Hilger, IOP, Philadelphia (1989).
12. P. Curie, *J. Phys.*, **3**, 393 (1894).
13. J. Simon, P. Bassoul and S. Norvez, *New J. Chem.*, **13**, 13 (1989).
14. J. Curie and P. Curie, *C. R. Acad. Sci. Paris*, **91**, 294 (1880).
15. T. Hahn (ed.), *International Tables for Crystallography*, Vol. A, *Space-Group Symmetry*, D. Reidel, Dordrecht (1983).
16. J. Simon, *C. R. Acad. Sci. Paris*, **324**, 47 (1997).
17. J. F. Nye, *Propriétés physiques des cristaux*, Dunod, Paris (1961).
18. A. W. Joshi, *Elements of Group Theory for Physicists*, Wiley Eastern, New Delhi (1988).
19. J. Sivardière, *La symétrie en mathématiques, physique et chimie*, Presses Universitaires de Grenoble (1995).

5 Interactions and Organization in Molecular Media

5.1 Directional and Nondirectional Forces	156
5.2 Charge/Charge (Coulomb) Forces	159
5.2.1 Lattice Energy of Ionic Crystals	159
5.2.2 Solubility of Ions	162
5.2.3 Hard Soft Acid Bases (HSAB Principle)	164
5.3 Molecular Polarization	167
5.4 Induced Dipole/Induced Dipole Interactions	171
5.4.1 Generalities	171
5.4.2 Packing of Aromatic Derivatives	172
5.4.3 Packing of Paraffinic Chains	176
5.5 Polar Molecular Units	180
5.6 Directional Nonbonded Interactions	182
5.7 Segregation in Mesophases	183
5.7.1 'Principe des Affinités Électives'	183
5.7.2 Limitations Due to Packing	187
5.8 Hydrogen Bonds	190
5.9 References	193

> *Si je n'ai plus d'argent pour acheter de la peinture, j'achèterais des pastels, sinon je prendrais une plume et de l'encre ou encore un simple crayon, et si on me jette en prison, je cracherais sur mon doigt et je peindrais sur les murs.*
>
> Pablo Picasso (1881–1973)

5.1 DIRECTIONAL AND NONDIRECTIONAL FORCES

Intermolecular forces may be classified into three main categories [1]:

(a) purely electrostatic forces (charge/charge, charge/permanent dipole, permanent dipole/permanent dipole, etc.);
(b) polarization forces (dipole moments induced by charges or dipole moments, etc.),
(c) repulsive forces which are quantum mechanical in nature (see Table 5.1).

It can be seen from Table 5.1 that intermolecular energies of interaction can be expressed as

$$E(r) = -\frac{c}{r^n}$$

where
r = distance
c = constant
n = integer

This expression is valid only if $r > \varnothing$ ($\varnothing$ = diameter of the molecular unit) [1].

When the molecular unit is part of a medium (solid, liquid, mesophase), it interacts with its neighbours and a 'cohesive energy' — the sum of its interactions with the surrounding molecular units — may be calculated. If d is the number of molecules per unit of volume, the number of molecular units in a region between r and $(r + dr)$ is $d(4\pi r^2 dr)$ [1]. The cohesive energy, E_{coh}, can be written as

$$E_{coh} = -\int_{\varnothing}^{L} \frac{c}{r^n} 4\pi r^2 d\, dr \quad (5.1)$$

$$= -\frac{4\pi c d}{(n-3)\varnothing^{n-3}} \left[1 - \left(\frac{\varnothing}{L}\right)^{n-3}\right] \quad (5.2)$$

where L is the size of the system. If $n > 3$ and $L \gg \varnothing$, this expression reduces to

$$E_{coh} = -\frac{4\pi c d}{(n-3)\varnothing^{n-3}} \quad (5.3)$$

Equation (5.2) shows that large distance contributions to the overall interaction vanish only when $n > 3$. For $n < 3$, the second term of equation (5.2) becomes non-negligible and the distant molecular unit contributions can dominate the cohesive energy.

In real cases, case (a) of Figure 5.1 never occurs since electroneutrality must be preserved and long-range intermolecular forces in a mixture of positive and negative charges rarely exceed 1000 Å [1].

All electrostatic interaction energies essentially derive from Coulomb's law:

$$E = \frac{e_1}{4\pi\varepsilon_0\varepsilon_r} \sum_i \frac{e_i}{r_i} \quad (5.4)$$

(see Chapter 4).

Table 5.1 Common types of interactions between atoms, ions and molecules in vacuum. $E(r)$ is the interaction energy (in J); e, electric charge (C); μ, electric dipole moment (C m); α, electric polarizability (C^2 m^2/J); r, distance between interacting atoms or molecules (m); k, Boltzmann's constant (1.38×10^{-23} J/K); T, absolute temperature (K); h, Planck's constant (6.626×10^{-34} J s); ν, electronic absorption (ionization) frequency (s^{-1}); ε_0, permittivity of the vacuum (8.854×10^{-12} C^2/J m). The force is obtained by differentiating the energy $E(r)$ with respect to distance r. (After Ref. [1])

Type of interaction		Interaction energy $E(r)$
Covalent		Short range
Charge/charge	$e_1 \quad r \quad e_2$	$e_1 e_2/(4\pi\varepsilon_0 r)$ (Coulomb energy)
Charge/dipole	Fixed dipole (μ, θ, r, e)	$-e\mu \cos\theta/(4\pi\varepsilon_0 r^2)$
	Freely rotating (μ, r, e)	$-e^2\mu^2/[6(4\pi\varepsilon_0)^2 kTr^4]$
Dipole/dipole	Fixed (μ_1, θ_1, r, Φ, μ_2, θ_2)	$-\mu_1\mu_2[2\cos\theta_1\cos\theta_2 - \sin\theta_1\sin\theta_2\cos\Phi]/(4\pi\varepsilon_0 r^3)$
	Freely rotating (μ_1, r, μ_2)	$-\mu_1^2\mu_2^2/[3(4\pi\varepsilon_0)^2 kTr^6]$ (Keesom energy)
Charge/nonpolar	$e \quad r \quad \alpha$	$-e^2\alpha/[2(4\pi\varepsilon_0)^2 r^4]$
Dipole/nonpolar	Fixed (μ, θ, r, α)	$-\mu^2\alpha(1 + 3\cos^2\theta)/[2(4\pi\varepsilon_0)^2 r^6]$
	Rotating (μ, r, α)	$-\mu^2\alpha/[(4\pi\varepsilon_0)^2 r^6]$ (Debye energy)
Two nonpolar molecules	$\alpha \quad r \quad \alpha$	$-\tfrac{3}{4}(h\nu\alpha^2/[(4\pi\varepsilon_0)^2 r^6])$ (London dispersion energy)
Hydrogen bond	(H–O–H ··· O–H structure)	Short range

Figure 5.1 Two cases of interaction energies: (a) coulomb interactions $(1/r)$, where long-range contributions dominate; (b) induced dipole/induced dipole interaction $(1/r^6)$, where only short-range contributions are significant

The corresponding numerical equation for two single positive charges is

$$E_i = \frac{14.43}{\varepsilon_r r_i} \quad (5.5)$$

where E is in electron volts for one mole, ε_r = relative dielectric constant (dimensionless) and r is in Å. For a distribution of charge, in general difficult to determine for a molecular unit, a sum of the various Coulomb contributions may be assumed. Charge/charge, charge/permanent multipole and permanent multipole/permanent multipole interactions are then taken into account.

The various types of interaction shown in Table 5.1 can be divided into directional and nondirectional forces:

(a) Charge/charge and induced dipole/induced dipole interactions are not directional. However, in the second case, although the van der Waals forces are always attractive whatever the relative orientation of the two molecular units, the magnitude of the interaction may change when unsymmetrical units are considered. The interaction of a charge with a polarizable molecular unit can also be considered as nondirectional, but also with the previous limitation.

(b) Covalent bonds are the forces in which the relative orientations intervene more importantly, but charge/dipole and dipole/dipole interactions and hydrogen bonds are also highly dependent upon the geometry of the system.

5.2 CHARGE/CHARGE (COULOMB) FORCES

5.2.1 Lattice Energy of Ionic Crystals

The electroneutrality of most media implies that the systems contain an equal number of positive and negative electrical charges. If they are regularly arranged within a periodic crystal, the energy of the set of ions may be calculated (lattice energy).

In a NaCl structure, the reference central Na^+ ion is surrounded by six nearest anions at a distance r, twelve cations at a distance $\sqrt{2}r$, eight more distant anions at $\sqrt{3}r$ and so on (Figure 5.2). The electrostatic energy of interaction of the Na^+

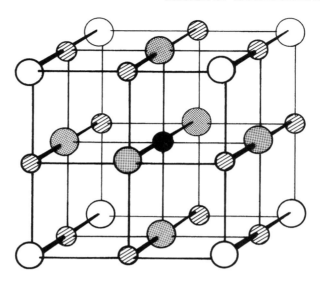

Figure 5.2 Interaction of a reference ion with the first and second neighbours in the rock salt (NaCl) structure

chosen as the origin and its positive and negative neighbours is [2]

$$E_{Na} = -\frac{e^2}{r}\left(6 - \frac{12}{\sqrt{2}} + \frac{8}{\sqrt{3}} + \cdots\right) \tag{5.6}$$

In the simple case of the NaCl structure:

$$E_{Na} = E_{Cl}$$

$$E_{TOT} = \frac{E_{Na} + E_{Cl}}{2}$$

The sum in parentheses in equation (5.6) is called the *Madelung constant*. This latter depends on the nature of the ions and on the geometry of the system (Figure 5.3). The various lattices may be seen as being formed of arrays of anions (the biggest ions) generating holes in which the cations can be incorporated (Figure 5.4).

Simple geometrical arguments allow the radius of the holes to be calculated as a function of the radius of the anions. The corresponding ratio is a characteristic of the lattice (Table 5.2).

The prediction — at least in a few simple and limited cases — of structures consisting of spherical cations and anions whose respective radii are known represents a first (and early known) supramolecular engineering attempt. In previous considerations, only attractive forces were described. However, short-range repulsive forces must also be taken into account. One can model them by using either an exponential or a power law equation (see Ref. [4] and references therein):

INTERACTIONS AND ORGANIZATION IN MOLECULAR MEDIA 161

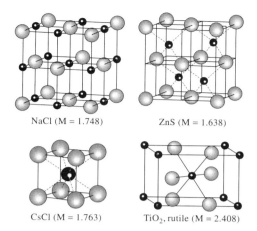

Figure 5.3 Madelung constants for a few crystalline lattices. (Modified from Ref. [3])

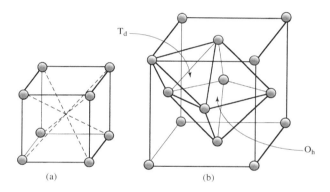

Figure 5.4 (a) Simple cubic lattice with a cubic 'hole'. (b) Face-centred cubic lattice showing one octahedral hole and one tetrahedral hole

Table 5.2 Theoretical and experimental (in parentheses) radius ratios (r_{hole}/r_{anion}) for various crystal structures. (Modified from Ref. [3])

Lattice	ZnS	NaCl	CsCl	CaF$_2$
Theoretical radius ratio	0.23	0.41	0.75	0.75
Experimental results	BeO (0.22)	MgSe (0.33)	CsCl (0.93)	SrCl$_2$ (0.62)
	MgTe (0.29)	MgS (0.35)	CsBr (0.87)	CaF$_2$ (0.73)
		CaTe (0.45)		CdF$_2$ (0.71)
		MgO (0.46)		HgF$_2$ (0.81)

$$E_{\text{rep}} = +a \exp(-br) \quad \text{with } a, b = \text{constants} \quad (5.7)$$

$$E_{\text{rep}} = +\frac{c}{r^{12}} \quad \left(\text{or } \frac{d}{r^n}\right) \quad \text{with } c, d = \text{constants} \quad (5.8)$$

The relative contributions of these forces is shown in Table 5.3 for a few ionic crystals.

Because of the presence of negative *and* positive charges in the lattices, the electrical field arising from a given ion is rapidly screened and decays more rapidly than the $1/r$ law. It has been shown that at large distances, the decay is exponential with distance [1].

Table 5.3 The various contributions to the lattice energy (in electron volts) (zero point energy is the energy of vibration of the ions at $T = 0$ K). (After Ref. [2])

Energy		Salt	
	LiF	NaCl	CsI
Coulomb	−12.4	−8.92	−6.4
Repulsion	+1.9	+1.03	+0.63
Induced dipole/induced dipole	−0.17	−0.13	−0.48
Zero point	+0.17	+0.08	+0.3

5.2.2 Solubility of Ions

In this section, the first example of the 'Principe des affinités électives' will be seen (like-as-like goes together) [5]. In the subsequent section, only alkali and alkaline earth cations will be considered. They are closed shell cations which can be seen as hard spheres with a small or negligible polarizability.

In a first step, one can consider the formation of ion pairs in solution [6, 7]:

$$M_1 + X \rightleftharpoons M_1, X \quad K_1, \Delta G_1 \quad (5.9)$$

$$M_2 + X \rightleftharpoons M_2, X \quad K_2, \Delta G_2 \quad (5.10)$$

where M = cation and X = anion. The selectivity of the association is given by

$$\log(K_1/K_2) = S \quad (5.11)$$

$$\Delta G = -RT \log K \quad (5.12)$$

$$\Delta G_1 - \Delta G_2 = \Delta(\Delta G) = -RTS \quad (5.13)$$

The free energy of formation may be expressed as a function of the various contributions:

$$\Delta G = \Delta G_{\text{MX}} - \Delta G_{\text{M}}^{\text{S}} - \Delta G_{\text{X}}^{\text{S}} + \Delta G_{\text{MX}}^{\text{S}} \quad (5.14)$$

where

ΔG_{MX} = Coulomb interaction anion/cation
ΔG^{S} = solvation energies of M, X or MX depending on the index

In a first approximation, it can be considered that ΔG^S_{MX} does not depend on the nature of the cation M_1 or M_2. Therefore:

$$\Delta(\Delta G) = (\Delta G_{M_1X} - \Delta G_{M_2X}) - (\Delta G^S_{M_1} - \Delta G^S_{M_2}) \tag{5.15}$$

The selectivity of ion pair formation depends on the difference between the driving energy (energy of the ion pair) and an antagonistic force (the solvation of the isolated ions):

$$\Delta G_{MX} = -\frac{14.43}{\varepsilon_r(r_M + r_X)} \tag{5.16}$$

where ΔG is in eV for one mole and r_M, r_X = ionic radii of M and X, respectively:

The same type of treatment may be applied by considering H_2O as a tripole (H, 0.32+; O, 0.64−):

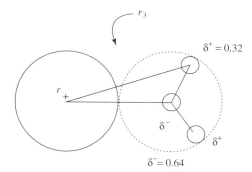

$$\Delta G^S_M = -\frac{14.43}{\varepsilon_r}\left(\frac{0.64}{r_M + r_O} - 2\frac{0.32}{r_3}\right) \tag{5.17}$$

where r_O = van der Waals radius of oxygen (1.5 Å)

The parameter r_3 can be calculated from the covalent distance of the OH bond (0.97 Å) and the H–O–H angle (104°).

One can then calculate ΔG when the size of the anion is varied. The result is shown in Figure 5.5.

It can be seen from Figure 5.5 that the biggest anions form ion pairs preferentially with the biggest cations ($Cs^+ > Rb^+ > K^+ > Na^+ > Li^+$) whereas the reverse order is found for small anions. The 'Principe des affinités électives' seems to apply in this case. It is noteworthy that this tendency is clear only at the two extremes (for very small or very large r values).

In the case of the complexation of ligands such as Cl^-, Br^-, I^-, RS^-, PR_3, etc., with second and third transition metal series different results may be obtained

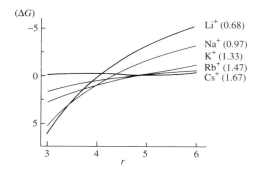

Figure 5.5 Eisenman's model results for the reaction: $X + M(H_2O) \rightleftharpoons MX + H_2O$. The size of the anion is varied on the x axis: $r = r_X + r_M$ (in Å). ΔG is shown on the y axis (in kcal/mole). The cation that forms the most stable complex is situated at the lowest part of the graph

Table 5.4 Solubility (in g/100 g) of various alkali halides in water (the temperature in °C is indicated as an index). (From Ref. [9])

	F^-	Cl^-	Br^-	I^-
Li^+	0.27^{18}	63.7^{0}	145^{4}	165^{20}
Na^+	4.22^{18}	35.7^{0}	116.0^{50}	184^{25}
K^+	92.3^{18}	34.4	53.48^{0}	127.5^{6}
Rb^+	130.6^{18}	77^{0}	98^{5}	152^{17}
Cs^+	367^{18}	$162.22^{0.7}$	124.3^{25}	44^{0}

[8], outlining the nongeneral validity of the 'Principe des affinités électives'. It is, however, a useful guideline. The 'Principe des affinités électives', although not general, can also be checked for the solubility of alkali salts in water (Table 5.4).

LiF (small cation/small anion) and CsI (large cation/large anion) are less soluble in water than the other salts. However, this apparently simple rule is not substantiated in the general case: the solubility of these same salts in ammonia, for example, does not follow this tendency. In liquid NH_3, the dielectric constant ($\varepsilon_r = 16.9$) is lower than in water ($\varepsilon_r = 78.5$) and ion pairs in solution can form.

5.2.3 Hard Soft Acid Bases (HSAB Principle)

In 1958 Ahrland, Chatt and Davies proposed a classification that could be used to understand which kind of metallic ions is more favourably bound to which type of ligands [2, 3]. Pearson, in 1963, proposed separating ligands and metallic ions according to their soft or hard character. Experimental observations have been carried out to determine the binding constants of complexes with a series of cations differing by their size (or hard/soft character) (Table 5.5).

Ahrland, Chatt and Davies divided the cations (metallic ions) into two classes (Table 5.6):

Table 5.5 Stability constants for the formation of ion pairs in aqueous solutions ($\log K$)

	Ionic radii (Å)	F^- 1.33	Cl^- 1.81	Br^- 1.96	I^- 2.20
Fe^{3+}	0.64	6.04	1.41	0.49	—
Zn^{2+}	0.74	0.77	-0.19	-0.6	-1.3
Cd^{2+}	0.97	0.57	1.59	1.76	2.09
Hg^{2+}	1.10	1.03	6.74	8.94	12.87

Table 5.6 Lewis acids and bases classified according to their hard and soft characters (Chatt–Pearson model). (After Ref. [10])

Acceptors (Lewis acids)	
Hard	Soft
H^+, Li^+, Na^+, K^+ Be^{2+}, Mg^{2+}, Ca^{2+}, Sr^{2+}, Mn^{2+} Al^{3+}, Sc^{3+}, Ga^{3+}, In^{3+} La^{3+}, Gd^{3+}, Lu^{3+}, Ce^{4+} Cr^{3+}, Co^{3+}, Fe^{3+} Ti^{4+}, Zr^{4+}, Th^{4+}, Pu^{4+} UO_2^{2+}, $(CH_3)_2Sn^{2+}$, VO^{2+}, MoO^{3+} HX (hydrogen-bonding molecules)	Cu^+, Ag^+, Au^+, Ti^+ Pd^{2+}, Cd^{2+}, Hg^+, Hg^{2+} Pt^{2+}, Pt^{4+} RS^+, RSe^+, RTe^+ I^+, Br^+, HO^+, RO^+ I_2, Br_2, ICN,... Trinitrobenzene,... Chloranil, quinones,... Tetracyanoethylene,... M^0 (metal atoms) Bulk metals
Border	
Fe^{2+}, Co^{2+}, Ni^{2+}, Cu^{2+}, Zn^{2+}, Pb^{2+}, Sn^{2+}, Sb^{3+}, Bi^{3+}, Rh^{3+}, Ir^{3+}, $B(CH_3)_3$, SO_2, NO^+, Ru^{3+}, Os^{2+}, R_3C^+, $C_6H_5^+$, GaH_3	
Donors (Lewis bases)	
Hard	Soft
H_2O, OH^-, F^- $MeCO_2^-$, PO_4^{3-}, SO_4^{2-} Cl^-, CO_3^{2-}, ClO_4^-, NO_3^- ROH, RO^-, R_2O NH_3, RNH_2, N_2H_4, NH_2^-	R_2S, RSH, RS^- I^-, SCN^-, $S_2O_3^{2-}$ R_3P, R_3As, $(RO)_3P$ CN^-, RNC, CO C_2H_4, C_6H_6 H^- R_3C^-
Border	
$C_6H_5NH_2$, Pyridine, N_3^-, Br^-, NO_2^-, SO_3^{2-}	

(a) These cations form the most stable complexes with the first elements of each column of the Mendeleiev table following a decrease of stability constants such as:

$$F > Cl > Br > I$$

$$N \gg P > As > Sb > Bi$$

$$O \gg S > Se > Te$$

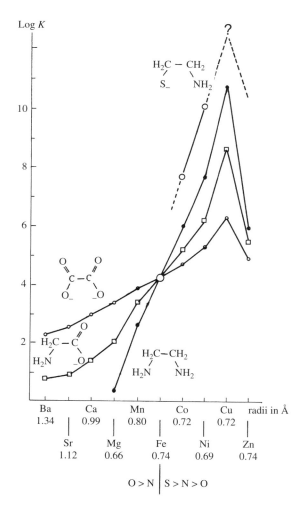

Figure 5.6 The Irving–Williams effect. Stability constants for the formation of complexes (log K) as a function of radius and hard–soft character of various divalent metallic ions [12]

(b) These cations form the most stable complexes in a reverse (or approximately reverse) sequence:

$$F < Cl < Br < I$$
$$N \ll P > As > Sb$$
$$O \ll S < Se \sim Te$$

The HSAB principle suffers many exceptions but it is however a useful guideline. This time, not only the Coulomb forces must be taken into account in the complexation process but also polarization effects. The HSAB principle seems to reflect the fact that large polarizable ligands associate preferentially with polarizable metallic ions (large radius, low degree of the oxidation number).

The determination of the hard or soft character can be in conflict with other factors. In the series from Ba to Zn (Figure 5.6), there is no simple relationship between the radius of the complexed cation and the nature of the heteroatoms of the ligands that form the most stable complexes. From Ba^{2+} to Fe^{2+}, O-containing ligands are preferred over N heteroatoms whereas the order $S > N > O$ is found from Fe^{2+} to Zn^{2+} (Irving–Williams effect) (Figure 5.6). [11, 12].

5.3 MOLECULAR POLARIZATION

An induced dipole is generated on a given molecular unit when it is plunged in an electrical field **E** [1]:

$$\mu_i = \alpha \mathbf{E} \tag{5.18}$$

where

α = polarizability in units of
$(4\pi\varepsilon_0) \text{ Å}^3 = (4\pi\varepsilon_0)10^{-30} \text{ m}^3 = 1.11 \times 10^{-40} \text{ C}^2\text{m}^2/\text{J}$
E = field in V/m
μ_i = induced dipole moment in C m ($1D = 3.336 \times 10^{-30}$ C m)

The direction of the induced dipole is shown in Figure 5.7 as conventionally adopted.

In the case where an oscillating electrical field is used, various polarization processes may be distinguished. The polarization of the molecular unit may arise from various contributions: dipolar (α_{dip}), vibrational (α_{vib}) or electronic (α_{e}) reorientations; they differ by their characteristic response times (Figure 5.8).

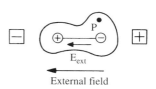

Figure 5.7 Charges inducing a dipole in a molecular unit: direction of the polarization. It must be stressed that the definition of the direction of the induced dipole and the external field is not the same relative to the charges

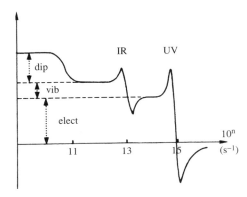

Figure 5.8 The frequency domains associated with the various contributions to the polarizability. The polarizability α as a function of the frequency (in s^{-1}) is represented. (After Ref. [13])

Electronic displacements in the optical domain take place at high frequencies ($10^{15}\,s^{-1}$); molecular vibration occurs at around $10^{13}\,s^{-1}$. Irradiation with Yag–Nd lasers (1.06 μm) corresponds approximately to the frontier between the two domains. The time needed for an electrical dipole to reorient in an oscillating electric field is in the range 10^{-11} s. Ion migration through organic media mainly depends on the state of order of the latter but starts to be important below $10^{2}\,s^{-1}$. In the subsequent section only electronic polarization will be considered.

The effect of an electrical field on conductive spheres was first studied by Lorenz [14]. It was demonstrated that, in this case, the induced electrical dipole moment is proportional to r^3, where r is the radius of the sphere. The refraction, which is proportional to the polarizability, does increase with the radius of the ion whatever its state of ionization. However, an $r^{4.5}$ relationship is approximately found instead of the expected r^3 behavior (Table 5.7). The refraction R is related

Table 5.7 Molar refractions (cm^3) of various ions as a function of their ionic radius (Å) (value in parentheses). (After Refs. [15] and [16])

Anion					
F^-	(1.33)	2.17	OH^-		4.42
Cl^-	(1.81)	8.22	NO_3^-		10.16
Br^-	(1.96)	11.60	SCN^-		16.54
I^-	(2.20)	17.53	ClO_4^-		12.66
Cation					
Li^+	(0.68)	0.12	Mg^{2+}	(0.66)	0.47
Na^+	(0.97)	0.65	Ca^{2+}	(0.99)	1.60
K^+	(1.33)	2.71	Sr^{2+}	(1.12)	2.56
Rb^+	(1.47)	4.10	Ba^{2+}	(1.34)	5.00
Cs^+	(1.67)	6.71			

INTERACTIONS AND ORGANIZATION IN MOLECULAR MEDIA

to the polarizability (α) through the Lorenz–Lorentz equation [14, 17]:

$$R = \frac{n^2 - 1}{n^2 + 2} \frac{M_w}{d} = \frac{4}{3}\pi\alpha \tag{5.19}$$

where

n = refractive index
M_w = molar mass
d = density

The polarizability of simple molecules may be estimated since, in most cases, it is an additive property. For alkanes of different lengthes, the molar refraction of the whole molecule may be calculated from individual bond refractions [18, 19]. The anisotropy of polarization is not taken into account. Many bond polarizabilities were determined (Table 5.8). The polarizability depends strongly on the size of the constitutive atoms and on the bond order but not on the polarity of the bond, C–H, C–O and C–N, which have very different bond polarities possess very close polarizabilities.

The sigma and pi contributions to the polarizability of bonds may also be estimated. In this case one must distinguish longitudinal (α_l) from transverse (α_t) polarizations (Table 5.9).

Polarizability arising from π electrons is of the same order of magnitude as for σ electrons in the longitudinal direction. The π contribution for a double bond is, for example, $30.2 - 18.2 = 12.0$, as compared with 18.2 for a single bond. However, π electrons are considerably more polarizable in the transverse direction: 9.6 against 0.2. The second difference between σ and π electrons is that π electrons may delocalize. The polarizability of polyenes (HC=CH)$_n$ [22] varies with the cube of the chain length while, for alkanes, the polarizability varies linearly with chain length. The bond additivity model is therefore not applicable whenever electron conjugation occurs.

Table 5.8 Molar bond refractions (cm^3) as referred to the sodium D-line. (After Refs. [18] to [20])

C–H	1.68	Si–H	3.32		
C–C	1.30	Si–C	2.47	Ge–C	2.9
C=C	4.17				
C≡C	5.87				
C–Cl	6.51				
C–Br	9.39				
C–I	14.61				
C–O	1.54	Si–O	1.87		
C=O	3.32				
C–S	4.61				
C=S	11.91				
C–N	1.57				
C=N	3.76				
C≡N	4.82				

Table 5.9 Longitudinal (α_l) and transverse (α_t) polarizabilities (cm³) for single, double and triple bonds. The value corresponding to an aromatic bond ($C_{ar}-C_{ar}$) is also indicated [21]

	C–C	$C_{ar}-C_{ar}$	C=C	C≡C
$\alpha_l \times 10^{25}$	18.2	23.2	30.2	36.8
$\alpha_t \times 10^{25}$	0.2	4.4	9.6	11.9

It is now necessary to consider polarization effects in condensed media. It will be seen in the corresponding section that charge carriers in van der Waals type crystals may be considered as plain molecular anions or cations. These charges polarize the medium around them via a conventional charge/induced dipole interaction.

The simplest approach is to consider the ion as a spherical cavity (radius a) in a continuous medium of permittivity ε (Figure 5.9). The polarization energy is then given by

$$P = -\frac{e^2(1-1/\varepsilon)}{2a} \quad (5.20)$$

with $\varepsilon = 3$, $a = 3$ Å, $P = -1.6$ eV [23]. This value is in good agreement with experimental determinations [24]. From naphthalene to tetracene, the polarization energy is in the range 1.7–1.8 eV. This energy is therefore considerable. It is now possible to detail the various contributions to the polarization energy [25]:

$$P = P_{c-id} + P_{id-id} + P_{c-q} + P_{c-iq} + \cdots \quad (5.21)$$

where

P_{c-id} = charge/induced dipole
P_{id-id} = induced dipole/induced dipole
P_{c-q} = charge/permanent quadrupole
P_{c-iq} = charge/induced quadrupole

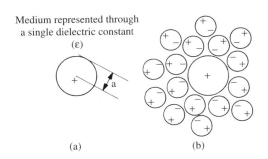

Figure 5.9 (a) Continuous and (b) discrete models of a charge in a polarizable condensed medium

The first term in equation (5.21) corresponding to a charge/induced dipole energy (P_{c-id}) is the most important one. The second term (P_{id-id}) is opposite in sign to P_{c-id} since the induced dipoles generated by the ion repell each other; its magnitude can be as high as 30–40% of the total energy for aromatic hydrocarbons [25]. P_{c-q} is significantly smaller. It has been calculated that it does not introduce an error exceeding 0.1 eV (6–7% of the total energy). Contributions from higher multipoles could also be considered, but in most cases they can be neglected [25].

5.4 INDUCED DIPOLE/INDUCED DIPOLE INTERACTIONS

5.4.1 Generalities

The induced dipole/induced dipole interaction energy is widely known as a 'van der Waals interaction' although some authors prefer to use the term 'London dispersion energy'. For a nonpolar atom or molecule, the time average of the electrical dipole moment is zero. However, at any instant, there is a finite dipole moment given by the instantaneous position of the electrons around the nucleus. This is not negligible, since for a small hydrogen atom one can calculate a value of 2.4D [1]. The instantaneous dipole moment generates an electrical field that polarizes all nearby molecular units. The resulting force between the nonpolar units is finite and the time average is not zero. Induced dipole/induced dipole interactions are weak and nondirectional.

Various formula are used to express these interactions [1]:

$$E_{id-id}(r) = -\frac{3}{2}\frac{\alpha_1\alpha_2}{(4\pi\varepsilon_0)^2 r^6}\frac{I_1 I_2}{I_1 + I_2} \tag{5.22}$$

where I = ionization potential. The molar lattice energy (cohesive energy) may be calculated by summing over all the molecular units contained in the medium. The comparison of theoretical and experimental results is given in Table 5.10.

The cohesive energies of molecular crystals may be estimated by considering the heat of vaporization, ΔH_v (transition liquid/gas), the heat of fusion, ΔH_{fus} (transition solid/liquid), and the heat of sublimation, ΔH_s (transition solid/gas):

$$\text{Solid} \xrightarrow{\Delta H_{fus}} \text{Liquid} \xrightarrow{\Delta H_v} \text{Gas}$$
$$\xrightarrow{\Delta H_s}$$

It is generally considered that (see Ref. [26] and references therein)

$$\Delta H_s \sim \Delta H_{fus} + \Delta H_v \tag{5.23}$$

A simple relationship has been proposed to estimate ΔH_v [26]:

$$\Delta H_v(298\,\text{K}) = 0.31\, n_Q + 1.12\, n_t + 0.71 \tag{5.24}$$

Table 5.10 Molar cohesive energies either calculated for close-packed structures (equation 5.22) or measured from the sum of the latent heat of melting plus the vaporization energy (approximately equal to the latent heat of sublimation) [1]

Interacting molecules	Molar cohesive energy (kJ/mole)	
	Theory	Measurement
Ne	2.0	2.1
Ar	7.7	7.7
Xe	15.6	14.9
CH_4	10.9	9.8
CCl_4	23.9	32.6

where

n_Q = number of quaternary carbon atoms
$n_t = n_{TOT} - n_Q$
n_{TOT} = total number of carbon atoms (ΔH in kcal/mole)

This equation can be used whenever experimental data are not available for ΔH_v. ΔH_{fus} is a measurement of the loss of cohesive energy when going from a well-ordered crystal to a liquid. This is also reflected in the difference ($\Delta H_s - \Delta H_v$).

In the case of aromatic derivatives (Table 5.11), the heat of vaporization increases, as expected, with the molar mass of the compound considered. On the contrary, there is not such a relationship for ΔH_{fus}: the increase is not directly related to the molar mass and isomers do not afford a constant value of ΔH_{fus}. This reflects the different arrangements (and subsequently different cohesive energies) within the various crystals.

The same tendencies are observed for the paraffinic derivatives (Table 5.12). The use of equation (5.24) implies that the same ΔH_v value is found for the same number of quaternary C atoms and total C atoms, independently of the other structural parameters of the molecular unit. ΔH_v values obtained in this way are used to calculate ΔH_s. By comparing the experimental and calculated values of ΔH_s, a satisfactory agreement can be see; this validates the use of equation (5.24).

An odd–even effect for the paraffinic chains is observed both on ΔH_{fus} and $(\Delta H_s)_{exp}$. This will be rationalized latter on by considering the structural differences between the two series of compounds.

5.4.2 Packing of Aromatic Derivatives

Since induced dipole/induced dipole interactions are nondirectional, the most stable condensed medium must also be one of the highest packing coefficient.

INTERACTIONS AND ORGANIZATION IN MOLECULAR MEDIA

Table 5.11 Heat of vaporization (ΔH_v), heat of fusion (ΔH_{fus}) and heat of sublimation (ΔH_s) for various aromatic derivatives. All units are in kcal/mole [26]

	ΔH_v calc. (see equation (5.24))	ΔH_{fus} exp.	ΔH_s calc. (see equation (5.23))	ΔH_s exp.
Benzene C_6: 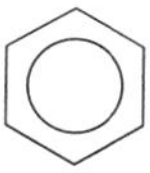	7.4	2.4	9.8	10.6
Naphthalene C_{10}: 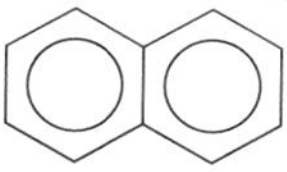	11.9	4.5	16.4	17.3
Anthracene C_{14}: 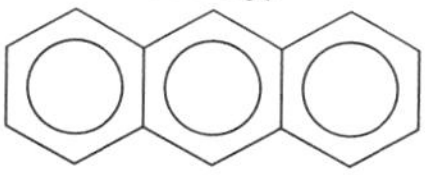	16.4	6.9	23.3	22.5
Phenanthrene C_{14}: 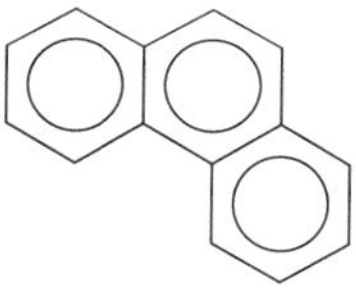	16.4	4.3	20.7	22.1
Triphenylene C_{18}: 	20.9	5.9	26.8	30.2
Perylene C_{26}: 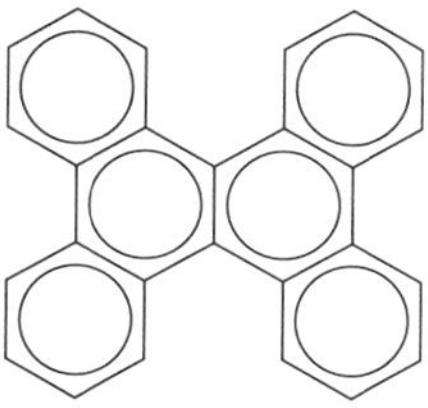	29.8	7.6	37.4	34.7

(continued overleaf)

Table 5.11 (*continued*)

	ΔH_v calc. (see equation (5.24))	ΔH_{fus} exp.	ΔH_s calc. (see equation (5.23))	ΔH_s exp.
Triptycene C_{20}: 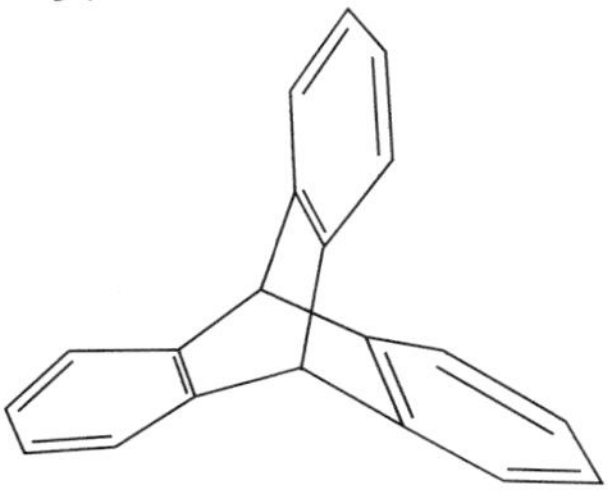	23.1	7.2	30.3	25.0

Table 5.12 Heat of vaporization (ΔH_v), heat of fusion (ΔH_{fus}) and heat of sublimation (ΔH_s) for various paraffinic derivatives. All units are in kcal/mole [26]

		ΔH_v calc. (see equation (5.24))	ΔH_{fus} exp.	ΔH_s calc. (see equation (5.23))	ΔH_s exp.
C_6H_{14}	n-Hexane	7.4	3.0	10.4	12.1
$C_{14}H_{30}$	n-Tetradecane	16.4	10.7	27.1	28.1
$C_{15}H_{32}$	n-Pentadecane	17.5	8.3	25.8	25.8
$C_{16}H_{34}$	n-Hexadecane	18.6	12.7	31.3	32.3
$C_{17}H_{36}$	n-Heptadecane	19.8	9.6	29.4	29.9
$C_{18}H_{38}$	n-Octadecane	20.9	14.7	35.6	36.5
$C_{20}H_{42}$	Eicosane	23.1	14.7	37.8	40.7

This last parameter is defined as [27]

$$k_p = \frac{ZV_0}{V} \qquad (5.25)$$

where

Z = number of molecular units in the unit cell
V = volume of the unit cell
V_0 = molecular volume

The packing coefficient generally varies from 0.6 to 0.8 for aromatic derivatives (Table 5.13).

As a comparison, spheres in a compact arrangement lead to $k_p = 0.741$. The packing coefficient significantly decreases when paraffinic chains are linked to an aromatic core (2,6-di-n-octylnaphthalene, $k_p = 0.595$); in many cases substituted derivatives form glasses. This is due to the poor adequacy of packing between the two subunits constituting the molecular unit. It seems that lower k_p values are obtained with unregular molecular shapes [27] (see, for instance, phenanthrene

Table 5.13 Packing coefficients k_p in crystals of various aromatic derivatives. (After Ref. [27])

Compound	k_p	Compound	k_p
Mononuclear:		α-Naphthylamine	0.705
Benzene	0.681	β-Naphthylamine	0.680
Resorcinol (α-form)	0.665	2,6-Diphenylnaphthalene	0.668
Resorcinol (β-form)	0.678	2,6-Dicyclohexylnaphthalene	0.690
p-Toluidine	0.677	2,6-Di-n-octylnaphthalene	0.595
Quinone	0.693		
p-Dibromobenzene	0.740	*Anthracene derivatives:*	
p-Bromochlorobenzene	0.714	Anthracene	0.722
p-Dichlorobenzene	0.687	9,10-Dibromoanthracene	0.773
Durene	0.704	9,10-Dichloroanthracene	0.800
		9,10-Anthraquinone	0.765
Polyphenyls:		1,2-Anthraquinone	0.781
Diphenyl	0.740	1,4-Anthraquinone (α form)	0.778
p-Diphenylbenzene	0.730	1,4-Anthraquinone (β form)	0.773
Quaterphenyl	0.746		
Triphenylmethane	0.638	*Other multinuclears:*	
sym-Triphenylbenzene	0.716		
Dibenzyl	0.705	Phenanthrene	0.684
Stilbene	0.720	Chrysene	0.737
Tolane	0.685	Naphthacene	0.735
		1,2-Benzpyrene	0.745
Naphthalene derivatives:		Perylene	0.805
		1,2-Benzanthracene	0.713
Naphthalene	0.702	1,2,5,6-Dibenzanthracene	0.708
1,2-Naphthoquinone	0.760	Coronene	0.726
1,4-Naphthoquinone	0.753	Graphite	0.887
2,6-Dimethylnaphthalene	0.740		
β-Methylnapthalene	0.712		
α-Naphthol	0.714		
β-Naphthol	0.710		

Resorcinol: [structure: benzene ring with two OH groups in meta positions]

p-Toluidine: [structure: benzene ring with NH₂ and CH₃ in para positions]

Durene: [structure: benzene ring with four CH₃ groups]

Dibenzyl: [structure: two phenyl rings connected by –CH₂–CH₂–]

Tolane: [structure: two phenyl rings connected by C≡C]

Chrysene: [structure: four fused benzene rings]

Naphthacene = tetracene

and anthracene). On the other hand, two different crystalline forms can afford almost the same k_p value (see, for example, resorcinol).

Most rigid aromatic molecules crystallize in the so-called 'herring-bone structure'. Columns are formed in which the molecular planes are tilted relative to the b-column axis (Figure 5.10). The same structure is found from naphthalene to hexacene: in all cases the molecular units conserve their center of symmetry in the crystal. The parameters a and b do not vary much whereas c increases to accommodate the long axis of the molecules. In the same time the packing coefficient k_p increases continuously from 0.704 (naphthalene) to 0.751 (hexacene).

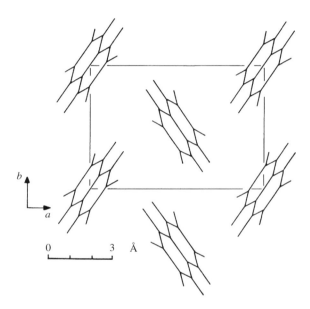

Figure 5.10 Projection on the (a, b) plane of the naphthalene structure (herring-bone arrangement). (Space group: $P2_1/a$, $z = 2$) (Reproduced by permission of Academic Press from S. C. Abrahams, D.W.I. Crnikshank cited in Ref. [29]

5.4.3 Packing of Paraffinic Chains

The paraffinic chains form condensed media (mesophases or crystals) with side-by-side stacking. The schematic representation shown in Figure 5.11 is usually used to represent the packing in a plane perpendicular to the long chain axis. In order to follow the Yin–Yang principle (bumps must fit hollows), two crystalline structures are possible with an oblique or rectangular cell (Figure 5.12). These arrangements arise when a hydrogen atom of one molecule fits into a hollow formed by three others of the adjacent molecule.

When the number of C atoms is odd, the paraffinic chains crystallize, with a rectangular two-dimensional cell leading to a rectangular lattice (Figure 5.12b). In

INTERACTIONS AND ORGANIZATION IN MOLECULAR MEDIA

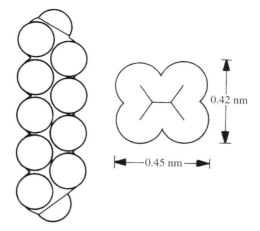

Figure 5.11 Molecular model of a paraffinic chain and corresponding projection perpendicular to the chain axis. (Modified from Ref. [28])

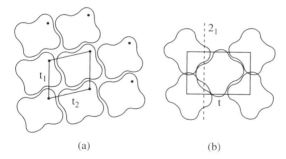

Figure 5.12 The two possible types of close packing for paraffinic molecules: (a) oblique and (b) rectangular. (After Ref. [29])

this case one of the chains is translated relative to the others (Figure 5.13b). The orthorhombic lattice may be schematically represented as shown in Figure 5.14.

When the paraffinic chains possess an even number of carbon atoms and with $n < 26$, the two-dimensional lattice is oblique and the tilting of the chains relative to the plane yields a triclinic cell (Figure 5.15). When $n > 26$ (n even), the phase is monoclinic with a rectangular two-dimensional lattice.

Another type of structure can be observed at temperatures close to the melting point, which is associated with a rotation along the chain axis. It is of hexagonal symmetry (Figure 5.16).

The projected area (on the plane normal to the chains) is around 18.5 Å^2 for the triclinic and orthorhombic phases. The length of rigid paraffinic chains may be readily calculated knowing the number of methylene groups in the chains [30]:

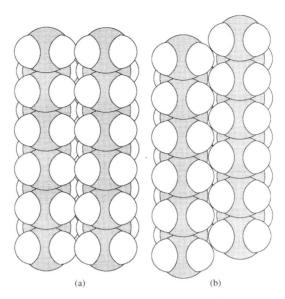

Figure 5.13 Translation of one paraffinic chain relative to the others. (a) Before, (b) after translation

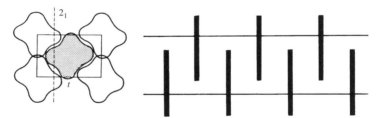

Figure 5.14 Schematic representation of the orthorhombic phase of paraffinic chains found for C_nH_{2n+2} (n odd) (in black the molecule translated along the chain axis). (After Ref. [29])

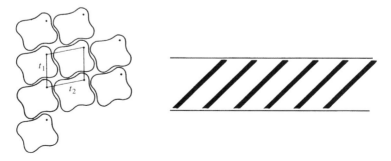

Figure 5.15 Schematic representation of the triclinic phase of paraffinic chains (n even, $n < 26$). (Reproduced by permission of Academic Press from Ref. [29])

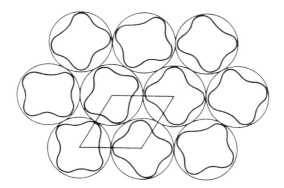

Figure 5.16 Two-dimensional lattice observed in the case where the chains can rotate along their long axis (hexagonal symmetry) (Reproduced by permission of Academic Press from [29])

$$l_{ext} = 1.265n + r_{CH_3} \qquad (5.26)$$

where

r_{CH_3} = van der Waals radius of the methyl group (2.0 Å) [31]
l_{ext} = length of the extended rigid paraffinic chains
n = number of CH_2 units

In the mesophases, the paraffinic chains are more or less disordered and a mean end-to-end distance may be roughly estimated by multiplying l_{ext} by a coefficient ρ_m which depends on the number of carbon atoms of the chain [30]:

$$l_{flex} = \rho_m l_{ext}$$

The coefficient ρ_m can be estimated from Flory's calculations (see Ref. [32]) (Table 5.14).

Table 5.14 Values of the coefficient ρ_m used to calculate the end-to-end distances of molten or quasi-molten paraffinic chains [32]

n	ρ_m	n	ρ_m
4	0.9	12	0.72
6	0.84	14	0.69
8	0.80	16	0.66
10	0.75	18	0.63

5.5 POLAR MOLECULAR UNITS

In the case where the molecular unit is both polar and polarizable, the relative importance of the various contributions may be questioned. The case of single crystals of nitrobenzene has been treated [33]. For the calculations, the cell parameters $a = 7.460$ Å, $b = 9.666$ Å, $c = 7.034$ Å in an orthorhombic lattice ($\alpha = \beta = \gamma = 90°$) were determined from the X-ray structure shown in Figure 5.17. Calculations of the total energy of the crystal give the contributions shown in Table 5.15.

The permanent multipole/permanent multipole contribution includes charge/charge, charge/dipole, dipole/dipole, etc., interactions. It can be seen that these are more than 10 times smaller than the induced dipole/induced dipole interactions. The electrostatic contributions are — except for the charge/charge interaction — highly directional and cancellation of various terms may occur within the crystal. Induced dipole/induced dipole interactions are not directional and are summed over all the various molecular units. The condensed phase of highest density (closest-packing principle) affords

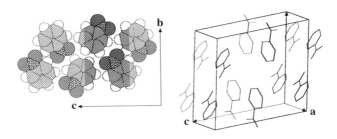

Figure 5.17 Arrangement of the molecules in the cell of the nitrobenzene crystal and orthogonal projection on the (b, c) plane. (After Ref. [33])

Table 5.15 Various contributions of the total energy of the crystal of nitrobenzene. (After Ref. [33])

	E (kcal/mole)
Permanent multipole[a] /permanent multipole	−1.95
Permanent multipole /induced dipole	−0.42
Induced dipole /induced dipole	−20.40
Repulsion	+6.66
Total	−16.11

[a]Multipole includes charges.

the largest induced dipole/induced dipole stabilization. A compromise must be found whenever directional and nondirectional contributions coexist in the crystal. Other packing energy calculations also concluded that dipole/dipole interactions contribute negligibly to the overall energy of molecular crystals [34].

To obtain noncentrosymmetric crystals is essential for many physicochemical properties such as the generation of second harmonics in optics [35]. An examination of the structures contained in the Cambridge database has been carried out to determine whether or not there is a correlation between the magnitude of the molecular dipole moment with the relative spatial arrangement of these dipoles within the structure [36, 37]. Three space groups have been selected:

(a) P1 (with $Z = 1$), in which all dipole moments point in the same direction;
(b) P$\bar{1}$ ($Z = 2$), where each polar unit has a counterpart with a cancellation of the dipole moments;
(c) P2_1 ($Z = 2$) (Figure 5.18).

The mean molecular dipole moments in these three space groups have been found to be [36]:

(a) P1: $\bar{\mu}_g = 3.41D$ (28 structures);
(b) P$\bar{1}$: $\bar{\mu}_g = 3.33D$ (161 structures);
(c) P2_1: $\bar{\mu}_g = 3.14D$ (179 structures).

There is almost no difference in the molecular ground state dipole moment in the three space groups considered, although polar and nonpolar arrangements are formed. This confirms that there is no straightforward correlation between the magnitude of the molecular dipole and the polar or nonpolar character of the crystal.

The frequency of occurrence of the space groups has also been studied [38, 39] (Table 5.16). Out of the 5926 compounds examined in Table 5.16, 25% of them (1532) crystallize in a noncentrosymmetric space group (669 in polar groups).

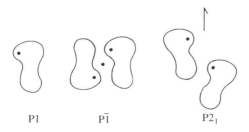

Figure 5.18 Possible relative orientations of the molecules in three of the space groups considered. (Modified from Ref. [36])

Table 5.16 Number of occurrences of the main space groups for 5926 organic crystals. (After Refs. [38] and [39])

Space group	Number of occurrences	Symmetry of crystal class		Major axes polar
		Chiral	Centrosymmetric	
$P2_1/c$	1783	(−)	(+)	(−)
$P2_12_12_1$	722	(+)	(−)	(−)
$P2_1$	458	(+)	(−)	(+)
$P\bar{1}$, $C2/c$, $Pbca$, $Pnma$	1026	(−)	(+)	(−)
$P2_12_12$	104	(+)	(−)	(−)
$Pna2_1$	100	(−)	(−)	(+)
$C2$	80	(+)	(−)	(+)
$P4_12_12$	37	(+)	(−)	(−)
Pc	31	(−)	(−)	(+)
Total compounds	5926			

The probability of obtaining a noncentrosymmetric structure is even lower if one disregards optically active molecular units.

5.6 DIRECTIONAL NONBONDED INTERACTIONS

Besides the formation of directional covalent bonds — which can be reversibly broken and formed as in disulfides RSSR — weaker directional interactions have

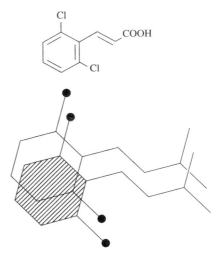

Figure 5.19 Intermolecular interactions arising in stacked molecules of 2,6-dichlorocinnamic acid. (Modified from Ref. [42])

been shown to occur between chlorine groups, ($-$Cl$\cdots$Cl$-$), cetone and chlorine, or cetone and phenyl moieties [40–42].

The term 'crystal engineering' was proposed by G. Schmidt during the early 1970s when he studied the solid state photopolymerization of *trans*-cinnamic acid derivatives [40]. It was realized, in particular, that chloro-substitution tends to 'steer' the structure in such a way that short (3.2–3.6 Å) nonbonded contacts between the chlorine groups exist within the crystal. (Those values must be compared with the expected van der Waals distance, which is in the range 3.4–3.8 Å.) In the cinnamic derivative series, chloro-substitution induces an arrangement that favours ($-$Cl$\cdots$Cl$-$) interactions as exemplified for 2,6-dichlorocinnamic acid in Figure 5.19 (for a review see Ref. [43]).

Oxygen-substituted aromatics have also been found to yield directional nonbonded interactions formally similar to hydrogen bonds: C$-$H$\cdots$O [42]. Sulfur–sulfur (S$\cdots$S) and sulfur–chlorine (S$\cdots$Cl) contacts have been described as tools to engineer crystal structures [44]. Fluoro-derivatives have also been studied for that purpose [45].

5.7 SEGREGATION IN MESOPHASES

5.7.1 'Principe des Affinités Électives'

The segregation of molecular fragments to form chemically homogeneous domains is very often encountered: rigid aromatic subunits pile up with each other; paraffinic chains do not mix with perfluoroalkyl chains; lipophilic groups separate from hydrophobic moieties; etc. [46, 47]. This effect has been already encountered for ions and the 'Principe des affinités électives' has been proposed to designate all these processes globally. A more precise insight into this phenomenon may now be gained for nonpolar condensed phases.

Let us consider N_i molecules of the compound i. In the condensed phase each molecule is surrounded by Z_c other molecular units [48, 49]. Z_c is considered to be unchanged with the type of molecules (regular solution case). If E_{ii} is the energy of interaction between two molecules i, the total energy within the condensed phase for pure phases is

$$Z_c \frac{N_i}{2} E_{ii} \qquad (5.27)$$

When two different molecules (1 and 2) are considered, each molecule 1 is surrounded by $Z_c N_1/(N_1 + N_2)$ molecules of 1 and $Z_c N_2/(N_1 + N_2)$ molecules of 2. Then the total energy of the interaction is [48]

$$\frac{1}{2} Z_c \frac{(N_1)^2}{N_1 + N_2} E_{11} + Z_c \frac{N_1 N_2}{N_1 + N_2} E_{12} + \frac{1}{2} Z_c \frac{(N_2)^2}{N_1 + N_2} E_{22} \qquad (5.28)$$

The difference in energy between the mixture and the segregated phase is the mixing enthalpy ΔH_M and is given by subtracting equation (5.28) from the two equations (5.27):

$$\Delta H_M = Z_c \frac{N_1 N_2}{N_1 + N_2} \Delta_{12} \quad (5.29)$$

with

$$\Delta_{12} = E_{12} - \frac{E_{22}}{2} - \frac{E_{11}}{2} \quad (5.30)$$

The nonbonded interactions between molecular units can be expressed by the equation (by ignoring the variation of the other parameters with distance):

$$E_{ij} \approx -\frac{1}{(d_{ij})^n} \quad (5.31)$$

where d_{ij} is the distance between the molecular units i and j and n is a parameter that depends on the type of interactions considered (see Table 5.1). Then Δ_{12} can be approximated by

$$\Delta_{12} \approx \frac{1}{(2r_1)^n} + \frac{1}{(2r_2)^n} - \frac{2}{(r_1 + r_2)^n} \quad (5.32)$$

where r_i are the corresponding radii of the spheres.

It can be readily seen that $\Delta_{12} = 0$ if $r_1 = r_2$ and that Δ_{12} is always positive if $r_1 \neq r_2$. The enthalpy of mixing is therefore always positive and unfavorable (Figure 5.20).

The overall tendency to segregate is given by the free energy of mixing ΔG_M:

$$\Delta G_M = \Delta H_M - T \Delta S_M \quad (5.33)$$

ΔS_M is always positive (the disorder increases by mixing), and segregation will occur whenever ΔH_M is larger than $T \Delta S_M$. From the equation found for Δ_{12}, it is clear that fragments of the same chemical nature (and therefore of the same size) will tend to form homogeneous domains. A more quantitative estimation

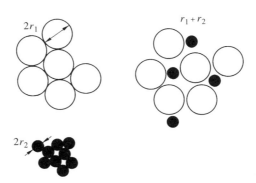

Figure 5.20 Segregated domains and mixture of two molecular units of different size

of the reciprocal 'antipathies' of chemically different fragments may be gained from the solubility parameter values established for polymers [51].

When apolar molecular units are considered, the main contribution to the interaction energy arises from induced dipole/induced dipole interactions:

$$E_{ij} = -\frac{3}{2} \frac{\alpha_i \alpha_j}{(d_{ij})^6} \frac{I_i I_j}{I_i + I_j} \quad (5.34)$$

where $d_{ij} = r_i + r_j$, I is the ionization potential and α_i is the polarizability coefficient of the molecular unit considered. In the case where the two molecular units have physicochemical properties that are not too different,

$$4I_1 I_2 \approx (I_1 + I_2)^2 \quad (5.35)$$

$$(4r_1 r_2) \approx (r_1 + r_2)^2 \quad (5.36)$$

Under these conditions,

$$|E_{12}| \approx \sqrt{E_{11} E_{22}} \quad (5.37)$$

and equation (5.29) may then be written as

$$\Delta H_M = \frac{N_1 N_2}{N_1 + N_2} \left[\left(\frac{E_{11} Z_c}{2}\right)^{1/2} - \left(\frac{E_{22} Z_c}{2}\right)^{1/2} \right]^2 \quad (5.38)$$

When the molecular units have different sizes, this expression becomes

$$\Delta H_M = V_M \left[\left(\frac{\Delta E_1}{V_1}\right)^{1/2} - \left(\frac{\Delta E_2}{V_2}\right)^{1/2} \right]^2 \Phi_1 \Phi_2 \quad (5.39)$$

where V_M is the total volume of the mixture, Φ_i the volume fraction of component 1 or 2 in the mixture, $\Phi_i = (N_i V_i / V_M)$, N_i is the number of moles of the species i and V_i the partial molar volume.

Models used to describe nonstoichiometric compounds are related to the previous description [50]. The quantity $(\Delta E/V)^{1/2}$ is often referred to as the solubility parameter δ and $\Delta E/V$ is the cohesive energy. Thus

$$\delta_i = \left(\frac{\Delta E_i}{V_i}\right)^{1/2} \quad (5.40)$$

Since the entropy ΔS_M is always positive, mixing will occur ($\Delta G_M < 0$) whenever ΔH_M is not too large; consequently $(\delta_1 - \delta_2)^2$ must be relative small. The solubility parameter δ of any molecule or molecular subunit may be estimated from their formula [51]. To each chemical group is associated a *molar-attraction constant* K; the solubility parameter is obtained by summing K over all the atoms and groupings in the molecules:

$$\delta = d \frac{\Sigma K}{M_w} \quad (5.41)$$

where d is the density and M_w the molar mass (molecular weight). The solubility parameters δ for linear paraffinic chains are almost independent of the molecular weight if chain ends are neglected. The values of the group molar attraction constants K are derived from the measurements of heats of vaporization or from vapor pressure determinations (see Ref. [51]) (Table 5.17).

The use of group molar attraction constants is of interest for predicting the type of mesophases that can be obtained. The molecular unit is first divided into subunits of a homogeneous chemical nature. The solubility parameters have been determined for high molecular weight polymers, and, therefore, fragments of sufficient size must be considered.

The antipathy character is not always related to moieties of different polarities. In the case of substituted phthalocyanines, the molecular subunit segregation properties arise from the *rigid* aromatic core and the *flexible* side chains (Figure 5.21).

Figure 5.21 demonstrates that segregation must occur between the phthalocyanine core and the paraffinic chains because of the difference in solubility parameters. The segregation is reinforced with high molecular weight side chains because of the decrease of the mixing entropy. In the case of phthalocyanine macrocycles substituted with long alkyl chains [52, 53], the segregation between the rigid aromatic moieties and the flexible alkyl chains leads to columnar mesophases (Figure 5.22) (see also Figure 1.48).

Biphenyl rigid cores have been substituted with alkyl and perfluoroalkyl chains, leading to mesogens consisting of three chemically different subunits [54–56]. The name 'polyphile' has been proposed to designate such compounds by analogy

Table 5.17 Group molar attraction constants derived from the measurements of vapor pressures. The solubility parameters, δ, calculated from K are in $cal^{1/2}/cm^{3/2}$. (After Ref. [51])

Group		K	Group		K
$-CH_3$		147.3	$-OH$		225.84
$-CH_2-$		131.5	$-H$	Acidic dimer	-50.47
$>CH$		85.99	$-OH$	Aromatic	170.99
$>C<$		32.03	$-NH_2$		226.56
			$-NH-$		180.03
$CH_2=$		126.54	$-N-$		61.08
$-CH=$		121.53	$-C\equiv N$		354.56
$>C=$		84.51	$-NCO$		358.66
$-CH=$	Aromatic	117.12	$-S-$		209.42
$-C=$	Aromatic	98.12	$-Cl_2$		342.67
$-O-$	Ether, acetal	114.98	$-Cl$	Primary	205.06
$-O-$	Epoxide	176.20	$-Cl$	Secondary	208.27
$-CO_2^-$		326.58	$-Cl$	Aromatic	161.0
$>C=O$		262.96	$-Br$		257.88
$-CHO$		292.64	$-Br$	Aromatic	205.60
$(CO)_2O$		567.29	$-F$		41.33

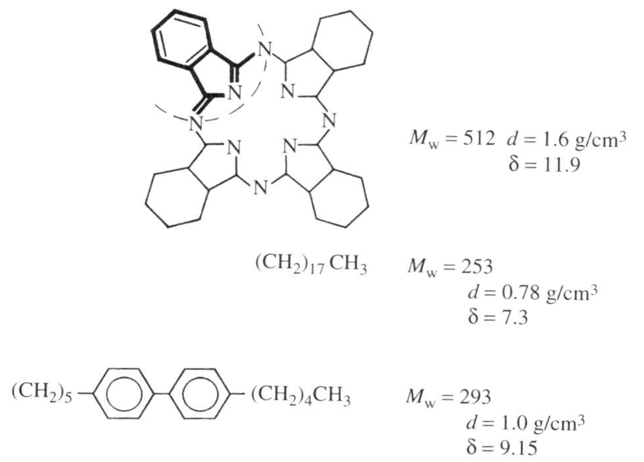

INTERACTIONS AND ORGANIZATION IN MOLECULAR MEDIA

$M_w = 512$ $d = 1.6 \text{ g/cm}^3$
$\delta = 11.9$

$(CH_2)_{17}CH_3$ $M_w = 253$
$d = 0.78 \text{ g/cm}^3$
$\delta = 7.3$

$(CH_2)_5$—⟨○⟩—⟨○⟩—$(CH_2)_4CH_3$ $M_w = 293$
$d = 1.0 \text{ g/cm}^3$
$\delta = 9.15$

Figure 5.21 Calculation of the solubility parameters of some molecular subunits

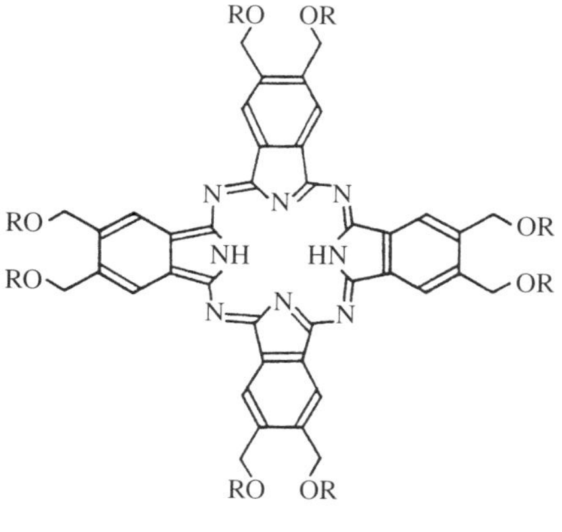

Figure 5.22 Octaalkoxymethyl-substituted phthalocyanines with $R = C_nH_{2n+1}$, $n = 8$, 12, 18. (After Ref. [52])

with amphiphilic derivatives like soaps in which two parts — one hydrophilic, the other hydrophobic — coexist in the same molecule [54]. The 'Principe des affinités électives' has been used in this case as a guideline to find a plausible structure of the smectic-type mesophases formed from these compounds [54–56].

5.7.2 Limitations Due to Packing

It has been shown in the above section that induced dipole/induced dipole interactions afford a strong tendency to segregation between molecular units of different sizes. However, the 'Principe des affinités électives' can fail for various reasons. The first one can be the presence of directional forces (dipolar interactions, hydrogen bonds, etc.). The other causes are related to the shape of the molecular units considered. When a hollow-to-bump fit (Yin–Yang principle) occurs between two different molecular units, the mixed phase may be favored over the segregated one (Figure 5.23).

As a matter of fact, even two spherical units with sufficiently different radii can produce a mixed phase when the holes generated by the packing of the large spheres can just be filled by the small ones. At least three cases may be distinguished depending on the ratio of the radii of the small and large spheres (Figure 5.24).

When the radius of the small spheres is such that they can be incorporated within the holes generated by the large spheres in contact (Figure 5.24a), large sphere/large sphere and small sphere/large sphere contacts are possible. In three dimensions for a sodium chloride lattice, this corresponds to [57]

$$R_{op} = \frac{r_1}{r_2} = \sqrt{2} + 1 = 2.41 \quad \text{(octahedral hole)} \tag{5.42}$$

For larger ratios (Figure 5.24b), the large spheres are in contact with each other but the small ones are not in contact with the lattice formed by the large spheres. For smaller ratios (Figure 5.24c), the small sphere size prevents large sphere/large sphere contacts.

The problem of determining the porosity coefficient P_{or} = (free volume/total volume) as a function of the ratio of the radii for the packing of two types of spheres has been studied for many years. Analytical models are difficult to find in the general case, but computer modelling and experimental determinations have been carried out. The porosity is related to the packing coefficient k_p via the relationship

$$k_p = 1 - P_{or} \tag{5.43}$$

It has been found experimentally [58] that disordered media made of identical spheres yield a value of k_p equal to 0.58. The compact disordered packing, which is obtained following a precise experimental protocol, affords $k_p = 0.62$–0.64. These values must be compared with the compactness of the close packing

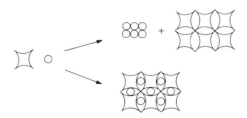

Figure 5.23 Nonapplicability of the 'Principe des affinités électives': the two different molecular units form a compact mixed phase

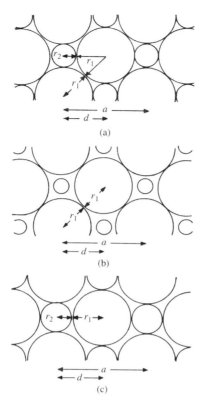

Figure 5.24 Two-dimensional packing of spheres having different diameters. (a) Optimal size ratio $(r_1/r_2) = R_{op}$. The large spheres are in contact with each other and the small spheres fill the holes. (b) Ratio $> R_{op}$. The large spheres are in contact but not the small ones. (c) Ratio $< R_{op}$. The small spheres are in contact with the large ones but not the latter with each other. (After Ref. [57])

($k_p = 0.741$). However, the lower values are closer to those expected in mesomorphic phases.

In the case of a mixture of two sets of spheres of different sizes, the porosity as a function of the volume fraction has also been experimentally determined [58] (Figure 5.25). For a given ratio of the diameters d_1/d_2, a maximum of compactness is found of

$$\frac{V_1}{V_1 + V_2} \approx 0.25$$

which must correspond to the maximum quantity of small spheres that can be incorporated within the holes generated by the packing of the large spheres.

In Figure 5.25, four ratios of d_1/d_2 have been studied in the range 0.115–0.536. It is known that periodic closest-packed spheres generate octahedral holes (ratio of 0.414) and tetrahedral holes (ratio of 0.225). The range studied is

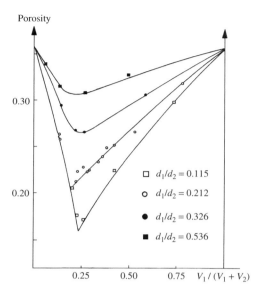

Figure 5.25 Porosity as a function of the volume fraction $V_1/(V_1 + V_2)$ for a mixture of spheres of diameter d_1 and d_2 [58]

therefore in this domain. However, a clearcut definition of the size of the hole cannot be given in disordered media. Figure 5.25 shows that, in the range studied, the porosity decreases whenever the size of the spheres becomes different. This is apparently true for all the volume fractions studied. The packing coefficient for identical spheres (0.64) is, however, far from that of a periodic compact arrangement of spheres (0.741).

5.8 HYDROGEN BONDS

Hydrogen bonds are established via a proton bound to two electronegative heteroatoms: X—H ⅠⅠⅠⅠⅠ Y (Table 5.18). In the case of KHF_2, the H bond energy can reach a value as high as 113 kJ/mole [2]. The symmetrical location of the hydrogen atom between the two fluorine atoms has also been demonstrated. H bonds are directional bonds. A deviation of 5° from colinearity yields a destabilization of 1 kJ/mole [59]. At 20 °C, the average bending due to thermal agitation in water is less than 8° [59].

In the solid state, H bonds are most often linear, although exceptions can be found. In the examination of neutron diffraction results for 74 different compounds containing O—H ⅠⅠⅠⅠⅠ H moieties, the mean angle has been found to be around 170° [60, 61]. In general, short (strong) H bonds tend to be more linear than long ones. Since H bonds are directional, strong and fairly selective, especially when cooperative effects are involved, it was tempting to try to use them for the design of crystals and mesophases.

Table 5.18 Bond energies corresponding to various H bonds. (van der Waals radii of F, 1.50–1.60 Å; O, 1.50 Å) [2]

Bond	Compound	Bond energy (kJ/mole)	AX⋯B distance (Å)
F—H—F	KHF$_2$	~113	
F—H⋯F	HF(l)	30	2.6–2.7
F—H⋯F	HF(g)	~28.6	
O—H⋯O	(HCOOH)$_2$	29.8	
O—H⋯O	H$_2$O (s)	~21	2.7–2.8
N—H⋯N	Melamine (H$_2$N–C$_3$N$_3$(NH$_2$)–NH$_2$, 1,3,5-triazine-2,4,6-triamine)	~25	3.4

Methods have been developed to find a systematic rationalization of hydrogen bonding patterns for engineering molecules to crystalllize in controlled and predictable arrangements [62]. At that time, co-crystals of *p*-aminobenzoic acid and 3,5-dinitrobenzoic acid were studied [62]. A classification was then proposed depending on the number of H bond patterns that were possible on each molecular unit [63] (Figure 5.26).

By examining the structure of H-bonded solids which can be found in the Cambridge structural database, three empirical rules could be proposed [63]:

(a) All good proton donors and acceptors can be used in hydrogen bonding.

(b) Six-membered-ring intramolecular hydrogen bonds form in preference to intermolecular hydrogen bonds.

(c) The best proton donors and acceptors remaining after intramolecular hydrogen bond formation form intermolecular hydrogen bonds.

More precise additional rules were also proposed [63].

Nanostructures have been studied in the same way, such as the 1:1 complex formed in aqueous solution between melamine and cyanuric acid. The cyclic hexameric structure has been proposed from powder diffraction studies (see references cited in Ref. [64]) (Figure 5.27). Other related structures have also been described [65, 66].

Amidine derivatives and various dicarboxylate compounds have been shown to form H-bonded patterns by X-ray crystallography [67] (Figure 5.28).

The formation of cooperative H bonds has also be employed to form mesogens. In this respect, interactions between uracil and 2,6-diaminopyridine have been used; the two molecular units are substituted with long paraffinic chains in order to induce a mesomorphic phase [68] (Figure 5.29).

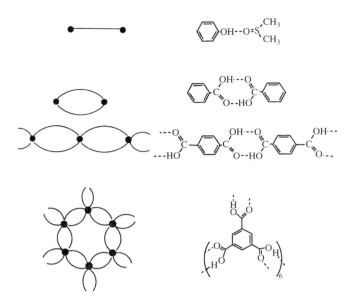

Figure 5.26 Some of the possibilities of H bond pattern formation. (Reprinted with permission from Ref. [63]. Copyright 1990 American Chemical Society)

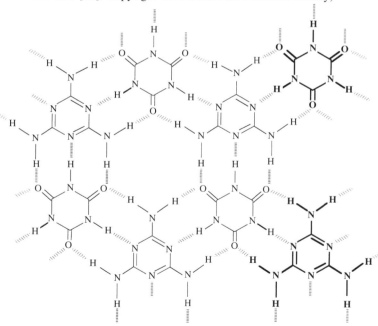

Figure 5.27 Cyclic hexameric structure proposed for the 1:1 complex of melamine and cyanuric acid from powder diffraction studies

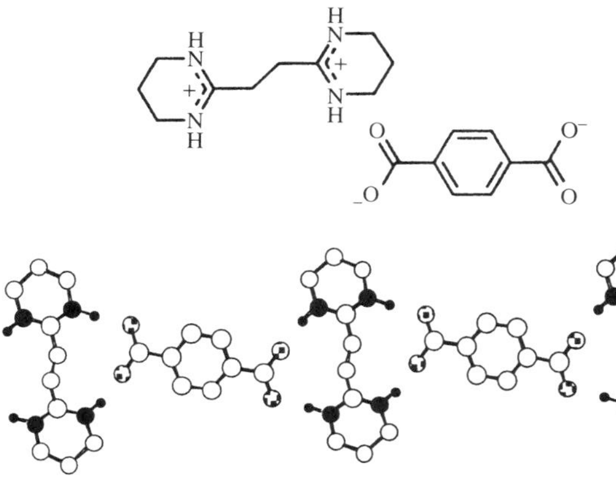

INTERACTIONS AND ORGANIZATION IN MOLECULAR MEDIA 193

Figure 5.28 H-bonded patterns which can be formed using bifunctional molecules: amidine derivatives and terephthalate dianion. (Modified from Ref. [67])

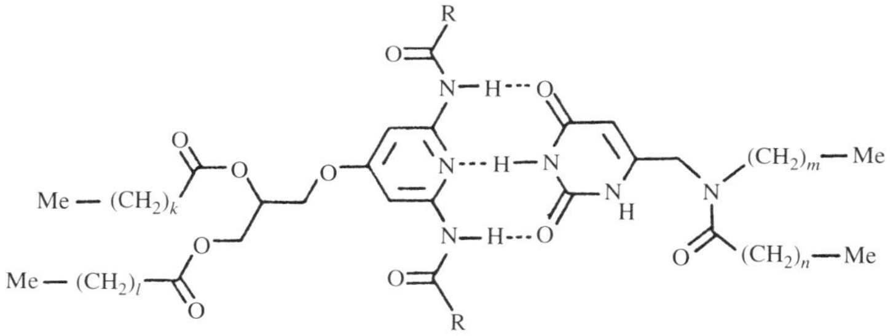

Figure 5.29 Type of interactions postulated for long-chain substituted uracil and 2,6-diaminopyridine derivatives. (Reproduced by permission of the Royal Society of Chemistry from [68])

Some of the previous compounds, depending on the number of C atoms of the side chains, show metastable mesomorphic phases. X-ray diffraction data agree with a hexagonal columnar mesophase with two molecules side by side forming a disc-like unit (Figure 5.29). Later, self-assembling discotic mesogens in which the discotic nature can be predicted from the shape of the molecular units were described [69]. Lyotropic mesophases have been reported with self-assembled rigid rods [70] (for a review see Ref. [71]).

5.9 REFERENCES

1. J. N. Israelachvili, *Intermolecular and Surface Forces*, Academic Press, London (1985).
2. F. A. Cotton and G. Wilkinson, *Advanced Inorganic Chemistry*, Interscience, New York (1972).

3. K. F. Purcell and J. C. Kotz, *Inorganic Chemistry*, W. B. Saunders, Philadelphia (1977).
4. J. Simon and J.-J. André, *Molecular Semiconductors*, Springer-Verlag, Berlin (1985).
5. From Goethe's novel, *Die Wahlverwandtschaften* (1809) in which the author applies to a social case the chemical principles related to 'affinity'.
6. Y. A. Ovchinikov, V. I. Ivanov and A. M. Shkrob, *Membrane Active Complexones*, Elsevier Scientific, New York (1974).
7. G. Eisenman, *Membranes*, Vol II, *Lipid Bilayers and Antibiotics*, Marcel Dekker, New York (1973).
8. J. D. Dunitz (ed.), *Structure and Bonding*, Vol. II, p. 1, Springer-Verlag, Berlin (1972).
9. R. C. Weast (ed.), *CRC Handbook of Chemistry and Physics*, 69th edn, Chemical Rubber Company, Boca Raton (1988–1989).
10. F. Basolo and R. G. Pearson, *Mechanisms of Inorganic Reactions*, John Wiley & Sons, New York (1958).
11. J. E. Huheey, *Inorganic Chemistry*, Harper & Row, Cambridge (1983).
12. Found in Ref. [11]: H. Sigel and D. B. McCormick, *Acc. Chem. Res.*, **3**, 201 (1970).
13. C. Kittel, *Introduction à la physique de l'etat solide*, Dunod, Paris (1958).
14. L. Lorenz, *Wied. Ann. Phys.*, **11**, 70 (1880).
15. A. Heydweiller, *Phys. Z.*, **26**, 526 (1925).
16. A. Dalgarno, *Adv. Phys.*, **11**, 281 (1962).
17. H. A. Lorentz, *Wied. Ann. Phys.*, **9**, 641 (1880).
18. A. I. Vogel, W. T. Cresswell, G. H. Jeffery and J. Leicester, *J. Chem. Soc.*, 514 (1952).
19. R. G. Gillis, *Rev. Pure Appl. Chem.*, **10**, 21 (1960).
20. R. J. W. Le Fèvre, *Advances in Physical Organic Chemistry* (ed. V. Gold), Academic Press, London (1965).
21. Sheng-Nien Wang, *J. Chem. Phys.*, **7**, 1012 (1939).
22. P. L. Davies, *Trans. Faraday Soc.*, **47**, 789 (1952).
23. F. Gutmann and L. E. Lyons, *Organic Semiconductors*, John Wiley & Sons, New York, 1967.
24. K. Seki, *Molec. Cryst. Liquid Cryst.*, **171**, 255 (1989).
25. E. A. Silinsh, *Organic Molecular Crystals*, Springer-Verlag, Berlin (1980).
26. J. S. Chickos, R. Annunziata, L. H. Ladon, A. S. Hyman and J. F. Liebman, *J. Org. Chem.*, **51**, 4311 (1986).
27. A. I. Kitaigorodskii, *Organic Chemical Crystallography*, Consultant Bureau, New York (1961).
28. L. H. Jensen, *J. Polym. Sci., Part C* (29), 47 (1970).
29. A. I. Kitaigorodsky, *Molecular Crystals and Molecules*, Academic Press, New York (1973).
30. J. Simon, J.-J. André and A. Skoulios, *Nouv. J. Chimie*, **10**, 295 (1986).
31. L. Pauling, *The Nature of the Chemical Bond*, 3rd edn, p. 260, Cornell University Press, Ithaca (1960).
32. C. Tanford, *J. Phys. Chem.*, **76**, 3020 (1972); **78**, 2469 (1974).
33. J. Caillet and P. Claverie, *Acta Cryst.*, **A31**, 448 (1975).
34. A. Gavezzotti, *J. Phys. Chem.*, **94**, 4319 (1990).
35. See, for example, D. S. Chemla and J. Zyss (eds.), *Nonlinear Optical Properties of Organic Molecules and Crystals* Academic Press, New York (1987).
36. J. K. Whitesell, R. E. Davis, L. L. Saunders, R. J. Wilson and J. P. Feagins, *J. Phys. D: Appl. Phys.*, **26**, B56 (1996).
37. J. K. Whitesell, R. E. Davis, L. L. Saunders, R. J. Wilson and J. P. Feagins, *J. Am. Chem. Soc.*, **113**, 3267 (1991).

38. D. Y. Curtin and I. C. Paul, *Chem. Rev.*, **81**, 525 (1981). The values have been taken in this article but this is mentioned in Ref. [39].
39. A. D. Mighell and H. M. Ondik, *J. Phys. Chem. ref. Data*, **6**, 675 (1977).
40. G. M. J. Schmidt, *Pure Appl. Chem.*, **27**, 647 (1971).
41. N. W. Thomas, S. Ramdas and J. M. Thomas, *Proc. Roy. Soc. Lond.*, **A400**, 219 (1985).
42. J. A. R. P. Sarma and G. R. Desiraju, *Acc. Chem. Res.*, **19**, 222 (1986).
43. G. R. Desiraju, *The Crystals as a Supramolecular Entity*, John Wiley & Sons, Chichester (1995).
44. V. Nalini and G. R. Desiraju, *J. Chem. Soc. Chem. Commun.*, 1030 (1986).
45. V. A. Kumar, N. S. Begum and K. Venkatesan, *J. Chem. Soc. Perkin Trans.*, **2**, 463 (1993).
46. C. Tanford, *Science*, **200**, 1012 (1978).
47. A. Skoulios and G. Finoaz, *J. Chim. Phys.*, **59**, 473 (1962).
48. C. Quivoron, in *Chimie macromoléculaire* (ed. G. Champetier) Vol. II, p. 3, Hermann, Paris (1972).
49. J. H. Hildebrand and R. L. Scott, *Regular Solutions*, Prentice Hall, Englewood Cliffs, New Jersey, (1962).
50. R. Collongues, *La Non-stoechiométrie*, Masson et cie, Paris (1971).
51. J. Brandrup and E. H. Immergut (eds.), *Polymer Handbook*, John Wiley & Sons, New York (1975).
52. C. Piechocki, J. Simon, A. Skoulios, D. Guillon and P. Weber, *J. Am. Chem. Soc.*, **104**, 5245 (1982).
53. C. Piechocki and J. Simon, *Nouv. J. Chimie*, **9**, 159 (1985).
54. F. Tournilhac, L. Bosio, J.-F. Nicoud and J. Simon, *Chem. Phys. Lett.*, **145**, 452 (1988).
55. F. Tournilhac, L. M. Blinov, J. Simon and S. V. Yablonsky, *Nature*, **359**, 621 (1992).
56. Yushan Shi, F. Tournilhac and S. Kumar, *Phys. Rev. E*, **55**, 4382 (1997).
57. N. W. Ashcroft and N. D. Mermin, *Solid State Physics*, Holt, Rinehart & Winston, New York (1976).
58. R. Ben Aïm, Thèse de Doctorat, Nancy (1970).
59. O. Beredsen, *Water Structure* (ed. O. Cole), Marcel Dekker, New York (1967).
60. F. H. Allen, O. Kennard and R. Taylor, *Acc. Chem. Res.*, **16**, 146 (1983).
61. C. B. Aakeröy and K. R. Seddon, *Chem. Soc. Rev.*, **22**, 397 (1993).
62. M. C. Etter and G. M. Frankenbach, *Chem. Mater.*, **1**, 10 (1989).
63. M. C. Etter, *Acc. Chem. Res.*, **23**, 120 (1990).
64. G. M. Whitesides, *Science*, **254**, 1312 (1991).
65. J. A. Zerkowski, J. C. MacDonald, C. T. Seto, D. A. Wierda and G. M. Whitesides, *J. Am. Chem. Soc.*, **116**, 2382 (1994).
66. V. A. Russel and M. D. Ward, *Chem. Mater.*, **8**, 1654 (1996).
67. G. Brand, M. W. Hosseini, R. Ruppert, A. de Cian, J. Fischer and N. Kyritsaka, *New. J. Chem.*, **19**, 9 (1995).
68. M.-J. Brienne, J. Gabard, J.-M. Lehn and I. Stibor, *J. Chem. Soc. Chem. Commun.*, 1868 (1989).
69. R. Kleppinger, C. P. Lillya and C. Yang, *Angew. Chem. Int. ed. Engl.*, **34**, 1637 (1995).
70. M. Kotera, J.-M. Lehn and J.-P. Vigneron, *J. Chem. Soc. Chem. Commun.*, **197**, (1994).
71. J. M. Lehn, *Supramolecular Chemistry*, VCH, Weinheim (1995).

6 Molecular Semiconductors: Properties and Applications

6.1 Introduction	197
6.2 Collective and Individual Approaches to Electronic Levels in Molecular Materials	199
6.2.1 Molecular Orbitals: A Qualitative Approach	199
6.2.2 Molecular Orbitals: Group Theory	200
6.2.3 One-Dimensional Band Structure	202
6.2.4 Bloch's Functions, Band Diagram	205
6.2.5 Fermi Level	208
6.2.6 Two- and Three-Dimensional Band Structures	209
6.2.7 Individual Approach: Supramolecular Orbitals	212
6.2.8 Collectivized Approach: An Insight into Band Theory	214
6.3 Molecular Semiconductors: Generalities	217
6.4 A Narrow-Band Molecular Semiconductor: Pc_2Lu	220
6.4.1 Synthesis	220
6.4.2 Electrochemical Properties	220
6.4.3 Crystal Structures	221
6.4.4 Conduction Properties	224
6.4.5 Relationship between ΔE_{redox} and E_{act}	227
6.4.6 Optical and Magnetic Properties	228
6.5 A Broad-Band Molecular Semiconductor: PcLi	229
6.5.1 Synthesis and Physicochemical Properties	229
6.5.2 Conduction Properties	230
6.5.3 Magnetic Properties	232
6.6 Band Structure of Metallophthalocyanines	233
6.6.1 Applicability of the Band Model	233
6.6.2 Band Structure of PcH_2	235
6.6.3 Band Structure of PcLi	237
6.6.4 Conclusions	239

MOLECULAR SEMICONDUCTORS: PROPERTIES AND APPLICATIONS

6.7 Liquid Crystalline Molecular Semiconductors	239
6.8 Junctions and Solar Cells	242
6.8.1 Generalities	242
6.8.2 Junction Studies in the Dark	245
6.8.3 Solar Cells and Photovoltaic Effect	248
6.8.4 Solar Cells: Classical Formulation	249
6.8.5 Molecular Solar Cells: Localized States Formulation	250
6.9 Conductivity-Based Gas Sensors	256
6.9.1 Generalities	256
6.9.2 Metallophthalocyanines: Thin-Film Morphology	258
6.9.3 Experimental Results: Monophthalocyanines	259
6.9.4 Experimental Results: Bisphthalocyanines	264
6.9.5 Conclusions	267
6.10 Field-Effect Transistors	269
6.10.1 Introduction	269
6.10.2 Thin-Film Transistors: Models and Applications in Industry	270
6.10.3 Molecular Field-Effect Transistors: A Chronology	274
6.10.4 Metallophthalocyanine-Based Field-Effect Transistors	276
6.11 References	291

> È già difficile per il chimico antivedere, all'infuori dell'esperienza, l'interazione fra due molecole semplici; del tutto impossibile predire cosa avverrà all'incontro di due molecole moderatamente complesse. Che predire sull'incontro di due esseri umani?
>
> Primo Levi (1919–1987)
> *La Chiave a Stella*

6.1 INTRODUCTION

The mechanisms of transport of charges in molecular materials has been obscured, for a very long time, because one applied band theory to the corresponding processes whereas it is not applicable, at least at room temperature. Persistance in this error came from the fact that two antagonistic models (electronic levels can be considered to be collectivized or not) yield the same macroscopic phenomenological equations (such as the expression of the logarithm of the conductivity as a function of the inverse of temperature). In many cases, no care was taken about the very general fact that the fit of experimental points to an equation does not confirm the hypotheses taken to establish the corresponding equation but only demonstrates that the model is *compatible* with the experiments.

A somehow arbitrary classification is generally taken that separates metals, semiconductors and metals (Figure 6.1). This classification corresponds, when pure compounds are considered, to basically different materials. There is not much in common between silicon, diamond and copper. One must, however, take care when considering materials containing impurities. Doped insulators may have conducting properties close to semiconductors or even metals although being fundamentally different in nature.

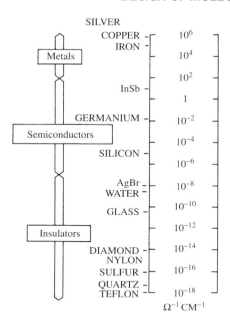

Figure 6.1 Conductivity domains of metals, semiconductors and insulators. A few inorganic compounds have been reported on the scale

In this chapter, molecular semiconductors are examined in detail. The term 'organic semiconductor' was used in 1948 by Vartanyan [1] and Eley [2]. However, it has been demonstrated [3] that all single-component materials consisting of small aromatic molecules or conjugated polymers are insulators, in the dark, at room temperature.

The energy necessary to generate charge carriers, may be derived by plotting $\log \sigma$ versus $1/T$ (σ = conductivity). In the case of covalent-type inorganic solids the forbidden energy gap separates the conduction and valence bands described in the framework of a conventional band theory (see Section 6.2.8). In this approach a one-electron approximation is used, in which, due to a strong delocalization of the charge carrier, possible polarization effects of the surrounding lattice may be ignored. A delocalized charge carrier in the conduction band can be regarded as a quasi-free particle possessing a relatively high mobility, often larger than $\mu_0 > 100 \, \text{cm}^2/\text{V s}$.

The situation is different in molecular crystals and other van der Waals solids. In molecular materials a large difference in energy exists between the *intramolecular* and *intermolecular* interactions. The energy of the intramolecular C–C bond is about 3.6 eV (83.1 kcal/mole) whereas the intermolecular interaction energy between C atoms of neighboring molecules is only about 4 meV (0.1 kcal/mole), three orders of magnitude smaller. The situation may be illustrated by comparing graphite and pyrene crystals (Figure 6.2).

MOLECULAR SEMICONDUCTORS: PROPERTIES AND APPLICATIONS

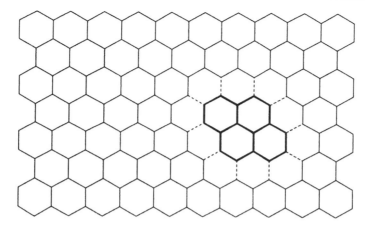

Figure 6.2 Collectivized (delocalized) or individual (localized) electronic levels (as illustrated by graphite and pyrene)

Graphite is a highly anisotropic material consisting of parallel planes of C atoms. A single layer may be considered as an enormous molecule formed by regular six-membered rings (Figure 6.2). Inside the layer the carbon atoms are tightly bound: the bond length is ~1.42 Å. Due to this strong covalent interaction the electrons within the layer are highly delocalized. As a result graphite is a fairly good conductor. In the case of pyrene crystals, the electrons are localized on individual molecular units. The van der Waals interaction between pyrene molecules in the crystal being weak, pyrene crystals exhibit a low conductivity and belong to the class of insulators [4].

The difference between localized and delocalized electronic levels may be understood from different ways; one of them is given in the following chapter. The analogies and the differences between band theory and the various hopping models will then be easily deduced.

6.2 COLLECTIVE AND INDIVIDUAL APPROACHES TO ELECTRONIC LEVELS IN MOLECULAR MATERIALS

6.2.1 Molecular Orbitals: A Qualitative Approach

The formation of molecular orbitals from atomic orbitals for describing the electronic energy levels of individual molecules may be extended to describe those of molecular materials [5]. In what follows, the procedure used to determine the electronic levels of condensed phases will be surveyed.

A molecular orbital may be approximated by a linear combination of atomic orbitals (LCAO):

$$\Psi_i = c_1\varphi_1 + c_2\varphi_2 + \cdots \tag{6.1}$$

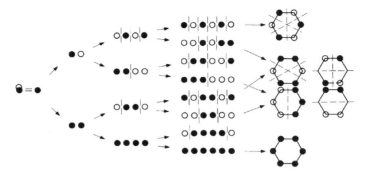

Figure 6.3 Stepwise construction of the molecular orbitals of benzene (in the polygons, the nodal plane must contain the centre of the polygon)

The relative signs of the coefficients of the atomic orbitals φ_i may be determined from simple symmetry arguments. As an example, the case of the π orbitals of benzene is described in Figure 6.3. It is possible to first build the orbitals corresponding to a linear chain by stepwise formation of bonding and antibonding pairs and then close the ring to get the π orbitals of benzene. The number of nodes gives the relative stabilities of the symmetry orbitals.

The apparent simplicity of the method employed to obtain the molecular orbitals of benzene, by merely assuming bonding and antibonding interactions between s (or equivalently p_z) orbitals, does not resist to a careful examination. Why are some of the orbitals disregarded? Why do some atomic orbital contributions vanish when one forms the benzene ring from the linear chains? These questions are treated in a next section.

6.2.2 Molecular Orbitals: Group Theory

A more satisfactory — but lengthier — way to establish molecular orbitals requires the use of group theory. In a first step, the nuclei are arranged in the geometry of the final compound. One-electron orbitals are then determined in the field generated by the nuclei. These orbitals are filled with as many electrons as there are in the molecule.

The D_{3h} symmetry with three s-type orbitals will be considered as an example:

$$\begin{array}{c} s_2 \\ \diagup \diagdown \\ s_1 \text{———} s_3 \end{array}$$

It is first necessary to find the representation associated with s_1, s_2, s_3 in the group D_{3h}. The number of orbitals unchanged by a symmetry operation gives the corresponding character:

D_{3h}	E	$2C_3$	$3C_2$	σ_h	$2S_3$	$3\sigma_v$
$\Gamma(s_1, s_2, s_3)$	3	0	1	3	0	1

It is then possible to decompose Γ into the irreducible representations of D_{3h}. If $a(\Gamma_i)$ is the number of times the irreducible representation Γ_i is contained in Γ:

$$a(\Gamma_i) = \frac{1}{h}\sum X(R)X_i(R) \qquad (6.2)$$

where

h = order of the group
$X(R)$ = character of the symmetry operation R of Γ
$X_i(R)$ = same for Γ_i

This equation has already been given (equation (2.7)). However, in its further use, the character can belong to imaginary numbers; in this case the product $X(R)X_i^*(R)$ must be considered:

D_{3h}	E	$2C_3$	$3C_2$	σ_h	$2S_3$	$3\sigma_v$
A_1'	1	1	1	1	1	1
E'	2	−1	0	2	−1	0
Γ	3	0	1	3	0	1

$a(E') = \frac{1}{12}[2 \times 3 + 2(-1) \times 0 + 3(0) \times 1 + 2 \times 3 + 2(-1) \times 0 + 3(0) \times 1]$

$a(E') = 1$

It can finally be found that

$$\Gamma = A_1' + E'$$

Knowing the irreducible representations associated with s_1, s_2, s_3, it is possible to find the linear combinations that are bases of the corresponding irreducible representation by using the projection operator:

$$P_{\Gamma_i}(\varphi) = \frac{l_i}{h}\sum_R [X_i(R)R(\varphi)] \qquad (6.3)$$

where

l_i = dimension of the irreducible representation
h = order of the group
$X(R)$ = character associated with the symmetry operation R
$R(\varphi)$ = application of the symmetry operation R on the function φ

By applying the previous formula, one can find

$$P_{A'_1}(s_1) = s_1 + s_2 + s_3$$

$$P_{E'}(s_1) = \tfrac{1}{6}(4s_1 - 2s_2 - 2s_3) = f_1$$

Two functions are necessary to define the basis of E'. The second one is obtained by applying the projection operator on s_2 and s_3:

$$P_{E'}(s_2) = \tfrac{1}{6}(4s_2 - 2s_1 - 2s_3) = f_2$$

$$P_{E'}(s_3) = \tfrac{1}{6}(4s_3 - 2s_1 - 2s_2) = f_3$$

By using the orthogonality condition, the second function $af_2 + bf_3$ is found:

$$f_1(af_2 + bf_3) = 0$$

Finally, the second function is found to be proportional to $(s_2 - s_3)$:

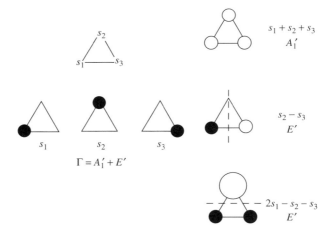

The previous formalism can be applied to any point symmetry group and, as a consequence, any number of atomic orbitals. However, it should not be deduced that any organization in condensed media could be described in this way since symmetry point groups cannot be employed instead of space groups. This point is more thoroughly examined in subsequent sections.

6.2.3 One-Dimensional Band Structure

Not all the symmetry elements of D_{3h} are necessary to describe the corresponding molecular orbitals derived from s-atomic orbitals. The symmetry group C_3 would have led to the same conclusions:

	C_3	E	C_3^1	C_3^2
A		1	1	1
E		1	ε	ε^*
		1	ε^*	ε

$$\varepsilon = \exp\left(\frac{2\pi i}{3}\right) = \cos\frac{2\pi}{3} + i\sin\frac{2\pi}{3}$$

It can be seen that each orbital is associated with only one symmetry element:

$$E(s_1) = s_1$$
$$C_3^1(s_1) = s_2$$
$$C_3^2(s_1) = s_3$$

When the projection operator is used, the coefficient of each orbital is therefore merely given by the character of the corresponding symmetry element:

$$P_{\Gamma_i}(s_1) = \tfrac{1}{3}[\chi(E)s_1 + \chi(C_3^1)s_2 + \chi(C_3^2)s_3]$$

$$A \longrightarrow \quad \Psi_1 = s_1 + s_2 + s_3$$

$$E \longrightarrow \begin{bmatrix} \Psi_2 = s_1 + \exp\left(\frac{2\pi i}{3}\right)s_2 + \exp\left(-\frac{2\pi i}{3}\right)s_3 \\ \Psi_3 = s_1 + \exp\left(-\frac{2\pi i}{3}\right)s_2 + \exp\left(\frac{2\pi i}{3}\right)s_3 \end{bmatrix}$$

These functions can easily be shown to be related to those previously found:

$$\Psi_2 + \Psi_3 = 2s_1 + 2\cos\left(\frac{2\pi}{3}\right)s_2 + 2\cos\left(\frac{2\pi}{3}\right)s_3$$

$$\cos\frac{2\pi}{3} = -\frac{1}{2}$$

$$\Psi_2 + \Psi_3 = 2s_1 - s_2 - s_3$$

and also

$$\Psi_2 - \Psi_3 = 2i\sin\left(\frac{2\pi}{3}\right)s_2 - 2i\sin\left(\frac{2\pi}{3}\right)s_3$$

$$\sin\frac{2\pi}{3} = \frac{\sqrt{3}}{2}$$

$$\Psi_2 - \Psi_3 = i\sqrt{3}(s_2 - s_3)$$

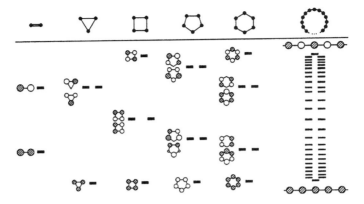

Figure 6.4 From a single atomic orbital to an infinite array of molecular orbitals (symmetry C_∞). (Reproduced by permission of Wiley-VCH from Ref. [5])

This calculation may be generalized for an infinite number of atomic orbitals disposed around a circle. In this case the group C_∞ must be considered (Figure 6.4).

In the symmetry group C_N (or C_∞ for an infinite number of atomic orbitals), the molecular orbitals will have the form:

$$\Psi_k = \sum_j \chi_k(R)\varphi_j \qquad (6.4)$$

with $\varphi_j = C_N^j(\varphi)$, where k depends on the corresponding irreducible representation. The table of characters of the symmetry group C_N is given in Table 6.1.

The angle of rotation Φ is given by

$$\Phi = \frac{2\pi}{N}$$

The character associated with the jth symmetry element is

$$\chi(C_j) = \exp ikj \left(\frac{2\pi}{N}\right)$$

and the molecular orbital associated with the irreducible representation Γ_k is

$$\Psi_k = \sum_{j=0}^{N-1} \left[\exp ikj \left(\frac{2\pi}{N}\right)\right] \varphi_j \qquad (6.5)$$

where Ψ_k can be obtained by applying the projection operator P_{E_k} on any function φ.

Whenever $N \to \infty$, the orbitals may be thought to represent a one-dimensional case. For an earlier slightly different approach see Ref. [6].

Table 6.1 Table of characters of the symmetry group C_N

C_N	E	C_N	C_N^2...	C_N^j...	C_N^{N-1}		
(E_0) A	1	1	1...	1...	1		
E_1	$\begin{cases}1\\1\end{cases}$	$\begin{array}{c}\varepsilon\\ \varepsilon^*\end{array}$	$\begin{array}{c}\varepsilon^2...\\ \varepsilon^{2*}...\end{array}$	$\begin{array}{c}\varepsilon^j...\\ \varepsilon^{j*}...\end{array}$	$\begin{array}{c}\varepsilon^{N-1}\\ [\varepsilon^{N-1}]^*\end{array}$	$\varepsilon = \exp\left(\dfrac{2\pi i}{N}\right)$	
E_k	$\begin{cases}1\\1\end{cases}$	$\begin{array}{c}\varepsilon^k\\ \varepsilon^{k*}\end{array}$	$\begin{array}{c}\varepsilon^{2k}...\\ \varepsilon^{2k*}...\end{array}$	$\begin{array}{c}\varepsilon^{jk}...\\ \varepsilon^{jk*}...\end{array}$	$\begin{array}{c}\varepsilon^{(N-1)k}\\ [\varepsilon^{(N-1)k}]^*\end{array}$	$\varepsilon^N = 1$	
N odd $E_{(N-1)/2}$ or N even		$\begin{array}{c}\varepsilon^{(N-1)/2}...\\ [\varepsilon^{(N-1)/2}]^*...\end{array}$	$\begin{array}{c}\varepsilon^{N-1}...\\ [\varepsilon^{N-1}]^*...\end{array}$	$\begin{array}{c}\varepsilon^{j(N-1)/2}...\\ [\varepsilon^{j(N-1)/2}]^*...\end{array}$	$\begin{array}{c}\varepsilon^{(N-1)^2/2}\\ [\varepsilon^{(N-1)^2/2}]^*\end{array}$		
$(E_{N/2})$ B	1	-1	1...	$e^{ij\pi}$...	-1		

The various functions Ψ_k can be extracted from the table of characters of C_N shown in Table 6.1. When $N \to \infty$, the irreducible representations $E_{(N-1)/2}$ or $B(E_{N/2})$ become indistinguishable. For all other irreducible representations, one of the basis functions corresponds to $+k$ value and the other one to $-k$ value. In total, N different functions are obtained.

6.2.4 Bloch's Functions, Band Diagram

The functions Ψ_k are called the *Bloch functions* in solid state physics. The functions shown in Figure 6.4 are usually employed in the calculations, but are real not imaginary. They can be obtained, as previously, by taking the sum and the difference of the imaginary functions of the same irreducible representations:

$$\Psi_k \pm \Psi_{-k}$$

The value of k, which is associated with a given irreducible representation, is also related to the number of nodes:
$k = 0$ (0 node):

$$\Psi_0 = \sum_j e^0 \varphi_j = \varphi_0 + \varphi_1 + \varphi_2 + \varphi_3 + \cdots$$

$k = N/2$ (N even):

$$\Psi_{N/2} = \sum_j (e^{\pi i j})\varphi_j = \sum_j (-1)^j \varphi_j \quad (6.6)$$

$$\Psi_{N/2} = \varphi_0 - \varphi_1 + \varphi_2 - \varphi_3 + \cdots$$

The functions Ψ_k for $k = 0$ and $k = N$ are identical.

It is possible to consider translations instead of rotations:

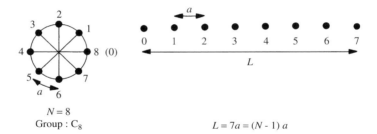

$N = 8$
Group : C_8

$L = 7a = (N - 1)\,a$

In this case, if there are N equivalent sites with a one-dimensional lattice parameter a, the overall length is given by $(N - 1)a$.

The functions Ψ_k may be written as

$$\Psi_k = \sum_j \exp(ikja)\varphi_j$$

with

$$a = \frac{2\pi R}{N}$$

where
$N =$ number of equivalent sites
$R =$ radius of the circle considered

Parameter a may be assimilated to a segment as long as R and N are large enough. In this case, however, in order to have a periodic function, the k values must be changed from $2\pi R/a$ (corresponding to N equivalent sites in the group C_N) to $2\pi R/(N-1)a$ (corresponding to the length L). By taking $R = 1$ and the notation shown in the previous scheme (with n an integer)

$$k \in \left(0, \ldots, \frac{n \times 2\pi}{(N-1)a}, \ldots, 2\pi/a\right) \qquad n \in [0, N-1]$$

or, equivalently:

$$k \in \left(0, \ldots, \pm\frac{n \times 2\pi}{(N-1)a}, \ldots, \pm\pi/a\right) \qquad n \in [0, (N-1)/2]$$

The corresponding zone of **k** will be called the *first Brillouin zone*.

However, it is easy to realize that

$$\Psi_{-\pi/a} = \sum_j (e^{-\pi i j})\varphi_j = \Psi_{\pi/a} = \sum_j (e^{\pi i j})\varphi_j \qquad (6.7)$$

The function may therefore be defined on the interval $[0, \pi/a]$.

MOLECULAR SEMICONDUCTORS: PROPERTIES AND APPLICATIONS

The functions Ψ_k are all associated with an energy E_k. Their classification may be made by considering the number of nodes: an increase of k (with s-type atomic orbitals) leads to a destabilization of the corresponding molecular orbital. By plotting E_k as a function of k, one obtains the classically called band diagram (Figure 6.5).

The difference in energy (ΔW) between the lowest and highest levels is called the bandwidth: its magnitude depends on the degree of interaction between the units and therefore to the *degree of collectivization* of the electronic wavefunctions. Calculations have been carried out for hydrogen atoms situated at 3, 2 or 1 Å; the corresponding bandwidths vary by a factor of almost 10 [5].

The band structure depends on the nature of the initial atomic orbital. In the case, for example, of p_z orbitals, the molecular orbitals at $k = 0$ will be less stable (higher energy) than for $k = \pi/a$ (Figure 6.6). The number of levels situated between the energies E and $E + dE$ is given by the density of states (DOS). It is noteworthy that the states are equally spaced along the k axis and therefore the density of states is higher when the band is nearly flat (Figure 6.7).

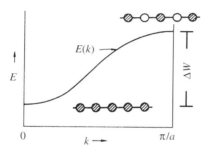

Figure 6.5 Plot of E_k versus k (band structure) for s-type orbitals. There are $N/2$ values of k between zero and π/a. When N is large the curve $E(k)$ is quasi-continuous. (Reproduced by permission of Wiley-VCH from Ref. [5])

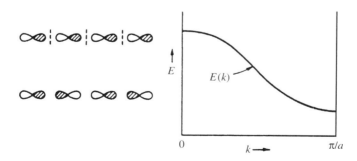

Figure 6.6 Band structure generated with p_z atomic orbitals. (Reproduced by permission of Wiley-VCH from [5])

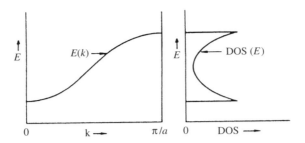

Figure 6.7 Band structure and corresponding density of states for s-type orbitals (Reproduced by permission of Wiley-VCH from Ref. [5])

6.2.5 Fermi Level

A Fermi–Dirac distribution is given by:

$$g_{FD} = \frac{1}{1 + \exp(E - E_F)/(kT)} \qquad (6.8)$$

where E_F is called the Fermi level. At $T = 0$ K all levels below E_F are occupied whereas the others remain empty. At higher temperatures E_F corresponds to an occupation density of 1/2:

$$g_{FD} = 1/2 \qquad E = E_F$$

It is worth pointing out that at high temperatures, the Fermi distribution tends to a Boltzmann distribution.

The same notion may be introduced starting from statistical thermodynamics. Conventional textbooks demonstrate that the entropy of a system is given by [7]:

$$S = k \ln C_{max} \qquad (6.9)$$

where

C_{max} = maximum number of possible combinations

It may then be demonsteted that

$$\overline{G}_i = \mu_i = G_i + RT \ln x_i \qquad (6.10)$$

where

$\overline{G}_i$ = partial free energy
G_i = free enthalpy
$R = \mathcal{N}k$, where $\mathcal{N}$ = Avogadro's number
x_i = molar fraction of the particle i in the mixture

The partial free energy $\overline{G}_i$ is also called the *chemical potential*. The term $R \ln x_i$ is the mixing entropy, the stabilization brought by mixing the various particles present.

The above formalism has been derived for dispersed media, for example an ideal gas. It can be shown (see, for instance, Ref. [8]) that

$$\mu_i = E_F$$

The concept of chemical potential used in chemistry (and electrochemistry) is equivalent to the concept of the Fermi level usually used in solid state physics.

6.2.6 Two- and Three-Dimensional Band Structures

Let us consider a lattice defined by the vectors $\mathbf{a}_x$ and $\mathbf{a}_y$:

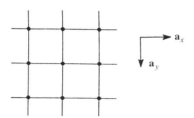

In one dimension, k could be considered as a scalar. However, in two dimensions the k values belong to different intervals since, in the general case, $|\mathbf{a}_x| \neq |\mathbf{a}_y|$. It will be considered that there are N_x sites along the Ox axis and N_y sites along Oy, the sites being considered to be arranged along a circle. The range of k values is given by

$$k_x \in [0, N_x - 1] \qquad N_x = \frac{2\pi}{a_x}$$

$$k_y \in [0, N_y - 1] \qquad N_y = \frac{2\pi}{a_y}$$

If translations are considered instead of rotations, it is found, as previously in one dimension:

$$k_x \in \left(0, \ldots, \pm n_x \frac{2\pi}{(N_x - 1)a_x}, \ldots, \pm \pi/a_x\right)$$

$$k_y \in \left(0, \ldots, \pm n_y \frac{2\pi}{(N_y - 1)a_y}, \ldots, \pm \pi/a_y\right)$$

The values of n_x and n_y on the interval $[0, \pi/a]$ are

$$n_x \in [0, (N_x - 1)/2]$$

$$n_y \in [0, (N_y - 1)/2]$$

The lattice is described by the functions $\varphi_{j_x + j_y}$ corresponding to the translations $j_x \mathbf{a}_x + j_y \mathbf{a}_y$. The Bloch functions corresponding to the two-dimensional square lattice are

$$\Psi_{k_x,k_y} = \sum_{j_x,j_y} [(\exp ik_x j_x a_x)(\exp ik_y j_y a_y)]\varphi_{j_x,j_y}$$

For special values of k_x and k_y, the following functions may be determined:

$$k_x = k_y = 0 \qquad \Psi_{0,0} = \sum_{j_x,j_y} \varphi_{j_x,j_y}$$

$$k_x = \frac{\pi}{a_x}, k_y = 0 \qquad \Psi_{\pi/a,0} = \sum_{j_x,j_y} (-1)^{j_x} \varphi_{j_x,j_y}$$

$$k_x = \frac{\pi}{a_x}, k_y = \frac{\pi}{a_y} \qquad \Psi_{\pi/a_x,\pi/a_y} = \sum_{j_x,j_y} (-1)^{j_x+j_y} \varphi_{j_x,j_y}$$

The k_x and k_y values may be easily associated with the irreducible representations shown previously for the symmetry group C_N (see Table 6.1). For $k = 0$, the wavefunction is associated with the irreducible representation A and therefore fully corresponds to the symmetry of the lattice. For other k values other irreducible representations are found and the Bloch functions have a periodicity $j_x a_x = (2\pi/k_x)$ along Ox and $j_y a_y = (2\pi/k_y)$ along Oy (Figure 6.8).

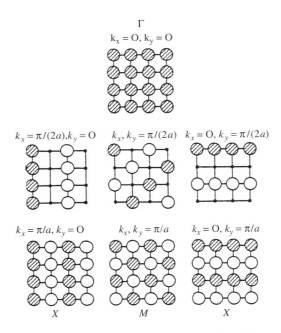

Figure 6.8 Representation of two-dimensional wavefunctions for various values of k. (The denomination corresponding to particular values of k is also given.) (Reproduced by permission of Wiley-VCH from Ref. [5])

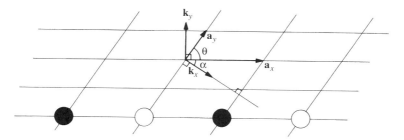

Figure 6.9 Translation vectors $\mathbf{a}_i$ and $\mathbf{k}_j$ in an oblique lattice

In the Bloch functions Ψ_{k_x,k_y}, the terms $k_i a_j (i \neq j)$ are not taken into account. This can be simply expressed in a vectorial form, the conditions needed being easily seen in an oblique lattice (Figure 6.9).

In a vectorial form, $\mathbf{k}_x$ may be taken perpendicular to $\mathbf{a}_y$ and $\mathbf{k}_y$ perpendicular to $\mathbf{a}_x$. Under these conditions, the scalar product $\mathbf{k}_x \cdot \mathbf{a}_x$ is given by

$$\mathbf{k}_x \cdot \mathbf{a}_x = |\mathbf{k}_x||\mathbf{a}_x|\cos\alpha$$

where
$\alpha = \pi/2 - \theta$
$\theta = (\mathbf{a}_x, \mathbf{a}_y)$

The moduli of the unit vectors must be expressed in the same reference frame:

$$|\mathbf{k}_x| = \frac{2\pi R_x}{a_x \cos\alpha}$$

where $a_x \cos\alpha$ is indeed the interplanar distance in the $\mathbf{k}_x$ direction. It can be taken as either $2\pi R_x = 1$ or $R_x = 1$, arbitrarily. Therefore, it becomes

$$\mathbf{k}_x \cdot \mathbf{a}_x = 2\pi$$

On the contrary, the terms $\mathbf{k}_i \cdot \mathbf{a}_j$ $(i \neq j)$ are null since by definition their relative angle is 90°.

The same type of expression may be found in three dimensions. In this case, the vectors $\mathbf{a}_i$ are generally written $\mathbf{R}$ and the wavefunction is written as

$$\boxed{\Psi_k = \sum_{j_x, j_y, j_z} [\exp(i\mathbf{k}\mathbf{R})]\varphi_{j_x, j_y, j_z}} \qquad (6.11)$$

The coordinates defining $\mathbf{k}$ are referred to as the *reciprocal space*. In vectorial form, this can be written as

$$\mathbf{k} = k_x \mathbf{a}_x^{-1} + k_y \mathbf{a}_y^{-1} + k_z \mathbf{a}_z^{-1}$$

where $\mathbf{a}_x^{-1}$, $\mathbf{a}_y^{-1}$ and $\mathbf{a}_z^{-1}$ are the unit vectors of the reciprocal space with

$$\mathbf{a}_x^{-1} = \mathbf{k}_x$$

The basic relationships relating the direct and reciprocal spaces are, as conventionally known,

$$\mathbf{a}_i \cdot \mathbf{a}_j^{-1} = 2\pi \quad i = j \ (i:\ x, y, z)$$

$$\mathbf{a}_i \cdot \mathbf{a}_j^{-1} = 0 \quad i \neq j$$

For the translations:

$$\mathbf{R} = j_x \mathbf{a}_x + j_y \mathbf{a}_y + j_z \mathbf{a}_z$$

6.2.7 Individual Approach: Supramolecular Orbitals

Until this point, the wavefunctions established are valid independent of the degree of interaction between the units constituting the lattice or the condensed phase. The elementary functions were — explicity or implicity — considered to be atomic orbitals. A slightly different model can now be made by postulating that the material consists of molecular units possessing their own molecular orbitals. The electronic wavefunctions describing the material must now be considered to be the sum of already formed molecular orbitals. We will name these orbitals *supramolecular orbitals*:

$$\Psi^{MO} = \sum_i C_i \varphi_i^{AO} \tag{6.12}$$

where
Ψ^{MO} = molecular orbital
φ^{AO} = atomic orbital

$$\Psi^{SO} = \sum_j C_j \Psi_j^{MO} \tag{6.13}$$

where
Ψ^{SO} = supramolecular orbital

The site symmetry must be considered at the very first stage of building the supramolecular orbitals. The order of the site symmetry group is equal to the order of the crystalline class of the material divided by the number of equivalent molecular units in the primitive cell.

The symmetry group of the molecular unit is equal to or is a supergroup of the site symmetry group. It is therefore necessary to find the irreducible representations associated with the molecular unit orbitals in the site symmetry group. In many cases, only the highest occupied molecular orbital (HOMO) and lowest unoccupied molecular orbital (LUMO) are considered.

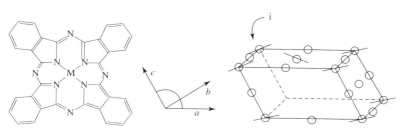

Figure 6.10 Chemical formula of a metallophthalocyanine and schematic representation of the unit cell (only the inversion centres have been represented) space group P2$_1$/a (Z = 2) (site symmetry $\bar{1}$)

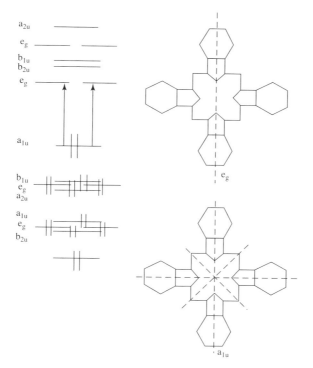

Figure 6.11 The HOMO (a$_{1u}$) and LUMO (e$_g$) of metallophthalocyanines belonging to the symmetry group D$_{4h}$

Metallophthalocyanines (symmetry D$_{4h}$) crystallize in most cases in the space group P2$_1$/a. In the lattice only the inversion centre is conserved (Figure 6.10). The HOMO and LUMO correspond to the irreducible representations a$_{1u}$ and e$_g$ in the symmetry group D$_{4h}$ (Figure 6.11).

The symmetry of the crystalline class is C$_{2h}$ (2/m). The correspondence between the D$_{4h}$ and C$_{2h}$ groups for the two frontier orbitals are shown in

Table 6.2 Correspondence between the irreducible representations of the orbitals of the individual molecular units (D_{4h}) or in the crystal (crystalline class C_{2h})

Molecular unit D_{4h}	Crystal C_{2h}				
	E	C_2	i	σ_h	
HOMO $2a_{1u}$	2	0	−2	0	
LUMO $2e_g$	4	0	4	0	
	1	1	1	1	A_g
	1	−1	1	−1	B_g
	1	1	−1	−1	A_u
	1	−1	−1	1	B_u

Table 6.2. It can be readily seen that

$$\text{HOMO: } 2\,a_{1u} \longrightarrow A_u + B_u$$
$$\text{LUMO: } 2\,e_g \longrightarrow 2(A_g + B_g)$$

In the symmetry group of the crystalline class (C_{2h}), it is now possible to build supramolecular orbitals by making linear combinations of molecular unit orbitals whose corresponding irreducible representations are determined knowing the site symmetry:

$$\Psi_k^{SO} = \frac{1}{\sqrt{N}} \sum_j (e^{i\mathbf{k}\cdot\mathbf{R}}) \Psi_j^{MO} \quad (6.14)$$

A more detailed description can be derived from examination of Refs. [9] and [10] and Sections 6.6.2 and 6.6.3.

The previous approach is valid only when intramolecular bonding is far larger than intermolecular interactions justifying a two-stage process. It can be considered that collectivization of the electronic levels occurs within the molecular unit in order to build molecular orbitals from atomic orbitals, whereas the integrity of the molecular units is more or less preserved within the crystal (the intermolecular interactions modify the energy levels of the molecular unit orbitals without changing their nature). The next step is to consider completely collectivized electronic levels.

6.2.8 Collectivized Approach: An Insight into Band Theory

Band theory is described in classical books (see, for instance, Refs. [11] and [12]. Since this approach is inappropriate for molecular materials (at least at room temperature), only the physical basis of this theory is given below.

The wavefunction associated with one electron in a potential well of the type:

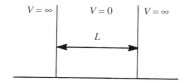

L = width of the well
V = potential

must be a solution of the Schrödinger equation with

$$H = -\frac{\hbar^2}{2m}\frac{\partial^2}{\partial x^2}$$

which contains only the kinetic energy part at a first approximation. A solution of the Schrödinger equation is

$$\Psi = \frac{1}{\sqrt{L}}\exp(ikx)$$

with

$$k = \frac{2\pi}{L}n \qquad n = 0, \pm 1, \pm 2, \ldots$$

The corresponding energy is given by

$$E = \frac{\hbar^2 k^2}{2m}$$

where
m = mass of the particle

By considering de Broglie's equation:

$$\lambda = \frac{h}{mv}$$

The kinetic energy is

$$E = \tfrac{1}{2}mv^2$$

$$p = mv$$

$$E = \frac{p^2}{2m} = \frac{\hbar^2 k^2}{2m}$$

$$k = \frac{p}{\hbar} = \frac{2\pi}{\lambda}$$

$$\boxed{k = \frac{2\pi}{\lambda}} \qquad (6.15)$$

There is therefore a relationship between the parameter k and the wavelength λ associated with the particle.

At three dimensions, the solution is given by

$$\Psi = \frac{1}{\sqrt{V}} e^{i\mathbf{k}\cdot\mathbf{R}} \tag{6.16}$$

and the energy is still given by

$$E = \frac{\hbar^2 k^2}{2m} \tag{6.17}$$

with

$$k_1 = \frac{2\pi}{L} n_1$$

$$k_2 = \frac{2\pi}{L} n_2$$

$$k_3 = \frac{2\pi}{L} n_3$$

for a cubic cell of edge L.

It is now necessary to consider the potential generated by the nuclei and the core electrons of the lattice (Figure 6.12). The corresponding hamiltonian is

$$H = -\frac{\hbar^2}{2m} \frac{\partial^2}{\partial x^2} + V(x) \tag{6.18}$$

where $V(x)$ has the periodicity of the lattice. The Bloch function is then

$$\Psi = e^{i\mathbf{k}\cdot\mathbf{R}} u(\mathbf{R}) \tag{6.19}$$

where $u(\mathbf{R})$ is a periodic function (where $u(\mathbf{R}) = c^{st}$ is the free electron approximation). Whenever $u(\mathbf{R}) \neq c^{st}$, zones of allowed and forbidden E_k appear (bands

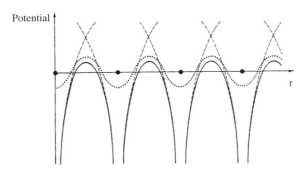

Figure 6.12 Periodic potential generated by a one-dimensional array of metallic cations: ———, potential along the line of the ions; ······, potential along a line between planes of ions; – – –, potential of single isolated ions. (From Ref. [12])

and forbidden gap). This can be qualitatively understood using Bragg's law: in a one-dimensional lattice of periodicity a,

$$2a \sin \theta = n\lambda$$

at one dimension $\theta = 90°$ for a wave propagating along the chain and so

$$2a = n\lambda$$

$$\lambda = \frac{2\pi}{k}$$

$$\boxed{k = n\frac{\pi}{a}} \qquad (6.20)$$

At the wall, the wave will be reflected, generating interferences between incident and reflected waves:

$$\Psi_+ = e^{i\pi x/a} + e^{-i\pi x/a} = 2\cos(\pi x/a)$$
$$\Psi_- = e^{i\pi x/a} - e^{-i\pi x/a} = 2i\sin(\pi x/a)$$

where Ψ_+ and Ψ_- are stationary waves. The corresponding probabilities of density are

$$\rho_- = |\Psi_-|^2$$
$$\rho_+ = |\Psi_+|^2$$

If ρ_- and ρ_+ lead to an energy difference E_g, we will have a forbidden gap [11]. The wavefunction just below the forbidden gap is Ψ_+ whereas the wavefunction just above the forbidden gap is Ψ_-.

In this approach a completely collectivized model is used: the electron (more precisely the electron of conduction) is considered as a wave interacting with a periodic potential generated by the nuclei and the core electrons.

6.3 MOLECULAR SEMICONDUCTORS: GENERALITIES

Since the intermolecular interaction energy is far smaller than the intramolecular bonding energy, the physicochemical properties of a molecular material may be derived from the characteristics of the constitutive molecular units. The molecular unit can be represented as A; its characteristics in condensed media may be derived from the ones in solution.

In the absence of any dopant and for a single component material, the generation of charge carriers is a disproportionation reaction [13–16]:

$$2A \rightleftharpoons A^+, A^- \quad \Delta G$$

where ΔG is the free energy of the reaction. The corresponding equilibrium constant K is

$$K = \frac{[A^+][A^-]}{[A]^2}$$

$$\Delta G^0 = -RT \ln K \qquad (6.21)$$

The redox potentials, which are usually determined in solution by cyclic voltammetry, correspond to the equilibria:

$$A \rightleftharpoons A^+ + e^- \quad E^0_{ox}$$
$$A + e^- \rightleftharpoons A^- \quad E^0_{red}$$

and then

$$\Delta G^0 = e(E^0_{ox} - E^0_{red}) \qquad (6.22)$$

The conductivity of a condensed phase is given by

$$\sigma = [A^+]e\mu_+ + [A^-]e\mu_-$$

where σ is expressed in $\Omega^{-1}\,\text{cm}^{-1}$, e, the electron charge, is equal to 1.6×10^{-19} C, and μ is the mobility in cm^2/V s.

The concentration of positive (or negative) charge carriers can be derived from previous equations:

$$[A^+] = [A]\exp\left(-e\frac{E^0_{ox} - E^0_{red}}{2kT}\right) \qquad (6.23)$$

The density of molecular units per cubic centimeter [A] may be estimated from the density d of the material and from the molar mass (M_w) of the constitutive molecular units:

$$[A] = \frac{d}{M_w}\mathcal{N} \qquad (6.24)$$

where $\mathcal{N}$ is Avogadro's number. For conventional molecular crystals $[A] \sim 10^{21}$ cm^{-3}.

The density of charge carriers A^+ and A^- in the solid state can therefore be estimated from the values of the redox potentials in solution. This seems to be an oversimplification since the polarity and the polarizability of the two media may be drastically different. However, it has been shown that for large conjugated molecules, the influence of the surroundings on the ionization equilibrium is small [8].

The concentration of charge carriers may therefore be estimated from simple experimental determinations. It is, on the contrary, difficult to find a simple model to estimate the mobility of charge carriers (the drift velocity of the charges in an electric field of unity) [3, 13]. An upper limit may, however, be found.

MOLECULAR SEMICONDUCTORS: PROPERTIES AND APPLICATIONS

For all molecular crystals known so far, the mobility of charge carriers does not significantly exceed $1\,\mathrm{cm^2/V\,s}$ at room temperature [3]. For polycrystalline or amorphous thin films, the mobility decreases by more than two orders of magnitude: mobilities are in the range 10^{-3}–$10^{-5}\,\mathrm{cm^2/V\,s}$.

It is now possible to classify the various types of molecular materials according to their conductivity and to determine which difference of redox potentials is necessary to have a high enough concentration of charge carriers (Table 6.3).

Organic crystals or conjugated polymeric systems have never been obtained in a pure enough state to be able to observe their intrinsic electrical properties at room temperature. In all cases $e\,\Delta E^0_{\mathrm{redox}}$ was typically more than 1.5 eV. The density of intrinsic charge carriers was therefore too low to be detectable and impurity states were predominant. The only exceptions are radical derivatives: bisphthalocyaninato lutetium Pc_2Lu and lithium phthalocyanine PcLi [14, 15] (Figure 6.13).

Table 6.3 Order-of-magnitude estimates of the conductivity (σ), the difference of redox potentials ($\Delta E^0_{\mathrm{redox}} = E^0_{\mathrm{ox}} - E^0_{\mathrm{red}}$) and the concentration of carriers (n) for insulators, semiconductors or metals

	Insulator	Semiconductor	Metal
$\sigma\,(\Omega^{-1}\,\mathrm{cm}^{-1})$	10^{-16}	10^{-7}	10^2
$e\,\Delta E^0_{\mathrm{redox}}$ (eV)	2	1	0.025
n (cm^{-3})	10^3	4×10^{12}	6×10^{20}

58	59	60	61	62	63	64	65	66	67	68	69	70	71
Ce	Pr	Nd	Pm	Sm	Eu	Gd	Tb	Dy	Ho	Er	Tm	Yb	Lu
140.12	140.91	144.24	(147)	150.4	151.96	157.25	158.93	162.50	164.93	167.26	168.93	173.04	174.97

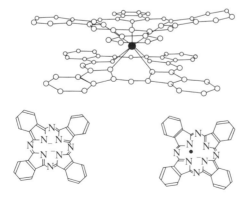

Figure 6.13 Representation of bisphthalocyaninato lutetium and of the two constitutive macrocycles. For the sake of simplicity the radical has been figured on one of the two macrocycles. The lithium derivative is derived from the radical macrocycle

6.4 A NARROW-BAND MOLECULAR SEMICONDUCTOR: Pc_2Lu

The width of a band in a solid, which must not be confused with the bandgap, is related to the magnitude of the interaction between the constitutive molecular units and to the mobility of charge carriers. It will be shown later that Pc_2Lu is a typical narrow-band intrinsic molecular semiconductor.

6.4.1 Synthesis

There are 15 different rare earths from lanthanum to lutetium with identical 5d and 6s electronic levels but differing by the number of electrons in the 4f level. Yttrium 39 is generally associated with the rare earths since it has the same electronic structure and possesses a comparable radius. Despite their name, rare earths are relatively abundant in Nature, but, in most cases, they are in low concentrations in many minerals.

Rare-earth phthalocyanines are obtained by reacting the corresponding acetates with o-phthalonitrile at 250–300 °C [17, 18]. In 1965, when the radical nature of the bisphthalocyanine had not yet been recognized, the formula Pc_2LnH was proposed. It took more than 15 years to find the correct formula from magnetic experiments [19].

Pc_2Sc has been found to be more easily synthesized than rare-earth phthalocyanines. In the previous syntheses, PcLnX (Ln, lanthanide ion) is formed beside the bisphthalocyanine derivatives. The reaction of heptanedionato complexes of rare-earth ions with $PcLi_2$ permitted the isolation of stable radical complexes $Pc^{\overline{\cdot}} M (dpm)_2$ (dpm, dipivaloylmethane) [20]. Supercomplexes of formula Pc_3Ln_2 were also suggested as side products in the formation reaction of Pc_2Ln [18]. Later experiments confirmed this assignment. Unsymmetrical naphthalophthalocyanines have been prepared in two steps via the formation of PcLuOAc [21].

6.4.2 Electrochemical Properties

Electrochromism of thin films of Pc_2Lu had already been demonstrated at the start of the 1970s [22]. Pc_2Lu deposited on electrodes was in contact with aqueous potassium chloride solutions. The initially green layer is transformed into a blue one by reduction and red by oxidation. In solution in dimethylformamide, four different colors have been seen: violet, blue, green and yellow-red. The spectra in CH_2Cl_2 for the oxidized and reduced forms are shown in Figure 6.14.

The molecular material films are very stable and after only $10^4 - 10^5$ reversals (green-red transition) color intensity decreases gradually [24]. The use of nonaqueous solutions (ethylene glycol) still increases the long-term stability under cycling [25].

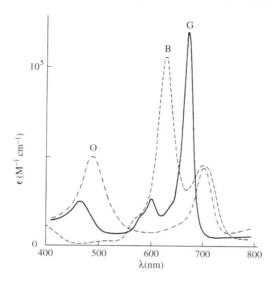

Figure 6.14 Absorption spectra in CH_2Cl_2 of $s\text{-}Pc_2Lu$(———), $s\text{-}Pc_2Lu^-$(– – –) and $s\text{-}Pc_2Lu^+$(– · – ·): $s\text{-}Pc = (C_{12}OCH_2)_8Pc$. (After Ref. [23])

Detailed electrochemical studies have been carried out [26], permitting the determination of the redox potentials associated with the color changes:

violet		dark blue		turquoise blue		green		yellow-red
$PcLu^{3-}$	$\rightleftharpoons$	Pc_2Lu^{2-}	$\rightleftharpoons$	Pc_2Lu^-	$\rightleftharpoons$	Pc_2Lu	$\rightleftharpoons$	Pc_2Lu^+
ΔE	-1.92		-1.54		-0.45		$+0.03$	
		0.38		1.09		0.48		1.20

$E_{1/2}$ versus ferrocene $\rightleftharpoons Pc_2Lu^{2+} \rightleftharpoons Pc_2Lu^{3+}$
$+1.23 \qquad +1.43$
0.20

The second and third reduction potentials do not significantly depend on the cation complexed (from Sm^{3+} to Y^{3+}). The first reduction and the first oxidation depend linearly on the ionic radii of the cations and the difference remains approximately constant and equal to 0.45 V [27].

Substitution of one of the two Pc macrocycles with napththalocyanine to give NPcPcLu leads to only small differences in the redox potentials ($E^0_{ox_1} = 0.08$ V, $E^0_{red_1} = -0.43$ V, $E^0_{red_2} = -1.46$ V, $E^0_{red_3} = -1.82$ V versus ferrocene) [28].

6.4.3 Crystal Structures

Various crystalline forms of rare-earth phthalocyanines have been obtained, depending on the methods of preparation. Crystallization from organic solutions frequently leads to solvates: the β-form $Pc_2Lu \cdot CH_2Cl_2$ is an example. The electrochemical oxidation of anions Pc$_2$Ln$^-$ permits solvent free materials to be obtained; in this way, three polymorphic forms of Pc$_2$Lu, α, β and γ have been characterized [29] (Table 6.4).

In the solvate β-Pc$_2$Lu · CH$_2$Cl$_2$, columns of molecules may be distinguished in the structure. The macrocycles make an angle of 36° with the c axis of the crystal (the axis of the columns). The interunit overlap may be estimated from the distance between carbon atoms belonging to two different molecular units. When the distances are smaller than the van der Waals value, strong interunit interaction energies are expected. The number of close contacts is also an important parameter for determining the overall interunit interaction energy. The shortest distances between the molecular units are in most cases greater than 0.34 nm. Only two contacts at 0.337 and 0.329 nm may be observed (Figure 6.15).

The γ-form of Pc$_2$Lu presents a different type of organization (Figure 6.16). In this case two-dimensional layers parallel to the (010) plane of the orthorhombic unit cell are formed. In the layers the planes of the macrocycles are parallel to each other. However, between two contiguous layers, the macrocyclic planes are perpendicular to one another. The shortest intermolecular intralayer distances are significantly smaller than the van der Waals values (from 0.328 nm to 0.338 nm).

The structure of α-Pc$_2$Lu is only known from studies on polycrystalline samples [34]. The cell is tetragonal with parameters very close to those known for Pc$_2$Nd and Pc$_2$Yb [29, 32] (Figure 6.17). The main structural characteristic influencing the transport properties is the columnar organization. The bisphthalocyanine macrocycles are perpendicular to the column axes.

At a molecular level, there is not a large difference between the β-solvate and the γ-form. In β-Pc$_2$Lu · CH$_2$Cl$_2$ the rings are rotated by 45° and the deviation of one benzene ring from the four coordinate nitrogen mean plane is 13.8° [30] (other phenyl rings deviate by 1.9–3.6°). In γ-Pc$_2$Lu the macrocycles are rotated by

Table 6.4 Structural characteristics of various bismetallophthalocyanines (see Refs. [32, 111])

	Form	a (nm)	b (nm)	c (nm)	a,c (°)	Z	Space group
Tetragonal							
Pc$_2$Lu	α	1.98	1.98	0.66	90	2	
Pc$_2$Yb	α	1.98	1.98	0.69	90	2	
Monoclinic							
Pc$_2$Nd	β	1.901	1.906	1.554	116.1	4	C2/c
Pc$_2$U	β	1.874	1.873	1.561	113.9	4	C2/c
Pc$_2$Th	β	1.876	1.876	1.581	113.2	4	C2/c
Orthorhombic							
Pc$_2$Sn	γ	1.055	5.074	0.890	90	4	P2$_1$2$_1$2$_1$
Pc$_2$Sc	γ	1.05	5.13	0.88	90		
Pc$_2$Nd	γ	1.067	5.157	0.847	90	4	P2$_1$2$_1$2$_1$
Pc$_2$Lu	γ	1.055	5.079	0.896	90	4	P2$_1$2$_1$2$_1$
Solvate orthorhombic							
Pc$_2$Lu · CH$_2$Cl$_2$	β	2.824	2.288	0.805	90	4	Pnma

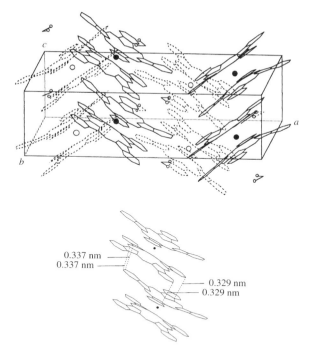

Figure 6.15 Structure of the β-form of the solvate Pc$_2$Lu · CH$_2$Cl$_2$. The structure is described in Ref. [30]

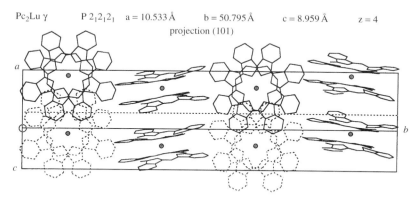

Figure 6.16 Structure of the γ-form of Pc$_2$Lu. This structure is described in Ref. [31]

41°. The benzene rings deviate through a smaller angle than in β-Pc$_2$Lu · CH$_2$Cl$_2$, but the two phthalocyanine rings are still not equivalent. The lutetium atom is displaced from the centre of the coordination polyhedron toward one of its bases [31]. It is supposed that the unpaired electron is located on one of the phthalocyanine ligands.

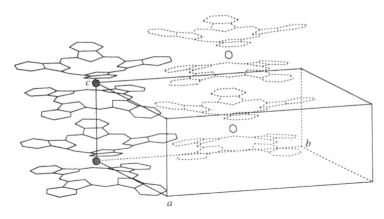

Figure 6.17 Schematic representation of α-Pc$_2$Lu (tetragonal form) [34]

Sublimation under vacuum of Pc$_2$Ln allows the preparation of thin solid films. The molecules can be deposited on glass substrates or on cleaved single crystals. High-resolution transmission electron microscopy of Pc$_2$Yb thin films indicates a square array of molecules in columns corresponding to the α-form [32]. The growth of thin films on glass substrates has been studied in detail for Pc$_2$Lu: the α-phase has been characterized for thick films [34].

Epitaxial growth on the (001) face of NaCl or KCl single crystals is also possible. In most cases Pc$_2$Ln form tetragonal phases with the columns perpendicular to the substrate. The interaction of the nitrogen atoms of the phthalocyanine with the Na$^+$ or K$^+$ ion of the substrate is thought to be the driving force for the flat arrangement of the macrocycle [188].

6.4.4 Conduction Properties

The conduction properties of Pc$_2$Nd were measured on compressed pellets [17]. A value of $4 \times 10^{-2}\,\Omega^{-1}\,\mathrm{cm}^{-1}$ for the conductivity at room temperature with an activation energy of 0.12 eV was found. These results, as shown later, cannot arise from the intrinsic properties of Pc$_2$Nd and demonstrate the presence of impurities. In 1985, the electrical properties of thin films (TF) and single crystals (SC) of Pc$_2$Lu were described [14]. The difference of redox potentials in Pc$_2$Lu is 0.48 eV: the density of intrinsic charge carriers is therefore exceptionally high in solid phases. On the contrary, the other metallophthalocyanines with $\Delta E_{\mathrm{redox}} \sim 2\,\mathrm{eV}$ [15] yield insulators.

The conductivity of single crystals of Pc$_2$Lu is $6 \times 10^{-5}\,\Omega^{-1}\,\mathrm{cm}^{-1}$ with a thermal activation energy of conduction of 0.64 eV, which is of the same order of magnitude as the difference of redox potentials. By comparison, the conductivities of divalent metallophthalocyanines (PcH$_2$, PcCu, PcNi) are extremely low — less than $10^{-12}\,\Omega^{-1}\,\mathrm{cm}^{-1}$ (Table 6.5).

Table 6.5 Electrical properties of single crystals (SC) or thin films (TF) of various metallophthalocyanines (PcM or Pc$_2$M)

		Solid state			Solution	
		σ_{RT} (Ω^{-1} cm^{-1})	E^e_{act} (eV)	μ_e (cm^2/V s)	$E^{ox}_{1/2}$ (V)	$E^{red}_{1/2}$ (V)
PcH$_2$	SC	$< 10^{-12}$	2.00	1.2	$+0.86^a$	-0.58^a
	TF			10^{-2}–10^{-3}		
PcCu	SC	$< 10^{-12}$	2.00	7	$+0.98^b$	-0.84^c
	TF	$< 10^{-10}$	1.98	10^{-2}		
PcNi	SC	$< 10^{-12}$	2.28		$+1.05^b$	-0.85^c
Pc$_2$Lu	SC	6×10^{-5} (β-form)	0.64		$+0.03^d$	-0.45^d
	TF	$\sim 10^{-5}$	0.52	1.3		

[a] versus NHE (normalized hydrogen electrode) in dimethylformamide, DMF ($\varepsilon = 36.7$).
[b] versus SCE (standard calomel electrode) in chloronaphthalene ($\varepsilon = 5.04$).
[c] versus SCE in DMF.
[d] versus ferrocene in CH$_2$Cl$_2$ ($\varepsilon = 9.08$).
[e] E_{act} in the equation $\sigma = \sigma_0 \exp[-E_{act}/(2kT)]$.

The conductivities of the materials are clearly related to the difference of redox potentials in solution. The thermal activation energies of the insulating materials are large — approximately 2 eV in agreement with ΔE_{redox}. The divalent metallophthalocyanines are representative of most, if not all, organic or organometallic derivatives. Charge transfer complexes in which several components are brought together to form a material are excluded from this classification. They cannot be sublimed to form thin films and they cannot be doped. Organic molecular crystals (from anthracene to pentacene) and conjugated polymeric systems are insulators when undoped (Table 6.6).

It can be seen from Table 6.5 that the conductivity of thin films of PcM (M = H$_2$, Cu) is higher than for single crystals. This means that the amount of impurities is higher in vacuum deposited layers. During their formation, the thin films can trap gas molecules which can act as dopants. A more systematic study

Table 6.6 Oxidation ($E^{ox}_{1/2}$) and reduction ($E^{red}_{1/2}$) potentials in solution of various aromatic derivatives or conjugated polymers compared with solid-state parameters: E_{act}, activation energy; E_{opt}, optical gap [13, 33]

	$E^{ox}_{1/2}$ (V)	$E^{red}_{1/2}$ (V)	ΔE_{redox} (eV)	E_{act} (eV)	E_{opt} (eV)
Anthracene	1.34^b	-1.96^b	3.3	4.10	4.40
Tetracene	1.04^b	-1.58^b	2.62	3.13	3.43
Pentacene	0.82^b	-1.34^b	2.16	2.47	2.83
Polyacetylene	0.2^b				
Polyphenylene	1^a	-2.6^a	3.6		

[a] versus Ag/Ag$^+$.
[b] versus SCE.

has been carried out with Pc_2Lu in which the intrinsically high density of charge carriers prevents the observation of such an artifact [34].

Thin films of Pc_2Lu were prepared by evaporation under vacuum: an increased crystallinity with thickness is noticed. For films thinner than 60.0 nm, no electron diffraction is observed and a smooth surface is seen by scanning electron microscopy. The material is in a quasi-amorphous state and the conductivity is $1.8 \times 10^{-5}\,\Omega^{-1}\,cm^{-1}$. In the range 60–100 nm a series of sharp diffraction rings belonging to the α-phase appears. Electron diffraction measurements as a function of the angle between the electron beam and the surface of the substrate show that the crystallites are not randomly oriented: the columns are approximately parallel to the plane of the substrate. The conductivity is of the order of $3.8 \times 10^{-5}\,\Omega^{-1}\,cm^{-1}$ and increases to $8.3 \times 10^{-5}\,\Omega^{-1}\,cm^{-1}$ at larger thicknesses (greater than 100 nm) for which large crystals of the α-phase are visible (size $\sim$300 nm) (Table 6.7).

Single-crystal and thin-film conductivity values do not differ by more than one order of magnitude. This demonstrates that impurities and grain boundaries do not play a dominant role in the charge transport process. This is corroborated by a.c. measurements at 10 GHz for β-$Pc_2Lu \cdot CH_2Cl_2$: good agreement with the d.c. value is found. The same measurement on the γ-phase indicates some extrinsic contribution.

The thermal activation energy of conduction of Pc_2Lu thin films is constant down to 200 K, where it starts to decrease. The concentration of impurity (from chemical or disorder origins) is therefore of the order of 10^{11}–$10^{14}\,cm^{-3}$ depending on the density of states taken into consideration.

The mobility of charge carriers has been measured by the space charge limited current (SCLC) technique on thin films of Pc_2Lu [15]. In the shallow trap

Table 6.7 Electrical properties of Pc_2Lu as a function of the organization state

	σ_{RT} ($\Omega^{-1}\,cm^{-1}$)	E_{act} (eV)
$Pc_2Lu \cdot CH_2Cl_2$ β-phase Orthorhombic $a = 28.24, b = 22.88, c = 8.05$ $D_{cal} = 1.64\,g/cm^3$	6×10^{-5} (d.c.)[a] 10^{-4} (10 GHz)	0.64
Pc_2Lu γ-phase Orthorhombic $a = 10.55, b = 50.79, c = 8.95$ $D_{cal} = 1.66\,g/cm^3$	5.3×10^{-5} (d.c.) 2×10^{-3} (10 GHZ)	0.26
Pc_2Lu mixture Amorphous + α-phase Pure amorphous Pure α (oriented; polycrystalline)	$\sim 10^{-5}$ 1.8×10^{-5} 8.3×10^{-5}	0.52

[a] Along the c axis.

approximation, the product $\theta\mu$ can be calculated, where θ is the proportion of free charge carriers and μ is the mobility. A value as high as $1.3\,\mathrm{cm^2/V\,s}$ is thus determined, which can be compared with those obtained with standard metallophthalocyanines: $\theta\mu = 10^{-2}$–10^{-3} for $\mathrm{PcH_2}$ or $\theta\mu = 10^{-2}$ for PcCu using the same method of measurement. SCLC measurements may give unreliable results unless the validity of Child's law is checked; this law is indeed observed for $\mathrm{Pc_2Lu}$ [35].

6.4.5 Relationship between $\Delta E_{\mathrm{redox}}$ and E_{act}

Experimentally, there is close agreement between the difference of redox potentials in solution and the thermal activation energy of conduction in the condensed phases. This agreement cannot be predicted from any model since the solvation energy of the various species in solution and the polarization energy in the solid state are extremely different. An equation has been proposed [33] to take into account the difference in the surroundings:

$$E_{\mathrm{act}} = (E_{\mathrm{ox}}^1 - E_{\mathrm{red}}^1) + (S_2^+ + S_2^-)\left(1 - \frac{1 - 1/\varepsilon_2}{1 - 1/\varepsilon_1}\right) \quad (6.25)$$

where E^1 is the redox potential in medium 1, S is the solvation energy of the ions less that of the neutral molecular units, and ε_1 and ε_2 are the dielectric constants of the media 1 and 2, respectively.

The correction introduced by the solvation energies S and the ratio of the dielectric constants is of the order of 0.4–0.5 [33]. An examination of the previous results shows that there is not such a large difference between E_{act} and $\Delta E_{\mathrm{redox}}$. The redox potentials of large aromatic molecules are rather insensitive to the media in which they are determined [36]. If μ_i^1 and μ_i^2 are the chemical potentials of the species i in the media 1 and 2, respectively, the energy change in going from one to the other is given by

$$\mu_i^1 - \mu_i^2 = RT \ln {}^1\gamma_i^2 \quad (6.26)$$

where ${}^1\gamma_i^2$ is the parameter characterizing the chemical potential change when going from medium 1 to 2. It is then assumed that, for large molecules,

$${}^1\gamma_{\mathrm{A}^+}^2 = {}^1\gamma_{\mathrm{A}}^2 = {}^1\gamma_{\mathrm{A}^-}^2$$
$$\mu_{\mathrm{A}^+}^1 - \mu_{\mathrm{A}^+}^2 = \mu_{\mathrm{A}}^1 - \mu_{\mathrm{A}}^2 = \mu_{\mathrm{A}^-}^1 - \mu_{\mathrm{A}^-}^2 \quad (6.27)$$

This means that the difference in chemical potential when going from 1 to 2 is the same for A, A^+ and A^- [8, 36].

These relations have important consequences for the creation of charge carriers:

$$2\mathrm{A} \rightleftharpoons \mathrm{A}^+ + \mathrm{A}^-$$
$$\Delta G = \mu_{\mathrm{A}^+} + \mu_{\mathrm{A}^-} - 2\mu_{\mathrm{A}}$$

Table 6.8 Redox potentials $\Delta E = (E^0_{ox} - E^0_{red})$ for Pc$_2$Lu and PcLuNPc as a function of the solvent (ε = dielectric constant) [8]

		ε	ΔE
Pc$_2$Lu	CH$_2$Cl$_2$	9.1	0.48
	DMF	36.7	0.48
PcLuNPc	DMF/THF 5:5	36.7/7.4	0.47
	CH$_2$Cl$_2$/THF 3:7	9.1/7.4	0.48

NPc = 2,3-naphthalocyanine

In solvents 1 and 2,

$$\Delta G_1 - \Delta G_2 = (\mu_{A^+} - \mu_A)^1 - (\mu_{A^+} - \mu_A)^2 + (\mu_{A^-} - \mu_A)^1 - (\mu_{A^-} - \mu_A)^2$$

we then have

$$\Delta(\Delta G) = \Delta G_1 - \Delta G_2 = 0$$

It is shown that the energy needed to create charge carriers is solvent (or medium) insensitive as long as the molecular units are sufficiently large. The physical origin of this assumption is still to be clarified [8, 37].

The redox potentials of Pc$_2$Lu and PcLuNPc have been measured in various solvents which differ in their polarity and dielectric constant (Table 6.8). No difference in $\Delta E = E^0_{ox} - E^0_{red}$ has been noticed.

6.4.6 Optical and Magnetic Properties

Rare-earth phthalocyanines all show a broad absorption band around 1380–1400 nm in the near-infrared region. This transition is associated with the radical nature of the subunit since oxidation to Pc$_2$Ln$^+$ or reduction to Pc$_2$Ln$^-$ causes its disappearance. It was tempting to associate this band to an intermacrocyclic transition between Pc$^{\bullet-}$ and Pc^{2-} subunits. However, a strong delocalization of the unpaired electron is expected since the intermacrocyclic interaction energy is large. A rough estimation has been made using models previously established for mixed valence compounds. It has been found that the molecular orbital describing the ground state can be formulated as [38]

$$\Psi_0 = 0.89\Psi_a + 0.46\Psi_b$$

where Ψ_a and Ψ_b are the molecular orbitals of the macrocycles a and b, respectively. The unpaired electron is therefore highly delocalized. The interaction energy $V_{ab} = \langle\Psi_a|H|\Psi_b\rangle$ is of the order of 0.40 eV. Calculations have been carried out on models of Pc$_2$Lu. Two orbitals result from the splitting of the $4a_u$ π-HOMO of the phthalocyanine monomer. They are separated by 0.83 eV, in good agreement with the energy of the near-infrared spectrum (0.90 eV) [39].

Besides optical absorption studies, electron paramagnetic resonance (EPR) measurements have been carried out. In solution [14, 40], Pc_2Lu shows a single broad band (width $11G$) with a g value close to the free electron value. The peak-to-peak linewidths are $6.0G$ and $6.3G$ for Pc_2Y and Pc_2La, respectively. The EPR spectrum of Pc_2Lu is therefore probably broadened by an unresolved hyperfin coupling with the nuclear spin 7/2 of lutetium (III).

In the solid state $\beta\text{-}Pc_2Lu \cdot CH_2Cl_2$ yields a narrower signal of $1.6G$. The paramagnetic susceptibility as a function of temperature follows a Curie–Weiss law with $T = -6\,\text{K}$ (antiferromagnetic interactions). In a simple Hubbard model knowing the disproportionation energy (U), it is possible to calculate the bandwidth of the semiconductor from the transfer integral t:

$$t = \langle \Psi_1 | H | \Psi_2 \rangle \tag{6.28}$$

where Ψ_i is the molecular orbital associated with site i.

Experimentally, magnetic measurements yield the value of J, the magnetic coupling constant:

$$J = \frac{3k\theta}{2ZS(S+1)} \tag{6.29}$$

where S is the spin ($S = \frac{1}{2}$), k is Boltzmann's constant, θ is the Curie–Weiss temperature and Z is the number of neighbours ($Z = 2$). Thus

$$J = \frac{2t^2}{U} \tag{6.30}$$

where U can be approximated by $e\,\Delta E_{\text{redox}}$. The bandwidth is given by $4t$ which, for Pc_2Lu, is of the order of 15–35 meV [14]. The anisotropy of the EPR linewidth shows that the system is highly one-dimensional [41]. The other crystalline form of Pc_2Lu (γ-form) reveals, on the contrary, a two-dimensional character [42].

6.5 A BROAD-BAND MOLECULAR SEMICONDUCTOR: PcLi

PcLi and Pc_2Lu are both molecular radicals. Many physical properties will reflect this similarity. The same type of electronic configuration could be attained by oxidation of divalent metallophthalocyanine but ionic species would be formed.

6.5.1 Synthesis and Physicochemical Properties

The first syntheses of PcLi have been made by chemical, electrochemical or photochemical oxidation of $PcLi_2$ [43]. A one-step electrosynthesis of PcLi starting from o-phthalonitrile has been described [44]. The overall yield is exceptionally high (70%) and the final product, which is fairly difficult to purify

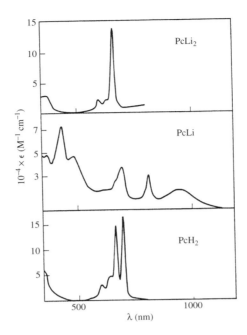

Figure 6.18 Comparison of the optical absorption spectra of $Pc^{2-}2Li^+$, PcLi and PcH_2. (From Ref. [45])

because of its sensitivity to water, dioxygen and acidic media, is of good purity. The optical absorption spectrum of PcLi is very different from other metallophthalocyanines (Figure 6.18). Oxidation of $PcLi_2$ leads to a partial filling of the HOMO ($a_{1u}(\pi)$) and to the appearance of new transitions coming from deeper levels.

Cyclic and stationary voltammetry was performed in a mixture of THF and chloronaphthalene [45]. The following reactions have been characterized:

$$PcLi^+ \underset{1.0}{\overset{-e^-}{\longleftarrow}} PcLi \underset{0.17}{\rightleftharpoons} PcLi^- \underset{-1.53}{\rightleftharpoons} PcLi^{2-} \quad \text{versus SCE}$$

The disproportionation reaction corresponds to $\Delta E_{\text{redox}} = 1.0 - 0.17 = 0.83$ V. This value is significantly larger than for Pc_2Lu (0.48 V).

Electrolysis of $PcLi_2$ at 0.15 V versus SCE in acetonitrile gives crystals of PcLi [46]. Their symmetry is tetragonal (space group P4/nnc) with $Z = 4$ (see below).

6.5.2 Conduction Properties

The disproportionation energy of PcLi is significantly larger than for Pc_2Lu: the intrinsic conductivity should be consequently smaller due to a lower density of charge carriers. This is not observed experimentally.

Single crystals of PcLi are two orders of magnitude more conductive than Pc$_2$Lu (Table 6.9). The intrinsic character of the electrical properties has been demonstrated by a measurement at 10 GHz: only a tenfold increase of the conductivity is observed compared with d.c. measurements [47]. Concomitantly, the thermal activation of conduction is four times lower than the difference of redox potentials obtained in solution. The characteristics of the molecular unit in solution cannot be used to rationalize the properties of the material: a large interaction energy between the constitutive units is effective. This is confirmed by the fact that the conductivity of thin films of PcLi is one or two orders of magnitude smaller than the conductivity of single crystals (Table 6.9).

The disorder in PcLi thin films creates deep traps. The depth of these traps is related to the magnitude of the interunit interaction energy. On the contrary, the conductivity difference between thin films and single crystals of Pc$_2$Lu is only a factor 6. This confirms the small intermolecular interaction energy in this case. The large intermolecular interaction energy in PcLi is reflected in its X-ray structure [46] (Figure 6.19).

In the tetragonal form of PcLi, the molecules form columns with the plane of the macrocycles perpendicular to the column axis. The intermolecular distances within a column are significantly smaller than the van der Waals values (Figure 6.20).

Table 6.9 Comparative conduction properties of thin films and single crystals of Pc$_2$Lu and PcLi

		σ_{RT} (Ω^{-1} cm^{-1})	E_{act} (eV)
Pc$_2$Lu	SC	6×10^{-5}	0.64
	TF	10^{-5}	
PcLi	SC	2×10^{-3}	0.2
	TF	10^{-4}–10^{-5}	

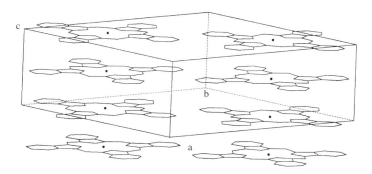

Figure 6.19 Structure of PcLi (P4/nnc: $a = 1.9575$ nm, $c = 0.6491$ nm, $Z = 4$). After Ref. [46]

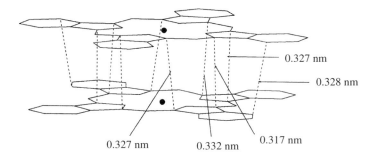

Figure 6.20 The shortest intermolecular C–C distances in PcLi tetragonal crystals. (After Ref. [46])

The lithium ion is situated at the centre of the cavity and the macrocyclic planes are rotated by 38°. The lithium–lithium distance (0.324 nm) is shorter than the van der Waals value of aromatic rings (0.34 nm), indicating a fairly important 'compression' of the π orbitals within the columns. A more direct insight into the magnitude of the intermolecular interaction energy may be gained from magnetic measurements.

6.5.3 Magnetic Properties

The magnetic properties of PcLi in solution (chloronaphthalene) may be affected by the formation of aggregates. The aggregates show a narrower line — close to the one observed for solids — superimposed on a broad line which can be associated with isolated molecules in solution [45]. In the last case, a Gaussian absorption line is observed with a peak-to-peak linewidth of $\Delta B_{pp} = 6.3$ G. The spectrum is similar to those reported for oxidized phthalocyanine derivatives. The broadening is due to unresolved hyperfine couplings with nitrogen atoms [189, 190].

Single crystals of PcLi show an extremely narrow ($\Delta B_{pp} = 50$ mG) Lorentzian-shaped single line centred at $g = 2.002$ [48]. An efficient spin exchange narrowing process is effective in the solid. Whereas the conduction properties are not affected by air exposure, the magnetic properties are importantly dependent on the presence of dioxygen [47, 49] (Figure 6.21).

The dioxygen effect is fully reversible and, almost instantaneously, without degradation of the sample, the O_2 molecules penetrate into channels present in the tetragonal structure of PcLi crystals. The presence of O_2 allows interchain coupling: the antiferromagnetic coupling between two PcLi molecules in a column competes with the antiferromagnetic coupling between the spin of $\frac{1}{2}$ of PcLi and the spin of 1 of O_2. At low temperatures the latter dominates, resulting in a ferromagnetic coupling between consecutive PcLi [50].

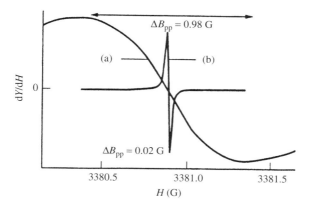

Figure 6.21 Modification of the EPR spectrum of a single crystal of PcLi when (a) exposed to air and (b) under vacuum (Y = intensity of the signal, H = magnetic field, B = magnetic induction, ΔB_{pp} = peak-to-peak width). (After Ref. [47])

Estimation of the coupling constant J from the variation of the susceptibility as a function of temperature in the absence of O_2 indicates a strong antiferromagnetic coupling (1000 K) between the spins and a bandwidth of approximately 1 eV [50]. PcLi is therefore a large bandwidth, intrinsic molecular semiconductor.

6.6 BAND STRUCTURE OF METALLOPHTHALOCYANINES

6.6.1 Applicability of the Band Model

More than 30 years ago, the applicability of the band model to explain charge transport properties in van der Waals crystals, such as metal-free phthalocyanine, has been examined [51]. At that time, the fact that the thermal activation energy of conduction, the optical absorption edge and the edge of photoconductivity are observed experimentally to be equal seemed to be good evidence that the band model was applicable. The corresponding bandwidths were, however, very small, of the order of kT, and associated with large uncertainties [52].

Discussions on the use of incoherent hopping or coherent model concepts gained in precision when ultrapure van der Waals crystals were obtained [53]. Two types of perylene single crystals have been studied: conventionally purified crystals and ultrapurified ones. The time-on-flight mobilities of charge carriers were then determined and the two kinds of crystals showed very different temperature dependences (Figure 6.22).

From the Hoesterey–Letson equation [54]:

$$\mu_{\text{eff}}(T) = \mu_0(T) \left[1 + \frac{N_t}{N_b} \exp\left(\frac{E_t}{kT}\right) \right]^{-1} \quad (6.31)$$

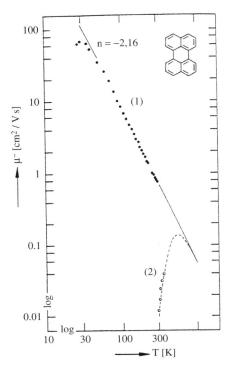

Figure 6.22 Electron mobility in ultrapurified perylene versus temperature for the electric field in the crystallographic $c' = a \times b$ direction: (1) ultrapurified perylene, (2) trap containing perylene (E_t = trap depth = 270 meV; N_t = density of traps = 0.17%). (Modified with permission (from Ref. [53] and [54])

where

μ_0 = microscopic mobility
μ_{eff} = effective mobility
N_t = density of traps
E_t = trap depth
N_b = density of molecular units (or band states)

it could be deduced that perylene, as representative of molecular crystals, reveals the presence of shallow traps [53]. This precludes the observation of the expected temperature dependence of the mobility around room temperature. Only ultrapure perylene crystals show the expected increase in mobility by decreasing the temperature down to 30 K where the mobility almost reaches $100 \, \text{cm}^2/\text{V s}$ [54]. The same type of studies has been carried out on napthalene [55].

In summary, for conventionally purified single crystals and thin films, the conduction processes in molecular materials are always dominated (or influenced) by traps and hopping models can be satisfactorily used to describe the charge transport phenomena. The exceptions found necessitate the use of ultrapure single crystals obtained after exceptional time consuming procedures (zone refining of potassium-treated material is, for instance, one of the purification steps [55]) as well as fairly low temperatures.

All calculations carried out on single crystals of phthalocyanines agree that the bandwidth is larger along the c^{-1} axis than along the two other directions [56–59]. However, a very unsatisfactory agreement is found in the magnitude of the bandwidth when different methods of calculations are used. It is worth pointing out that in many cases the effect of vibronic overlap on the energy bands, which should reduce the bandwidth, is not considered [60].

6.6.2 Band Structure of PcH$_2$

As an example, we will detail the method of determining the band diagram of single crystals of monoclinic PcH$_2$ β-phase (P2$_1$/a, Z = 2). For the hole band structure arising from the a_u molecular orbital (symmetry group D$_{2h}$), the configuration interaction can be neglected since it is separated from the nearest level by about 0.5 eV [57] (Figure 6.23).

The highest occupied molecular orbital (HOMO) of PcH$_2$ corresponds to the irreducible representation a_u in the point symmetry group D$_{2h}$. The site symmetry is $\bar{1}$, only the inversion centre i is conserved. The orbitals of the unit cell (crystalline class C$_{2h}$) which must be considered are:

C$_{2h}$	E	C$_2$	i	σ_h
(2a$_u$)	2	0	−2	0
(2a$_u$)		A$_u$ + B$_u$		

In the space group P2$_1$/a, the C$_2$ axis corresponds to the screw axis and σ_h to the glide plane. The irreducible representations of the *orbitals of the unit cell* are shown below:

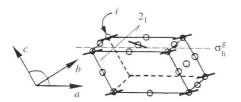

	E	C$_2$(2$_1$)	i	$\sigma_h(\sigma_h^g)$
A$_u$	1	1	$\bar{1}$	$\bar{1}$
B$_u$	1	$\bar{1}$	$\bar{1}$	1

In the irreducible representations A$_u$ and B$_u$, the effect of the symmetry operations containing a translation is different. This is illustrated for a simple case in Figure 6.24.

It is now possible to determine the supramolecular orbitals in the a^{-1}, b^{-1} and c^{-1} directions. Both a^{-1} and c^{-1} belong to the plane (a, c) while b^{-1} is parallel to b. It can be seen from Figure 6.25 (Plate 3) that, at $k = 0$, the

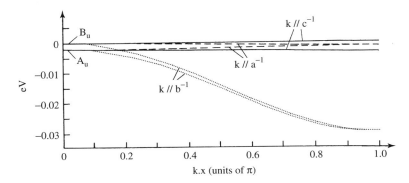

Figure 6.23 Hole band structures of β-metal-free phthalocyanine. (From Ref. [57])

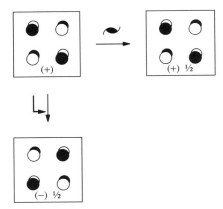

Figure 6.24 Illustration of the effect of a screw axis or a glide plane on orbitals corresponding to the irreducible representation A_u

two supramolecular orbitals derived from A_u and B_u are different. However, at $k = \pi/b$, the two supramolecular orbitals are similar: they will therefore be degenerated at this point.

An identical process may be applied in the (a, c) plane which contains the vectors a^{-1} and c^{-1}. This time again the supramolecular orbitals at $k = \pi/a$ are similar and therefore degenerated (Figure 6.26, Plate 4).

Finally, the c^{-1} direction may be considered. In this case (Figure 6.27, Plate 4), the two supramolecular orbitals derived from A_u and B_u are not degenerated and will have different energies whatever the value of k in the direction considered.

Another view of the arrangement of the phthalocyanine molecules in the crystal as viewed along the b axis is shown in Figure 6.28.

The bandwidth which reflects the degree of interaction between the molecular units may be calculated in the three directions of the reciprocal space. As expected

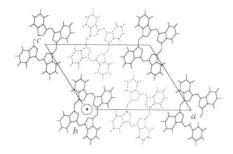

Figure 6.28 Geometrical arrangement of the phthalocyanine molecules in the crystal. The b axis is perpendicular to the plane. The dotted lines show the phthalocyanine at $|b| = \frac{1}{2}$.

Table 6.10 The bandwidths of β-PcH$_2$ [57]

Band	k	PcH$_2$
Hole	a^{-1}	1.05 meV
(a$_u$)	b^{-1}	29.76
	c^{-1}	0.86
Electron	a^{-1}	3.57
(b$_{2g}$)	b^{-1}	101.55
	c^{-1}	1.56

from the geometrical arrangement of the molecules in the crystal, the bandwidth for k values parallel to the b^{-1} direction is by far the largest one [57] (Table 6.10).

Other calculations carried out on metallophthalocyanines show, at least qualitatively, the same tendencies. However, calculations give results differing by orders of magnitude for the bandwidths [61–63].

6.6.3 Band Structure of PcLi

The band structure of crystals of PcLi has been calculated by the valence effective hamiltonian method [39]. One can start from the molecular orbitals of the isolated molecular unit (Figure 6.29).

As usual for metallophthalocyanines, the HOMO is of symmetry a$_{1u}$ (with a single electron) whereas the LUMO is degenerated (irreducible representation e$_g$). For calculating the band structure, the tetragonal form (P4/nnc) was chosen. The symmetry characteristics of the cell are shown in Figure 6.30.

In the structure, the staggering angle is 38.7°, a value of 40° having been chosen in the calculation. The lithium atom is not explicitly included. The valence band arising from a$_{1u}$ is well above the 92 fully occupied bands. The conduction band arising from e$_g$ is degenerated (Figure 6.31).

The width of the valence band is 1.08 eV, in good agreement with a previous estimation based of magnetic measurements [14]. Since this band is only half

238   *DESIGN OF MOLECULAR MATERIALS*

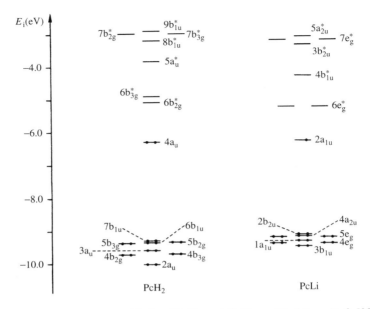

Figure 6.29 Main molecular orbitals of PcH$_2$ and PcLi modified from Ref. [39]

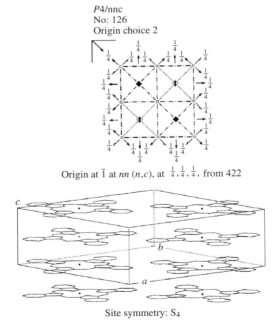

Figure 6.30 The various symmetry operations associated with the space group P4/nnc and the structure of PcLi ($Z = 4$). (After Refs. [16] and [64])

MOLECULAR SEMICONDUCTORS: PROPERTIES AND APPLICATIONS

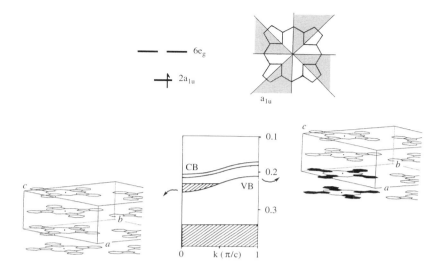

Figure 6.31 Band structure of PcLi with corresponding supramolecular orbitals (units in atomic unit = 13.6 eV) (CB, conduction band; VB, valence band). The valence band is only half filled. (Modified from Ref. [39])

filled, no energy gap for electrical conduction along the stacking axis should be measured. This is not the case since PcLi single crystals lead to $E_{act} = 0.2$ eV [15]. On the other hand, a total width of 0.99 eV is obtained for the almost degenerate conduction bands [39].

6.6.4 Conclusions

In summary, both experimental conductivity results and calculations demonstrate that PcLi affords an intrinsic molecular semiconductor with large bandwidth (and a small forbidden energy gap). This is undoutedly due to the radical nature of the molecular unit which favours the $\pi-\pi$ overlaps between the phthalocyanine macrocycles. On the contrary, divalent ion monophthalocyanines or metal-free monophthalocyanine are insulators when undoped and the intermolecular interaction energy is fairly small, with bandwidths of the order of kT. In PcLi one can consider that the electronic levels are collectivized whereas in PcM (M = H_2, Cu, Zn, etc.) the wavefunctions are restricted to a single molecular unit.

6.7 LIQUID CRYSTALLINE MOLECULAR SEMICONDUCTORS

Supramolecular engineering has been encountered in the domain of molecular semiconductors at only one level: the design of a molecular unit suitable to yield predictable macroscopic electrical properties. However, mastering the relative geometrical arrangement of the molecular units has not yet been considered. As it has been seen previously, the prediction of the structure of three-dimensional periodic assemblies is still out of reach of chemists. In this respect only tendencies, guesses or hopes can be given. The problem is slightly easier for mesophases in which a higher degree of disorder is introduced. In this case supramolecular engineering can be envisaged. It has been shown [65] that phthalocyanine macrocycles substituted with long paraffinic side chains may form columnar liquid crystals (Figure 6.32).

Columnar liquid crystalline phases have been first observed with hexa-substituted benzene derivatives [66] and subsequently with many rigid aromatic cores surrounded by flexible paraffinic chains. The same studies have been carried out with a series of phthalocyanine derivatives. In the first stage only complexes of monophthalocyanines were studied, but the use of the Pc_2Lu and PcLi subunits was tempting because of their peculiar redox and electronic properties. Lutetium bisphthalocyanine and lithium phthalocyanine macrocycles substituted with long paraffinic chains have thus been synthesized [67, 68] (Figure 6.33).

The phthalocyanine moieties stack to form molecular spines which are separated from each other by approximately 1.5 nm of a medium made of molten paraffinic chains. The molecular spines may be used as one-dimensional conducting wires at a submicrometer scale. The domain of stability of the columnar (discotic) mesophase depends on the nature of the connecting link between the ring and the side chains ($-OCH_2-$, $-CH_2O$, $-CH_2CH_2-$). However, in all cases, X-ray diffractions at small angle show an hexagonal order typically associated with columnar liquid crystals [68] (Table 6.11).

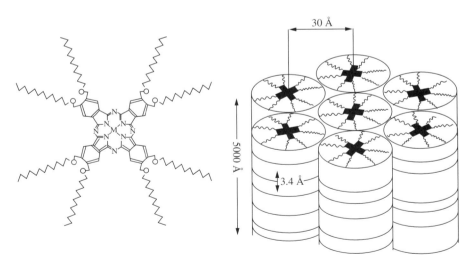

Figure 6.32 A long-chain substituted phthalocyanine and the corresponding type of mesophases. The rigid cores form a submicronic wire. (After Ref. [65])

MOLECULAR SEMICONDUCTORS: PROPERTIES AND APPLICATIONS

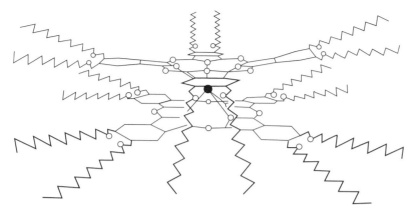

Figure 6.33 Molecular structure of lutetium bisphthalocyanine substituted with dodecyloxymethyl side chains [67]

Table 6.11 Structural parameters determined by X-ray for the substituted lutetium and lithium phthalocyanine derivatives (D = intercolumnar distance, L = correlation length of disordered side chains; h = stacking period along the columns). The distances are in nanometers [68]

	K	M
$[(C_{18}OCH_2)_8Pc]_2Lu$		Hexagonal $D = 3.7$ $L = 0.43, 0.41$ $h = 0.73$
$[(C_{12}O)_8Pc]_2Lu$	Rectangular $a = 2.9$ $b = 2.46$	Hexagonal $D = 3.46$ $L = 0.46$ $h = 0.33$
$(C_{12}O)_8PcLi$	Square $a = b = 2.55$	Hexagonal $D = 3.49$ $L = 0.46$ $h = 0.335$

The spectroscopic and redox properties of the substituted lutetium and lithium phthalocyanines in solution are identical to those of the unsubstituted compounds. Consequently, the density of intrinsic charge carriers in the solid state should be comparable. Electrical measurements as a function of frequency (10^{-3} Hz–10 GHz) do not clearly demonstrate the intrinsic nature of the charge carriers. This is due to the difficulty of purifying these materials which cannot be recrystallized or sublimed. However, it could be estimated that intracolumnar conduction processes are detected for frequencies higher than 10^2–10^4 Hz. The corresponding mobilities are rather small: 10^{-4}–10^{-2} cm^2/V s [68].

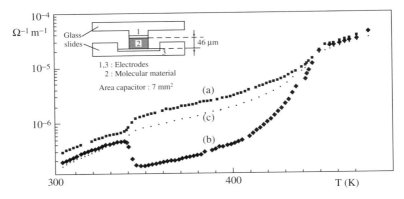

Figure 6.34 Variation of the conductivity σ as a function of temperature measured at 10 kHz. An external magnetic field ($H = 0.9\,\text{T}$) was applied (a) perpendicular and (b) parallel to the direction of the a.c measurement electrical field; (c) is obtained in the absence of magnetic field [69]. For the fabrication of the cell see Ref. [70]

$[(C_{12}O)_8Pc]_2Lu$ has been partially orientated in a magnetic field and an anisotropy of conductivity of approximately 10 has been measured [69] (Figure 6.34):

$$K_1 \xrightarrow[(314\,\text{K})]{41\,°\text{C}} K_2 \xrightarrow[(358\,\text{K})]{85\,°\text{C}} M \xrightarrow[(462\,\text{K})]{189\,°\text{C}} I$$

The orientation is achieved by heating the sample up to about 20 °C beyond the mesophase–isotropic phase transition and measurements were recorded on cooling down the sample slowly. It was previously known that the columns tended to be aligned perpendicularly to the magnetic field. In consequence, in *experiment* (a) (Figure 6.34) the columns are parallel to the electrical field ($\sigma_\parallel$) applied for measuring the conductivity, whereas the columns are perpendicular ($\sigma_\perp$) in experiment (b). Since conduction is very reasonably thought to be easier along the column axis, one expects $\sigma_\parallel > \sigma_\perp$. This is indeed experimentally observed [69].

6.8 JUNCTIONS AND SOLAR CELLS

6.8.1 Generalities

The formation of molecular material-based junctions and their use for converting solar photons into electricity was described 40 years ago [71, 72]. However, after a rapid initial development, the poor performances of the devices led to their abandonment. Most of the results on junctions and solar cells have already been described [3], but it is useful to understand the molecular mechanisms occurring in related devices (transistors, sensors).

The description of a p–n junction or Schottky contact can be made using molecular terms. Let us consider a material composed of molecules A and

MOLECULAR SEMICONDUCTORS: PROPERTIES AND APPLICATIONS 243

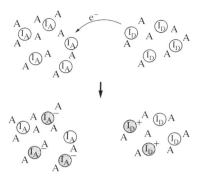

Figure 6.35 Schematic representation of the formation of a p–n junction in a molecular material composed of molecules of type A, doped with acceptor, I_A, or donor, I_D, impurities

containing either electron donor impurities I_D or electron acceptor impurities I_A. A junction is formed by joining the two counterparts. A transfer of electrons from the donors to the acceptors occurs (Figure 6.35).

$$I_A + I_D \rightleftharpoons I_A^- + I_D^+$$

The amount of charge transfer may be determined if the solid-state redox potentials of the impurities, E_A and E_D, can be estimated:

$$I_A + e^- \rightleftharpoons I_A^- \quad E_A$$
$$I_D^+ + e^- \rightleftharpoons I_D \quad E_D$$

The densities of ionized acceptors and donors are obtained from Nernst's law in the case of a monoelectronic transfer:

$$\log K_{eq} = \frac{E_A - E_D}{0.059}$$

with

$$K_{eq} = \frac{[I_A^-][I_D^+]}{[I_A][I_D]}$$

where $[I_A]$, $[I_D]$ are the densities of impurities.

It has been previously seen that the reducing or oxidizing properties of a molecule in the solid state may be approximated by the redox potentials in solution. Consequently, E_A and E_D may be estimated from standard electrochemical measurements (cyclic voltammetry or polarography). The classical image of junction formation by equalization of the Fermi levels is formally equivalent to writing the previous mass action law. If the absolute difference between the oxidation and reduction potentials is less than 0.25 V, partial charge transfer between the two kinds of impurities occurs ($K_{eq} \leq 10^4$). This approach, however, rapidly reaches its limitations. For example, the probability of electron transfer depends on the distance of the impurity from the interface. This is not presently taken into account. Such a model must therefore be used with care; it must only be taken as a way of visualizing the chemical mechanisms arising during the formation of the junction.

A more satisfactory treatment has been proposed that considers only electrostatic interactions [73]. Two antagonistic forces govern the amount of charge transfer: (a) the driving force is given by the difference in redox potentials between the electron donor and electron acceptor; (b) the opposite force is due to ion–ion repulsion between ionized impurities. With this model, an order-of-magnitude estimate of the extent of the space charge regions may be made. For a molecular size of A of 14 Å, a dielectric constant $\varepsilon_r = 4$ and an impurity concentration of 10^{-3} (mol/mol), the energy required to ionize a new couple of donor–acceptor impurities, is 0.5 eV for the very first layer [73]. It is already apparent that doped insulators, in which the concentration of dopants is high enough to ensure a conductivity in the semiconductor range, can only yield very narrow space-charge regions.

In the space-charge regions, the concentration of charge carriers is lower than in the bulk since the proportion of ionized impurities is higher:

$$I_A/I_D \rightleftharpoons I_A^-/I_D^+$$
$$A, I_A \rightleftharpoons A^+, I_A^-$$

and $$[A^+][I_A^-] = \text{constant}$$

Whenever $[I_A^-]$ increases, the density of A^+ decreases.

Consequently, the resistances of the space-charge regions are higher than the bulk ones. When an external voltage is applied, the space-charge region can increase or decrease its extent depending on the sign of the voltage. One therefore obtains a diode effect (Figure 6.36).

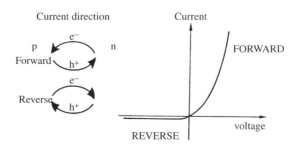

Figure 6.36 Current–voltage plot for a p–n junction: effect of a reverse or a forward bias

6.8.2 Junction Studies in the Dark

Most of the studies on molecular materials have been made using Schottky-type devices. Studies on polycrystalline films of PcM, M = H_2, Cu, Ni, Mn, showed that a small amount of a liquid polar impurity at the interface was essential to obtain a rectification effect in the dark [74]. It was postulated at that time that an ionic space-charge barrier in the vinicity of the least noble electrode was necessary in order to obtain a significant rectification ratio. Various chemicals were tested for their ability to give rise to this ionic space-charge barrier.

A huge influence of air on the junction properties of PcMs was observed. Figure 6.37 shows a current–voltage plot of an Au/PcZn/Al cell made without breaking the vacuum at any stage (10^{-7}–10^{-8} torr) [75]. No rectification effect is observed. The I–V curve is almost perfectly symmetrical, although the two metallic electrodes have different work functions ($\Phi_{Au} = 5.1$ eV, $\Phi_{Al} = 4.28$ eV) [76]. In the low-voltage range studies (± 2 V), the plots of $\log I$ versus $\log V$ give a straight line with a slope of 1.2–1.5, i.e.

$$I \sim V^n \qquad n = 1.2-1.5 \qquad (6.32)$$

The current is proportional to the voltage when the number of charges injected from the electrodes is less than the density of the charge carriers generated within the molecular material. This is roughly the case at low voltages.

At higher voltages ($|V| > 3-5$ V), the currents flowing through the Au/PcZn/Al device become space-charge limited; the number of charges injected from the electrodes is higher than the bulk ones. The $\log I$– $\log V$ plots remain linear but the parameter n rises to approximately 3.

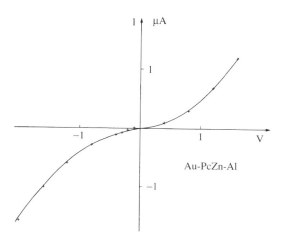

Figure 6.37 Current–voltage plots for an organic solar cell of Au/PcZn/Al made and studied entirely under vacuum (10^{-7}–10^{-8} torr). The thickness of PcM is 2.0 μm. (After Ref. [75])

$M_1/PcH_2/M_2$ devices (M = Al, Au, Pb) made under ultra-high vacuum conditions have previously been reported [77]. In this case, contrary to the above results, a rectification effect was observed. The dark current is lower for a given voltage when the nonsubstrate electrode, M_2, is positive. This probably indicates a nonsymmetrical distribution of traps in the vicinity of the two metallic electrodes. M_1, which is in contact with the glass substrate, must present less structural irregularities than M_2, which is deposited on to the PcM polycrystalline film.

The behavior of the cells is drastically different in the presence of air (Figure 6.38). The doping is achieved by leaving the PcZn thin films in ambient air for 10 mn before a semitransparent overlayer of aluminum is deposited. All subsequent studies are carried out without breaking the vacuum (10^{-7} torr). This brief exposure to air is sufficient to give rise to a large rectifying effect ($r_{rf} = 16\,000$ at ± 1.6 V) as compared with the cell made entirely under high vacuum ($r_{rf} = 1.2$). The sensitivity to air has been found to be highly dependent upon the nature of the central metal ion. For PcNi, for example, doping must be carried out at 150 °C under an atmosphere of pure O_2 (15 h) to demonstrate the same rectifying effect (Figure 6.39). At room temperature, noticeable effects are obtained only after several weeks of exposure. For comparison, a PcNi thin film was treated at 150 °C under argon, left 10 mn in air and then studied under vacuum (Figure 6.39). Although the 'apparent' resistance of the thin layer is markedly lowered, no rectification effect is observed. It was once assumed that rectification is due to the presence of a thin layer of Al_2O_3 at the interface between PcM and the aluminum electrode. To test this hypothesis, Al was deposited first and left 5 mn in air to be superficially

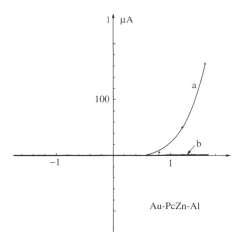

Figure 6.38 Current–voltage plots for Au/PcZn/Al devices. (a) PcZn is first sublimed on Au and left 10 mn in air before the layer of Al is deposited. All subsequent studies are made under vacuum (10^{-7} torr). (b) The depositions of PcZn and Al and determination of the characteristics are all carried out without breaking the vacuum at any stage. (After Ref. [75])

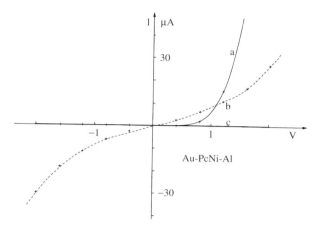

Figure 6.39 Current–voltage characteristics for Au/PcNi/Al devices. (a) PcNi O_2-doped at 150 °C, 15 h, 1 atm of O_2. (b) PcNi treated under argon at 150 °C, 15 h. In cases (a) and (b), after the deposition of the aluminum electrode, the $I-V$ plots are drawn under vacuum (10^{-7} torr). (c) Device made and studied entirely under vacuum. In all cases the thickness of the organic layer is 2.0 μm. (After Ref. [75])

oxidized. Then overlayers of PcZn and Au were successively evaporated under vacuum. Without breaking the vacuum, the $I-V$ characteristic in the dark was measured. No rectification was observed ($r_{rf} = 0.9$ at ±1.6 V) [75]. The same results have been obtained for Al/PcCu/Au devices [78, 79]. To show how the rectification really arises, a sandwich cell of Au/PcZn/Au was prepared where the organic layer had been left 5 mn in air before deposition of the gold overlayer [75]. Although symmetrical and nonoxidizable electrodes were used, a strong rectification effect was observed ($r_{rf} = 6100$ at ±1.5 V). The superficial metal oxide layers do not directly intervene in the diode behavior of the sandwich cell.

From these $I-V$ determinations in the dark, several conclusions may be drawn. First, air is essential for the establishment of a space-charge region; the rectification ratio is directly related to the exposure to air. While aluminum electrodes are superficially oxidized in air, the insulating Al_2O_3 is not responsible for the rectification effect. Devices with two similar noble metallic electrodes, such as Au/PcM/Au, can give rise to a rectification effect when an unsymmetrical distribution of dopants (or traps) within the organic material is generated. Second, a space-charge layer is definitely formed *within the molecular material*. The depletion region is probably formed by ionization of PcM–O_x complexes through the equilibrium:

$$PcM-O_x/metal \rightleftharpoons PcM-O_x^-/metal (+)$$

where

$$O_x = oxidant(O_2, O_3 \ldots)$$

Capacitance measurements have been carried out to elucidate the mechanisms of the space-charge region formation.

The capacitance varies with the applied voltage with a $1/C^2$ versus V dependency (C = capacitance). The width of the space-charge region, l_{sc}, is classically given for n-type materials by [3]

$$l_{sc} = \sqrt{\frac{2\varepsilon_s(V_{bi} \pm V)}{qN_d}} \tag{6.33}$$

where

N_d = density of ionized donors
ε_s = dielectric constant of the semiconductor
V_{bi} = built-in potential
V = applied voltage

However, electrical characterizations of metallophthalocyanine–metal contacts in a large frequency range have demonstrated that the previous conventional model cannot be used [80]. Three different contributions to the electrical a.c. response may be distinguished. Over the very few angstroms of the molecular material extends a surface-charge layer associated with a high capacitive term. The surface-charge capacitance is only slightly dependent upon superimposed d.c. voltages. The so-called space-charge region extends below the surface-charge layer over approximately 2000–4000 Å. However, the corresponding capacity is almost independent of any d.c. superimposed voltage. At the same time, the *resistance* of the space-charge region is highly dependent on externally applied d.c. voltages probably indicating a Frenkel–Poole mechanism [3]. Finally, the rest of the metallo-organic layer shows the properties of the bulk material, as expected. In consequence, apparent variations of the overall capacitance do not correspond to properties of the space-charge region. The $1/C^2$ versus V relationship cannot be used without close examination of the measurements over an extended range of frequencies (10^{-3}–10^5 Hz) [80].

6.8.3 Solar Cells and Photovoltaic Effect

Becquerel in 1839 discovered that the potential of an electrode immersed in an electrolyte varies when this electrode is exposed to light. It was thus shown that the conversion of light into electricity was possible. However, it took more than a century to produce a solar cell converting sunlight into electrical power efficiently (∼6%) [81]. Single-crystal silicon cells can now exhibit conversion efficiencies as high as 24% [82], one photon out of four being converted into an electron flowing in the external circuit. The state-of-the-art in the domain of molecular semiconductors is not at that level. Power conversion efficiencies are more in the order of 10^{-1}–10^{-5}%. Even though molecular solar cells can hardly compete nowadays with their silicon-based counterpart, the understanding of the mechanisms involved in the photoelectric phenomena is important.

6.8.4 Solar Cells: Classical Formulation

Figure 6.40 shows a p–n junction illuminated by photons with an energy greater than the bandgap. The photons absorbed on both the p and n sides generate an e^-/h^+ pair. Because of the field ξ_{bi}, the minority carriers are swept down the energy barrier. The separated photogenerated carriers set up an electrical field which is opposite to the built-in electric field ξ_{bi}. The difference in potential between the two sides of the junction is reduced from V_{bi} to a smaller value $V_{bi}-V_f$. The effect would be the same if an external forward-bias voltage of value V_f were applied across the junction. The maximum open-circuit voltage under illumination V_{oc} is therefore less than or equal to the built-in potential V_{bi}.

A photovoltaic effect may also arise in Schottky junctions if the contact is irradiated through a metallic electrode thin enough to be semitransparent (Figure 6.41). Photoeffects may arise from three main processes. First, light may be absorbed in the metal and excite electrons from the metal to the semiconductor over the energy barrier ΔE_{MS}. This contribution is generally fairly small and is usually negligible compared to the other mechanisms. Second, light may be absorbed in the space-charge layer of the semiconductor: the e^-/h^+ generated is separated by the built-in electric field of the junction before recombination may occur. Third, light may be absorbed in the bulk semiconductor; in this case, the

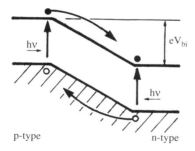

Figure 6.40 The p–n junction irradiated with photons whose energy $h\nu$ is greater than the bandgap E_g. The minority carriers generated by illumination are swept down the energy barrier. (After Ref. [83])

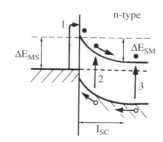

Figure 6.41 Energy-band diagram of a Schottky contact with an n-type semiconductor. Under illumination, three mechanisms giving a photocurrent are possible. (1) The photons excite electrons from the metal to the semiconductor over the barrier ΔE_{MS}. (2) Light is absorbed within the space-charge layer of the semiconductor, the e^-/h^+ pair generated is separated by the built-in electric field before recombination can occur. (3) Light is absorbed in the bulk semiconductor. The minority carriers (holes) diffuse to the junction where they are collected. (After Ref. [84])

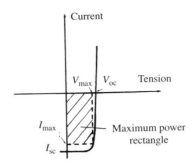

Figure 6.42 Typical $I-V$ curve for a solar cell under illumination (I_{sc} = short-circuit current, V_{oc} = open-circuit voltage, V_{max}, I_{max} = voltage and intensity corresponding to the maximum power output, respectively)

minority carriers (the holes in Figure 6.41) must diffuse up to the junction to be collected [84]. As for p–n junctions, irradiation is equivalent to the application of a voltage.

The maximum power extractable from the cell is a major parameter characterizing a solar cell:

$$P_{out}(\max) = I_{max} V_{max} \qquad (6.34)$$

where I_{max} and V_{max} are the current and voltage corresponding to the maximum power output. The fill factor (FF) is used to characterize the maximum power output:

$$\text{FF} = \frac{V_{max} I_{max}}{I_{sc} V_{oc}} \qquad (6.35)$$

A typical current–voltage plot under irradiation is shown in Figure 6.42.

6.8.5 Molecular Solar Cells: Localized States Formulation

In this section only metal/molecular material junctions will be considered since only a very few cases of molecular p–n devices have been described.

It is fruitful to divide the photovoltaic effect arising in molecular materials into a succession of elementary steps. First, a photon excites a given molecule to an upper electronically excited state. The probability of excitation is related to the absorption coefficient of the material which, in turn, determines the penetration length of the photon l_a. The excitation migrates from molecule to molecule over a distance l_{ex} without dissociation to free charge carriers. It is assumed that the extent of the space-charge region within the semiconductor (l_{sc}) is negligible compared to the other characteristic distances. A fraction $[l_{ex}/(l_a + l_{ex})]$ of the 'excitons' can diffuse up to the front surface where they undergo chemical changes (Figure 6.43). Near the electrode, the excited state may dissociate by ejecting one electron (or hole) into the metallic electrode. This charge-transfer

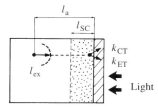

Figure 6.43 Schematic representation of the photovoltaic effect for a Schottky contact (localized states picture) (l_{sc} = space-charge region depth, l_a = light penetration depth, l_{ex} = exciton diffusion length, k_{CT} = charge transfer to the electrode (rate constant), k_{ET} = energy transfer to the electrode (rate constant))

process may be characterized by the rate constant k_{CT}. The excited state may alternatively exchange the excess energy with the metal via a nonradiative energy transfer with rate constant k_{ET}. The rates of these two quenching processes do not have the same dependence on the distance between the molecule in the excited state and the electrode. Energy transfer arises through a long-range dipolar mechanism, whereas the charge-transfer process mostly occurs through a short-range tunneling mechanism.

The space-charge region may arise from the ionization of impurities (usual Schottky barrier), from the carriers injected from the electrode (insulator–metal contact) or from a high trap density in the vicinity of the interface. In all these cases, the mechanisms of exciton dissociation near the interface are complicated and various processes can intervene. Consider a molecular crystal 'doped' with an acceptor impurity I_A. The classical space-charge region may be written schematically as

$$A, I_A^-/\text{metal} (+)$$

The photovoltaic effect under irradiation may then be written as

$$A, I_A^-/\text{metal} (+) \xrightarrow{h\nu} A, I_A/\text{metal}$$

There is a question as to whether the photon directly excites the electron from I_A to the metallic electrode or whether some intermediates are formed:

$$A, I_A^- \xrightarrow{h\nu} A^*, I_A^-$$

$$A^*, I_A^- \longrightarrow A^+, I_A^{2-}/\text{metal} \longrightarrow A^+, I_A^-/\text{metal} (+e^-)$$

This latter mechanism has been postulated [85] in the case of metallophthalocyanines doped with chloranil. Absorption may also occur in the charge-transfer band of the pair A, I_A such that

$$A, I_A \xrightarrow{h\nu_{CT}} A^+, I_A^-$$

or

$$A^+, I_A^- \xrightarrow{h\nu_{CT}} A, I_A$$

Electron transfer from or to the metallic electrode via these photoexcited charge-transfer complexes is then possible:

$$A^+, I_A^-/\text{metal} \longrightarrow A^+, I_A/\text{metal}\ (+e^-)$$

The ability of the photons to detrap the charge carriers is strongly dependent upon the electric field effective in the space-charge region. Thus, the probability of photostimulated electron transfer should show a strong dependence on the distance from the interface.

Figure 6.44 shows the approximate energy level diagram for typical PcMs relative to the work functions of the most usual metallic electrodes.

The effect of air on the photovoltaic behavior of $M_1/\text{PcM}/M_2$ devices is drastic, as expected from the dark junction properties. In the absence of air, an Au/PcNi/Al system shows an ohmic behavior in the dark and no photovoltaic effect ($V_{oc} = 0$, $I_{sc} \sim 0$). When the metallo-organic layer is treated at 150 °C in an atmosphere of pure dioxygen, a strong rectification is observed in the dark and a large photovoltaic effect arises ($V_{oc} = 680$ mV, $I_{sc} = 5.4\ \mu\text{A/cm}^2$, 100 mW/cm^2 white light) (Figure 6.45).

Further experiments have been reported to elucidate the chemical mechanisms underlying the photovoltaic effect [75]. Two series of studies have been described. In the first one, Au/PcZn/M sandwiches have been realized by successive vacuum sublimations. The current–tension (I–V) characteristics in the dark and under illumination are measured *in situ*. The vacuum (10^{-7}–10^{-8} torr) is not broken at any stage of the experiments. Irradiation is carried out through the last deposited semitransparent electrode M. The nature of M is varied so as to span a wide range of work junctions from gold ($\Phi_{Au} = 5.47$ eV) to samarium ($\Phi_{Sm} = 2.7$ eV). The results are compared with those obtained for air-exposed molecular solar cells. The conditions of doping have been strictly defined and standardized to all the cells. The organic layer is left 1 h in air and then treated at 150 °C for 15 h at 10^{-2} torr. This treatment is necessary to standardize the amount of dopant

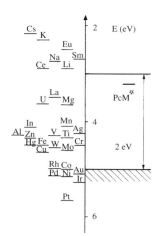

Figure 6.44 Energy of the donor and acceptor levels of PcH$_2$ relative to the workfunctions of metals: metal workfunctions from Ref. [76] and surface ionization potentials of PcMs from Refs. [86] and [87] (M = H$_2$, 5.2; Zn, 5.0; Cu, 5.0; Ni, 4.95; Fe, 4.95; PcM*, Q-band energy)

MOLECULAR SEMICONDUCTORS: PROPERTIES AND APPLICATIONS

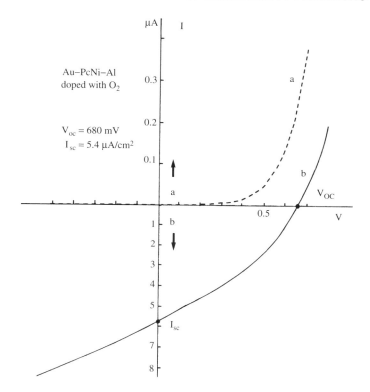

Figure 6.45 Electrical properties of O_2-doped Au/PcNi/Al devices: (a) in the dark, (b) under illumination (100 mW/cm² white light). Doping is achieved by treating the PcNi layer at 150 °C for 15 h in a pure atmosphere of O_2. The electrical characterizations are made under vacuum (10^{-5} torr)

remaining in the organic layer. The cell is again exposed to air for 10 mn and the last metallic overlayer is sublimed. The cells are then transferred into a jar where further studies are made under vacuum (10^{-4}–10^{-5} torr).

The electrical properties of the Au/PcZn/M devices entirely made under vacuum (conditions a) or exposed to air (conditions b) are shown in Table 6.12. The rectification ratio (r_{rf}) at ±0.5 V has been used to characterize the dark electrical behavior of the sandwiches. The open-circuit photovoltage V_{oc} and the short-circuit photocurrent I_{sc} are utilized to characterize the devices under illumination.

In a classical scheme, the space-charge region is formed by ionization of impurities present in the semiconductor near one of the interfaces. The space-charge region is detected in the I–V plots by giving rise to a rectification effect. For all air-exposed devices, rectification ratios, varying from 1.2 to 80 at ±0.5 V, are observed; r_{rf} rises rapidly with the voltage considered and at ±1 V, it reaches several thousands. On the contrary, no rectification effect in

Table 6.12 Electrical properties of Au/PcZn/M devices in the dark and under illumination. (from Ref. [75])

Metal	Φ_M^c (eV)	Conditions a[a]			Conditions b[b]		
		r_{rf} (± 0.5 V)	V_{oc} (mV)	I_{sc} (μA/cm^2)	r_{rf} (± 0.5 V)	V_{oc} (mV)	I_{sc} (μA/cm^2)
Au	5.47	1.0	450	0.6	1.2	20	3.2
Cu	4.59	1.0	0	0	3	80	4.7
Cr	4.50	1.1	450	0.3	19.0	228	51
Al	4.41	1.0	420	0.3	82	640	144
In	4.12	0.5	100	0.015	39	300	57
Sm	2.7	1.0	230	0.2	?[d]	1640	72

Thickness used: PcZn 2 μm; Au and M, 300 Å with the exception of In and Sm ($e = 600$ Å). Abbreviations: r_{rf} = rectification ratio in the dark (± 0.5 V); V_{oc} = open-circuit photovoltage; I_{sc} = short-circuit photocurrent.
[a] Conditions a: cells made and studied entirely under vacuum, without breaking the vacuum at any stage (10^{-7}–10^{-8} torr); irradiation conditions, 100 mW/cm^2, white light.
[b] Conditions b: the PcZn layer is first deposited on the gold counterelectrode, left 1 h in air and treated at 150 °C under vacuum (10^{-2} torr) for 15 h. The organic layer is again left 10 mn in air and the second metallic electrode is deposited. The studies are made under vacuum (10^{-4}–10^{-5} torr) with irradiation conditions of 100 mW/cm^2 white light.
[c] From Ref. [76].
[d] Too instable to be measured accurately.

the dark is noticed for the cells made entirely under vacuum ($r_{rf} \sim 1$). From examination of Table 6.12, it is clear that there is no straightforward relationship between the dark junction properties and the magnitude of the photovoltaic effect. In the absence of air (conditions a), a large photovoltage is found, while the rectification ratios are close to unity. Additionally, there is no obvious relationship between the workfunction of the semitransparent metallic electrode Φ_M and V_{oc}. In the presence of air (conditions b), both a strong rectifying effect and a strong photovoltaic effect are observed; V_{oc} increases when the workfunction of the metallic electrode is lowered. This behavior is expected for Schottky junctions between p-type materials and metals. It therefore seems that at least two different mechanisms give rise to the photovoltaic effect. One of them is correlated to Φ_M and to the presence of air; the second one is independent of both of these parameters.

It is possible to tentatively assign these two contributions to chemical mechanisms. The semitransparent electrode through which irradiation is performed is always negative; in consequence, the air-independent mechanism must involve some electronic transfer process between the metallophthalocyanine photochemically excited PcM* and the electrode:

$$\text{PcZn*/electrode} \longrightarrow \text{PcZn}^{\overset{+}{\bullet}}/\text{electrode}(-)$$

On the other hand, the air-dependent contribution should occur through a two-step mechanism. In the first step, the ionization of O_2 gives rise to a space-charge region near one of the interfaces:

$$\text{PcZn}, O_2/\text{electrode} \longrightarrow \text{PcZn}, \overline{O_2}/\text{electrode} (+)$$

In the second step, the reverse reaction takes place under illumination:

$$\text{PcZn}^*, \overline{O_2}/\text{electrode} (+) \longrightarrow \text{PcZn}, O_2/\text{electrode}$$

Under illumination, the $\log I - \log V$ plots are linear both under reverse- and forward-bias conditions. For air-doped Au/PcNi/Al devices in the forward direction, the relationship $I \sim V^n$ is found with $n \sim 4-5$. The same value of n was found in the dark, indicating a space-charge-limited current. Under reverse-bias conditions, the parameter n lies between 1 and 2, the trapping levels are probably all filled and the conduction is possibly governed by the trap-filled limit regime.

Many attempts have been made to use additional doping agents to improve the electrical performances of molecular solar cells. Once again, it must be emphasized that, in most cases, the doping was made in addition to air doping, no precautions being taken to avoid the presence of O_2 and related gases (such as ozone) in the materials.

PcMg pressed powders have been coated with a film of air-oxidized tetramethyl-p-phenylenediamine by evaporating an acetone solution of the amine [71, 72]. In the absence of dopant, no photovoltage under irradiation was observed. Surface doping induced a photovoltage of 200 mV under strong illumination conditions (500 W lamp). Based on the same principle, solar cells consisting of two different molecular materials have been described [88–90]. Sandwiches such as NESA/PcM/organic dye/In have been realized (Figure 6.46). The organic dye is deposited on the PcM thin film by the spin-coating method. High open-circuit photovoltages and high short-circuit photocurrents are obtained when the dye has a higher oxidation potential than the corresponding PcM. The energy conversion efficiencies of these devices are remarkably good and reach 0.5% for white light under natural conditions of irradiation (70 mW/cm^2) [88] (Table 6.13). It has been assumed that the organic dye yields a rectifying interface with the metallophthalocyanine: however, the mechanism involved has not been further characterized.

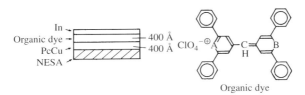

Figure 6.46 Molecular solar cells made up from two different dyes (NESA is the commercial trade mark for indium tin oxide). (After Ref. [88])

Table 6.13 Electrical properties and conversion efficiencies of the molecular solar cells shown in Figure 6.46

Substituents		V_{oc}^a (V)	I_{sc}^a (mA/cm^2)	Conversion efficiency (%)[a]
A	B			
O	O	0.55	0.9	0.29
O	S	0.45	1.6	0.43
S	Se	0.31	1.5	0.28
Se	Se	0.24	2.0	0.28
O	Se	0.42	0.4	0.10

[a] Air mass 2 irradiation, Kodak 600 slide projector with glass and water filters (70 mW/cm^2).

6.9 CONDUCTIVITY-BASED GAS SENSORS

6.9.1 Generalities

Even a very simple and incomplete description of the interaction between gases and molecular materials necessitates taking into account many elementary physicochemical processes (Figure 6.47).

The first step in the detection of a gas is the *adsorption* of the molecule on the surface of the material. In most cases, the surface is already recovered by another type of molecule less firmly bound — usually dioxygen — and this latter must be removed to allow the adsorption of a new molecule. The basis of such processes was given by Langmuir in 1916 [91].

If there are n molecules per unit of volume, in a given time t, only the ones situated at the distance $l \leq v_x t$ of a surface can reach it. If the corresponding surface area is $\mathcal{A}$, the total number of collisions is

$$v_x t \mathcal{A} n \text{ (or } v_x \mathcal{A} n \text{ per second)}$$

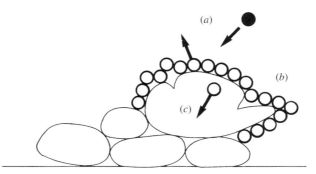

Figure 6.47 Schematic representation of some of the physicochemical processes involved in the interaction of gases with a polycrystalline molecular material [(a), displacement; (b), defects (structural and chemical); (c), intracrystallite diffusion]

The pressure is equivalent to a force per surface unit, which is in turn related to a momentum mv_x per unit of time. The pressure is therefore given by the product of momentum × number of collisions divided by the surface:

$$F = (2mv_x)(nv_x \mathcal{A}) \tag{6.36}$$

$$p = F/\mathcal{A} = 2nmv_x^2$$

By considering rate averages [92],

$$\langle v_x^2 \rangle = \langle v_y^2 \rangle = \langle v_z^2 \rangle = \tfrac{1}{3}\langle v^2 \rangle$$

$$p = \frac{4}{3}n\left\langle \frac{mv^2}{2} \right\rangle \tag{6.37}$$

$$\left\langle \frac{mv^2}{2} \right\rangle = \frac{3}{2}kT \tag{6.38}$$

Therefore the number of collisions per unit of surface is given by

$$p/2\sqrt{mkT} \tag{6.39}$$

The rate of adsorption is then

$$v_{ad}^+ = \frac{pS}{\sqrt{mkT}} \tag{6.40}$$

with S proportional to

$$f(\theta)\exp -\frac{E}{kT} \tag{6.41}$$

where

E = energy needed for adsorption
$f(\theta)$ = probability that the collision arises at a vacant site
θ = number of adsorbed molecules divided by the total number of sites

At equilibrium, for a single species, the rates of adsorption and desorption are equal and, under these conditions, the Langmuir isotherm can be obtained (Figure 6.48):

$$\theta = \frac{ap}{1+ap} \tag{6.42}$$

The term a depends on temperature. At low pressures $ap \ll 1$ and Henri's law is obtained, where an increase in temperature will decrease the number of molecules adsorbed at the surface of the molecular material.

A second process — the diffusion within the molecular material — must now be considered. The diffusion rate is given by the number of molecules crossing a reference area in a time Δt. It has been shown that it is related to the number of molecules in the volume extending over a distance $v_x \Delta t$ from the surface.

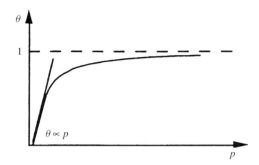

Figure 6.48 Percentage of occupied sites as a function of the pressure. (Reproduced from Ref. [93])

However, this time, back-crossing must also be considered. If n_a is the density of molecules as a function of the position,

$$J_x = -D \frac{dn_a}{dx} \quad (6.43)$$

where
D = diffusion coefficient ($= \mu kT$, μ = mobility)
J = flux of molecules per unit of time and surface

An increase in temperature leads to a higher diffusion coefficient.

The previous equations are not given in an attempt to formulate a model explaining experimental data on a quantitative basis. Many approximations have been implicitly taken. How can a surface of a polycrystalline material be defined rigorously? Are the sites identical? What are the effects of grain boundaries or structural defects on diffusion processes? However, if only the qualitative aspects of the processes are considered, the previous equations may be helpful.

6.9.2 Metallophthalocyanines: Thin-Film Morphology

Since 1935, the structures of most PcMs have been determined from X-ray diffraction measurements [94–97]. Three polymorphic forms are known, designated by the letters α, β and x (Figure 6.49).

Large size single crystals are, in most cases, of the β-type. They are generally grown by sublimation under a stream of nitrogen (7 torr) at a temperature of 400–500 °C. The crystals are needle-shaped, typically 1 cm long, 0.1 cm wide and 0.01 cm thick. PcMs crystallize in a monoclinic lattice. The large area surfaces are (001) faces and the needle direction is the b axis (Figure 6.50).

PcMs form polycrystalline films of the α-type [101–105] when evaporated under vacuum (10^{-5}–10^{-6} torr) on to a substrate maintained at room temperature.

More detailed studies have been carried out in the case of vacuum deposited films of PcZn [106]. The rate of deposition was 0.5 Å/s on to an amorphous

MOLECULAR SEMICONDUCTORS: PROPERTIES AND APPLICATIONS

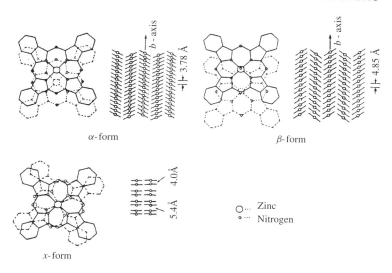

Figure 6.49 Schematic representation of the three main molecular stackings found for metallophthalocyanines. (After Refs. [98] to [100])

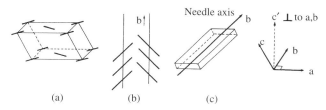

Figure 6.50 Basic parameters of the crystal of β-PcM: (a) unit cell, (b) stacking in the (a, b) plane, (c) needle axis

carbon substrate maintained at room temperature and the film thickness was around 50 Å. The corresponding pressure was 5×10^{-7} torr. High-resolution electron micrographs of PcZn films show that they are composed of crystallites whose sizes do not exceed 300 Å (Figure 6.51).

6.9.3 Experimental Results: Monophthalocyanines

The variation of a resistance by exposure to a chemical substance leads to a 'chemiresistor', as it was called in 1960 [85]. The conductivity detection of gases using molecular materials can occur whenever the redox properties of the gaseous molecules are suitable to form charge carriers.

It has been shown previously that all molecular materials — with only two exceptions — are insulators when undoped. The thermally generated charge

Figure 6.51 Electron micrograph of a PcZn thin film, 1 cm = 100 Å (see text for the conditions). (Reproduced by permission of the International Union of Crystallography from Ref. [106])

carrier concentration is given by

$$[A^-] = [A^+] = [A] \exp -\frac{E^0_{ox} - E^0_{red}}{2kT} \quad (6.44)$$

In usual cases $E^0_{ox} - E^0_{red} \approx 2\,\text{eV}$ and then (with $[A] \sim 10^{21}\,\text{cm}^{-3}$), the intrinsic concentrations of anions or cations are $[A^+] = [A^-] = 10^4\,\text{cm}^{-3}$ (out of 10^{21} molecules per cm³). Minute amounts of impurities, susceptible to generate around 10^4 charges per cm³, are sufficient to be — at least theoretically — detected.

Even very poor oxidizing and reducing agents can therefore act as dopants. These calculations are, however, not realistic since the molecular material is usually polluted by dopants fortuitously present in the thin molecular layers [3].

MOLECULAR SEMICONDUCTORS: PROPERTIES AND APPLICATIONS

Most of the studies concerning PcM-based gas sensors are related to the detection of NO_2. This is the result of chance (NO_2 is a fairly good oxidizing agent) and necessity (environmental problems necessitate accurate measurement of the NO_2 content in air).

The importance of the surface on the conductivity of thin films of PcM in the presence of gases has been revealed from the thickness dependence measurements of the sheet conductance [107]. The surface conductivity is defined as

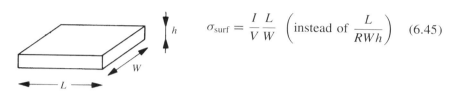

$$\sigma_{\text{surf}} = \frac{I}{V}\frac{L}{W} \quad \left(\text{instead of } \frac{L}{RWh}\right) \quad (6.45)$$

where

I = intensity of the current (in A) corresponding to the voltage (V)
σ_{surf} in A/V

Gas-sensing properties are influenced by various factors such as [107] film heterogeneities, differences of degree of crystallinity, crystallite size, relative orientation of the crystallites, grain boundaries, etc. Depending on the adsorption on edges, corners or different crystalline faces of structural defects, the gas molecules can be more or less strongly bound to the material.

Thin films of PcPb ($h < 1000$ nm) have been deposited on SiO_2 or Al_2O_3 in high vacuum (10^{-8} torr). First exposure to O_2 induces orders of magnitude differences in the conductivity of the thin film [108]. Subsequently, reversible changes upon O_2 exposure in the range 150–200 °C have been found [108]. By comparing the changes of conductivity induced by an exposure to small amounts of NO_2 upon PcCu and PcPb, it is possible to gain an insight into the influence of the morphology of the thin films on gas sensitivity (Figure 6.52).

It can be seen from Figure 6.52 that the time response of the PcPb device is approximately 10 times lower than for PcCu. In the same time, the reversibility is more easily obtained in the case of PcPb. The structures of PcCu and PcPb thin films are very different: PcCu is planar and yields α-type thin films, whereas PcPb is an out-of-plane complex and affords a monoclinic structure by deposition under vacuum (Figure 6.53). The protuberance brought by the voluminous Pb^{2+} ion probably facilitates the diffusion of gases within the molecular material layer.

The variation of conductance of PcPb films with time was found to be fairly fast (90% of the full response time in 90 s) and reversible (recovery time 140 s) [112] (Figure 6.54). However, these characteristics are only obtained at fairly high temperatures (100–170 °C) and for moderate concentrations of NO_2.

It has been found that at least two different types of binding sites can be distinguished [113]. At low concentrations of NO_2, adsorption occurs on sites that are weakly bound to the species already present (N_2 or O_2 in most cases).

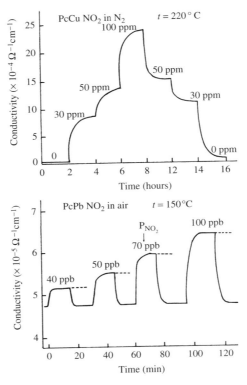

Figure 6.52 Conductivity as a function of time of thin films of PcCu and PcPb in the presence of various concentrations of NO_2. (After Refs. [93], [108] and [109])

Their replacement, which does not take much time or much energy, can occur at low concentrations of NO_2. This process is satisfactorily reversible.

At high NO_2 contents, NO_2 can displace molecules at sites where they are firmly bound. These sites are responsible for the slow processes that can intervene in the detection. A way to get ride — at least partially — of the second sites is to treat the PcPb layers in air at 360 °C where destructive oxidation of them is thought to occur [113, 114]. The molecular layer thin films then lead to a smaller but faster response to NO_2.

The simultaneaous measurement of the weight change (quartz microbalance) and of the conductivity has been carried out for thin films of PcAlF ($h = 3000$ Å) submitted to 100 ppm of NO_2 in N_2 at room temperature (Figure 6.55). The mass variations detected by the quartz microbalance strictly follow the conductivity changes, NO_2 replacing a lighter molecule (N_2 or O_2) [109].

The effect of O_2 on PcAlF thin films was also studied using a quartz microbalance. Within the uncertainties, no weight difference of the molecular material layer was observed with 100 ppm of O_2 in N_2 or pure O_2 compared

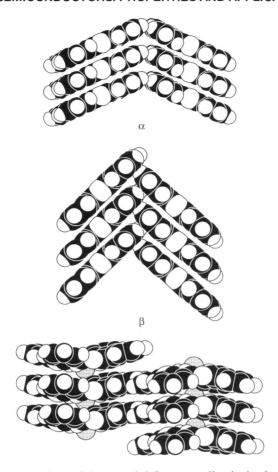

Figure 6.53 Representations of the α- and β-forms usually obtained with planar metal-lophthalocyanine and the monoclinic modification obtained with PcPb [110, 111]. The figures have been drawn using MOLDRAW-1989. (J. M. Cense, 1989)

to pure N_2. However, in the same time, an important conductivity change was noticed [109]. This probably means that N_2 is indeed displaced by O_2 but the weight change cannot be detected.

Phthalocyanine derivatives substituted with crown-ether macrocycles [115] are soluble enough in organic solvents to be deposited from solutions. At room temperature, significant conductivity increases are noticed with satisfactory reversibility when the layers are exposed to NO_2 in the range 1–5 ppm. However, after treatment of the molecular thin films with aqueous solutions of KCl, an addition of NO_2 leads to a decrease of the conductivity (Figure 6.56). KCl treatment has been shown to yield important changes in the film morphology. No satisfactorily model has, however, been given so far to rationalize these results.

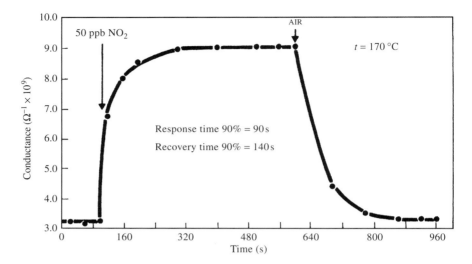

Figure 6.54 Variation of the conductance of a PcPb film at 170 °C ($h = 8000$ Å) under 50 ppb of NO_2 and return to air. (Reprinted from Ref. [112] with permission from Elsevier Science)

Ozone (O_3) is also an important oxidizing agent present in the atmosphere either naturally (photochemical generation) or because of pollution (with synergistic effects with NO_2). Depending on the season, the irradiance or the time of day, its concentration can vary in the range 5–50 ppb. The conductivity of thin films of PcCu has been shown to be influenced by the presence of O_3 [116, 117]. Satisfactory reproducibility and lifetime were obtained by switching between clean and sample air and by measuring the slope of the resistance during, for example, 1.5 s–1 min, depending on the ozone concentration. Under those conditions, the accuracy was of the order of 5% and the range of concentration of O_3 measurable was in the interval 10–5000 ppb [116, 117].

6.9.4 Experimental Results: Bisphthalocyanines

As previously seen, the density of intrinsically generated cations and anions is higher for Pc_2Lu than for conventional metallophthalocyanines:

$$2Pc_2Lu \rightleftharpoons Pc_2Lu^+, Pc_2Lu^-$$

$$[Pc_2Lu^+] = [Pc_2Lu^-] \approx 9.1 \times 10^{16} \text{ carriers/cm}^3$$

A second difference is due to the morphology of the thin films obtained by vacuum sublimation which is different for Pc_2Lu and PcM [34].

The effect of NO_2 on thin films of Pc_2Lu is indeed quite different compared to monophthalocyanines [118] (Figure 6.57). In the first stage, NO_2 acts as a

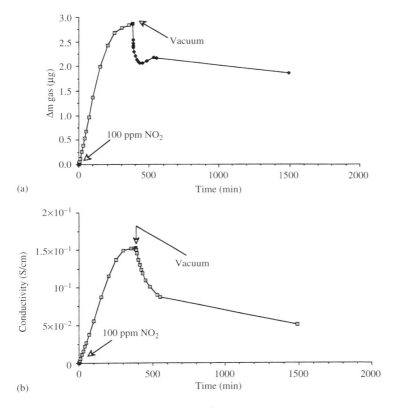

Figure 6.55 Thin film of PcAlF ($h = 3000\,\text{Å}$) submitted to 100 ppm of NO_2 (RT) in N_2 and then under vacuum: (a) mass variation as a function of time, (b) conductivity variation. (Reprinted from Ref. [109] with permission of Elsevier Science)

dopant and Pc_2Lu is partly transformed into Pc_2Lu^+. For usual metallophthalocyanines the amount oxidized remains extremely small. In the case of Pc_2Lu, on the contrary, NO_2 is sufficiently oxidizing to transform gradually all the molecular material into Pc_2Lu^+. The charge transport should then be given by the disproportionation reaction:

$$2Pc_2Lu^+ \rightleftharpoons Pc_2Lu, Pc_2Lu^{2+}$$

instead of the usual one:

$$2Pc_2Lu \rightleftharpoons Pc_2Lu^+, Pc_2Lu^-$$

The former requires more energy than the second and the density of charge carriers generated in this way is extremely small: pure Pc_2Lu^+, X^- material should be insulating.

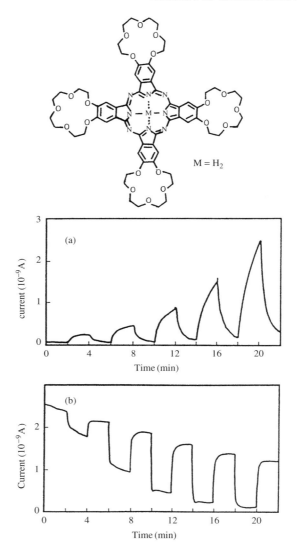

Figure 6.56 Chemical formula of the 15-crown-5 derivative of metal-free phthalocyanine and conductivity changes upon exposure of 1,2,3,4 and 5 ppm of NO_2 in dry air for 2 min with reversal in clean air: (a) before treatment, (b) after KCl treatment. (Reproduced by permission of the Royal Society of Chemistry from Ref. [115])

The effect of the crystallinity of the thin film may also be determined by comparing the properties of Pc_2Lu and $PN_nNN_{8-n}Lu$ ($n \approx 4$). The latter compound is obtained by reacting equimolecular concentrations of phthalonitrile and 2,3-naphthalonitrile with lutetium acetate [191]. This compound affords only quasi-amorphous thin films because it consists of several isomers and products.

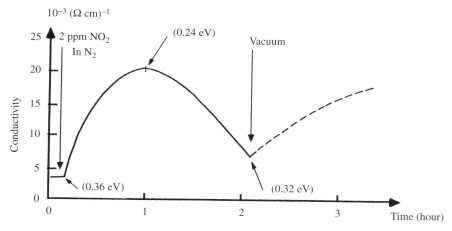

Figure 6.57 Conductivity as a function of time for Pc_2Lu thin films in the presence of 2 ppm of NO_2 at room temperature (in parentheses are the thermal activation energies of conduction E in $\sigma = \sigma_0 \exp[-E/(2kT)]$)

The time response to NO_2 exposure is characterized by t_{max} — the time necessary to obtain the maximum of conductivity σ_{max} (Table 6.14). Amorphous Pc_2Lu and $PN_n NN_{8-n} Lu$ lead to approximately the same time responses, whereas crystalline Pc_2Lu leads to 20 times slower kinetic constants [93, 119]

The oxidization of Pc_2Lu may be more easily completed by using a stronger oxidizing agent like Br_2.

$$Pc_2Lu \xrightarrow{Br_2} Pc_2Lu^+, Br^-$$

The detection of reducing agent, like ammonia, may then be achieved [119].

In the same way, H_2O vapour may be detected. In molecular materials, the effective dielectric constant is usually fairly low and ion pairs are easily formed. The incorporation of H_2O molecules within the layer significantly increases the dielectric constant (and reinforces ions solvation) facilitating charge migration [120] (Figure 6.58).

6.9.5 Conclusions

Gas sensors based on molecular materials and more particularly on metallophthalocyanines were described a long time ago. Whereas their performances are not evidently inferior to their mineral counterparts such as SnO_2, only a few technological investments have been made to transform the devices fabricated in laboratories into industrial sensors. Because of their intrinsic qualities and irreplaceable physicochemical characteristics their use in industry is a question of time.

Table 6.14 Time (t_{max}) necessary to reach the conductivity maximum for thin films of lutetium complexes upon exposure to 2 ppm of NO_2 at room temperature [93]

Product	t_{max}
Pc_2Lu	
300 Å (quasi-amorphous)[a]	32 min
3000 Å (crystalline)[b]	24 h 30 min
$PN_nNN_{8-n}Lu$	
1000 Å (amorphous)	50 min

[a] No electron diffraction.
[b] Deposited under vacuum on a substrate heated at 200 °C. Approximately the same results are obtained for films deposited at 80 °C: amorphous and crystalline phases then coexist.

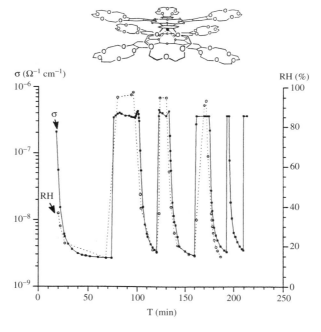

Figure 6.58 Conductivity of a $[(15C5)_4Pc]_2Lu$ thin film (thickness 150 Å) successively exposed to a stream of wet and dry nitrogen. The thin film was previously treated with Br_2. The relative humidity (RH) was measured with a capacitive humidity sensor [120]

6.10 FIELD-EFFECT TRANSISTORS

6.10.1 Introduction

In 1947 Bardeen, Brattain and Shockley produced the first transistor (trans-resistor) which was ten times smaller in dimension that the vacuum lamps used at that time. Since then, only mineral derivatives were employed for making the electronically active layer of the device: the semiconductor.

The mechanisms involved in field-effect transistors (FET) can be easily described in molecular terms. When a potential is applied on a molecular material using two metallic electrodes, which can be called the *source* and the *drain*, a current arises from the cations and anions thermally generated (ohmic domain) in the molecular material. For most divalent ion metallophthalocyanines this current is negligible.

Another mechanism of charge carrier generation may compete or totally replace the previous one when the molecular material is deposited on to a thin dielectric layer recovering a conductive electrode (the *grid*). By applying a potential between the source and the grid, charges are created on the two sides of the insulating layer following the equation $Q = CV$ (where Q = number of charges in C, C = capacitance, V = voltage in V) (Figure 6.59).

An understanding of classical silicon-based FET requires the use of band theory. A schematic picture of the device is shown in Figure 6.60.

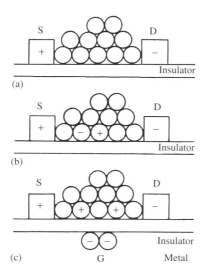

Figure 6.59 Three cases that can occur when a potential is applied between two electrodes (S, source; D, drain). (a) The molecular material is a pure insulator. (b) The thermal energy is sufficient to create ion pairs (charge carriers) in the material (intrinsic case). (c) By applying a source-to-grid voltage, charges are generated following $Q = CV$ ($Q_+ = Q_-$)

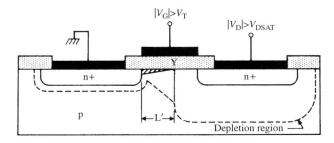

Figure 6.60 Schematic representation of a field-effect transistor using silicon-type materials. (Reproduced by permission of John Wiley & Sons, Inc. from Ref. [121])

Above a critical grid voltage (threshold voltage V_T), the grid can modulate the source-to-drain current. In the absence of grid voltage (V_{GS}), no source-to-drain current can flow because two n^+/p diodes are effective at the two electrode–semiconductor interfaces. Near these interfaces, depletion regions are generated in the semiconductor in which the concentration of charge carriers is lower than in the bulk. Such depletion regions do not occur when molecular materials are used. They are insulating in the absence of a grid voltage and this mechanism is therefore not necessary.

6.10.2 Thin-Film Transistors: Models and Applications in Industry

Silicon is by far the most widely used mineral material for making thin-film transistors used, in particular, in liquid crystal displays [122]. The performances of a FET are determined by two main parameters:

(a) The source-to-drain intensities of the current in the presence (I_{ON}) or in the absence (I_{OFF}) of a given grid voltage. This parameter is expressed as I_{ON}/I_{OFF}.
- (b) The switching time is the time needed to ensure the OFF/ON transition. This parameter is related to the mobility of charge carriers.

These two parameters are given for two typical examples of materials, amorphous silicon (a-SiH) and polycrystalline silicon (poly-Si), in Table 6.15.

The nonlinear response of the FET can be visualized by plotting the source-to-drain current I_{DS} as a function of the source-to-grid voltage V_{GS} (Figure 6.61). By varying V_{GS} from 1 to 3 V, the corresponding current I_{DS} increases by three orders of magnitude.

It is worth pointing out that the electrical properties of thin films of amorphous silicon greatly depend on the conditions in which they are prepared. a-Si is, in most cases, prepared by thermal decomposition of SiH_4. The undoped thin-film conductivity varies from 10^{-11} to $10^{-5}\,\Omega^{-1}\,cm^{-1}$ depending on the substrate temperature used for the deposition (100 and 550 °C, respectively) [123].

Table 6.15 Electrical parameters of thin-film transistors (TFT) based on amorphous (a-SiH) or polycrystalline (poly-Si) silicon [122, 123]

	a-SiH[a]	poly-Si[b]
μ (cm^2/V s)	0.3	15–20
V_T (V)	6–8	10
I_{ON} (A)	10^{-6}	10^{-4}
I_{OFF} (A)	10^{-13}	10^{-6}

μ = mobility of charge carriers; I_{ON}, I_{OFF}, see text (I_{ON} is measured at $V_{GS} = 20$ V and $V_{DS} = 10$ V).
[a]Glow discharge ($T = 300\,°C$; 10–15% H content).
[b]0.1 µm grain size.

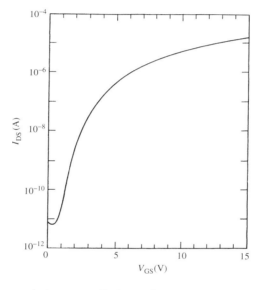

Figure 6.61 Source-to-drain current (I_{DS}) as a function of source-to-grid voltage (V_{GS}) for a TFT based on amorphous silicon [124]

The switching time of the FET is given by

$$T = \frac{L^2}{\mu V_{DS}} \quad (6.46)$$

where

μ = mobility of charge carriers
L = interelectrode spacing

This is the time necessary for a charge carrier to go from one electrode to the other, and is inversely proportional to the mobility of the charge carriers.

The current–voltage characteristics of the FET must be readily deduced from basic equations [121]:

$$\sigma = ne\mu \quad \text{and} \quad Q = CV$$

The charge generated in the semiconductor at a distance y from the source is given by

$$e\Delta n(y) = \frac{C_i}{h}[V_{GS} - V(y)] \qquad (6.47)$$

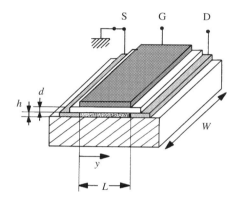

where

$\Delta n(y)$ = excess charge density induced by the grid voltage V_{GS}
C_i = capacitance of the insulator
d = dielectric thickness
h = thickness of the semiconducting layer (or channel)
$V(y)$ = potential due to the source-to-drain voltage

$\Delta n(y)$ is the density of charges induced by the field effect in excess of n_0, the density of charge carriers in the absence of grid voltage. In the last case the corresponding conductivity is σ_0. The source-to-drain current is then given by

$$I_{DS} = Wh[\sigma_0 + \Delta\sigma(y)]E(y) \qquad (6.48)$$

where

W = width of the electrode
$\Delta\sigma(y) = e\Delta n(y)\mu$
$E(y)$ = field at abcissa y

and

$$I_{DS} = Wh e\mu[n_0 + \Delta n(y)]\frac{dV(y)}{dy} \qquad (6.49)$$

In the case where the mobility of charge carriers is the same in the absence or in the presence of field and whenever Ohm's law is applicable (no injection of charge carriers),

$$I_{DS} = W\mu C_i \left[\frac{ehn_0}{C_i} + V_{GS} - V(y)\right]\frac{dV(y)}{dy} \quad (6.50)$$

By integrating between $y = 0$ and $y = L$, the interelectrode spacing, this becomes

$$I_{DS} = \frac{W\mu C_i}{L} V_{DS}\left(V_{GS} - V_T - \frac{V_{DS}}{2}\right) \quad (6.51)$$

with

$$V_T = -\frac{ehn_0}{C_i} \quad \text{(threshold voltage)}$$

When the source-to-drain voltage reaches a certain value, the associated corresponding current does not increase anymore (saturation current). This corresponds to

$$\frac{dI_{DS}}{dV_{DS}} = 0$$

$$\frac{dI_{DS}}{dV_{DS}} = \frac{W\mu C_i}{L}(V_{GS} - V_T - V_{DS})$$

$$\frac{dI_{DS}}{dV_{DS}} = 0 \quad \text{if } V_{DS} = V_{GS} - V_T$$

Then

$$\boxed{(I_{DS})_{sat} = \frac{W\mu C_i}{2L}(V_{GS} - V_T)^2} \quad (6.52)$$

Three other parameters are important for characterizing FETs:

(a) The transconductance g_m

$$g_m = \left(\frac{\partial I_{DS}}{\partial V_{GS}}\right)_{V_{DS}} \quad (6.53)$$

(b) The drain resistance r_d

$$r_d^{-1} = \left(\frac{\partial I_{DS}}{\partial V_{DS}}\right)_{V_{GS}} \quad (6.54)$$

(c) The amplification ratio A:

$$A = g_m r_d \quad (6.55)$$

Equations 6.51 and 6.52 allow an interpretation to be made of the experimental curves whose typical form is shown in Figure 6.62.

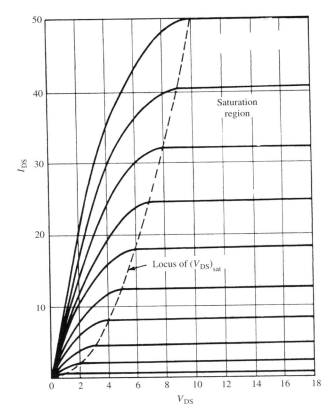

Figure 6.62 Source-to-drain current (I_{DS}) as a function of the source-to-drain voltage (V_{DS}) for various grid voltages (V_{GS}). (Reproduced by permission of John Wiley & Sons, Inc. from Ref. [121])

6.10.3 Molecular Field-Effect Transistors: A Chronology

The history of field-effect transistors is relatively brief since most of the studies only appeared around 1990.

In 1970 [125], it was shown on a single crystal of chloranil that the conductivity could be modulated by a transverse voltage. A few years latter [126, 127], the same authors described similar studies with tetrathiotetracene. These early publications, however, did not mention the expected saturation effect for a given source-to-drain voltage.

In 1983, Ebisawa *et al.* [128] described an attempt to fabricate an FET based on polyacetylene. The I_{DS} versus V_{DS} plots did indeed show a modulation with V_{GS} but no saturation.

In 1987, a group of researchers under the name GRIMM (Groupe de Recherches Interdisciplinaires sur les Matériaux Moléculaires) described for the first time the fabrication of a field-effect transistor based on lutetium bisphthalocyanine (Pc_2Lu) and zinc monophthalocyanine [129]. Approximately at the same time, an FET based on other metallophthalocyanines was described [130]. In the same year, a conjugated polymeric material — polythiophene — was used as the active part of an FET [131]. Very satisfactory electrical characteristics could be obtained already at that time, I_{ON}/I_{OFF} being of the order of 10^2-10^3. Similar studies were then carried out with poly (p-bisphenol) [132]. In this case, the mobility deduced from the FET characteristics was 4×10^{-4} cm^2/V s whereas it was around 10^{-3} cm^2/V s for polythiophene [133]. Another bispthhalocyanine/monophthalocyanine couple has been used for fabricating FETs [134]. An I_{ON}/I_{OFF} ratio of the order of 10^3 with a mobility of 10^{-4} cm^2/V s has been described with a polyacetylene-based transistor [135]. More detailed FET characterizations were subsequently reported for the couple $Pc_2Lu/PcZn$ [136–138] or for conjugated polymers [139].

In 1989, the use of an oligomeric form of thiophene (sexithiophene) for making an FET was described [140]. At that time, the I_{ON}/I_{OFF} ratio for this device was of the order of 2 together with low mobilities of charge carriers (3×10^{-4} cm^2/V s).

In all previous cases, the source and drain electrodes were metallic (in general gold). One publication mentions the use of a conjugated polymer — polypyrrole — to fabricate these conducting electrodes [141]. However, this change did not importantly affect the performances of the FET.

Around 1990 many papers were published concerning molecular FET [142–178]. It is beyond the scope of this book to detail all of them. In a further section metallophthalocyanine-based FET will be thoroughly described. The key points of the published discoveries will be given below for the other compounds.

Langmuir–Blodgett thin films prepared from a mixture of quinquethiophene or poly(3-hexylthiophene) and arachidic acid have been used to fabricate FETs [144]. However, very small mobilities were thus obtained ($\mu = 10^{-5}-7 \times 10^{-7}$ cm^2/V s). Alkyl-substituted polythiophene derivatives soluble in organic solvents and which can be deposited by casting, evaporation or spin coating have been used for the same purpose [162–164]. The insulating layer has been replaced by an organic polymeric film [146, 147] and all-organic thin film transistors have been described [146]. Following this line, printing techniques have been used to fabricate FETs [170].

Willander *et al.* demonstrated in 1993 [148] that the mobility of charge carriers for a poly(3-alkylthiophene)-based FET increases with the gate voltage (in contrast to inorganic semiconductors) from 1.2×10^{-5} cm^2/V s for $V_{GS} = -2$ V to 9.2×10^{-5} cm^2/V s at -4 V. This has been attributed to the creation of new hopping sites. It could also be due to some field-assisted detrapping process of charge carriers at the molecular material–insulating layer interface. A more precise study has been carried out with PcNi, indicating the same effects [171] (see the following section for other details). The same authors described a Schottky gated FET in which two gold electrodes are deposited on each side of an aluminum electrode; the contact between the deposited poly(3-alkylthiophene) and aluminum is rectifying. The source-to-drain current is then modulated by a voltage applied on the aluminum electrode [148, 149].

Thin films of diamond deposited by chemical vapor deposition and doped with boron have been used as an active layer in the FET [166]. Molecular ionic derivatives were also employed [167, 168] and it has been demonstrated that the mobility can reach 1.8×10^{-1} cm^2/V s after doping with iodine [167].

Fullerene compounds C_{60} [177] and C_{70} [172] also lead to good FET characteristics, with $\mu = 0.08$ cm^2/V s and $I_{ON}/I_{OFF} = 10^6$ for C_{60}.

Electronics circuits have also been fabricated using pentacene or poly(thienylenevinylene) for an electrically active layer [179]. In particular, an NOR gate has been constructed from three FETs [179].

6.10.4 Metallophthalocyanine-Based Field-Effect Transistors

The first metallophthalocyanine-based FET was composed of two layers, one with a fairly high intrinsic conductivity (Pc$_2$Lu) and the other insulating when undoped (PcZn) [129, 136]. While not expected at the time of the discovery, both layers play an important role in the electrical characteristics of the FET. It is therefore easier to try to understand the FET behaviour for single layers, in a first step.

The exposure of PcM thin films in ambient air yields a drastic effect on the FET characteristics as compared to those determined in vacuum. In air, the currents are multiplied by approximately 10 for the same gate voltage and source-to-drain potential. Whereas a negligible current is observed at $V_{GS} = 0$ under vacuum, in air an important current can be observed [180, 181] (Figure 6.63).

Two mechanisms compete at $V_{GS} = 0$:

(a) Intrinsic generation of charge carriers. This process is negligible at room temperature for the experiments carried out under vacuum for PcM. On the contrary, doping in air (with O$_2$ or chemically related species) can yield charge carriers following:

$$\text{PcM}, \text{O}_2 \rightleftharpoons \text{PcM}^+, \text{O}_2^{\cdot -} \underset{\text{PcM}}{\rightleftharpoons} \text{PcM}, \text{O}_2^{\cdot -} + \text{PcM}^+$$

Note that O$_2$ can be replaced by another oxidizing molecule.

(b) At sufficiently high V_{DS}, injection of charge carriers from the electrodes is possible. The importance of this contribution has been outlined in the very first publications concerning molecular FETs [129, 136]. The probability of electron transfer from (or to) the electrode to (or from) the molecular material depends on the energy difference between the Fermi level of the metal and the electronic levels (mainly HOMO and LUMO) of the metallophthalocyanines. However, to estimate this difference, both the source-to-drain and the source-to-gate differences of potential must be taken into account.

MOLECULAR SEMICONDUCTORS: PROPERTIES AND APPLICATIONS

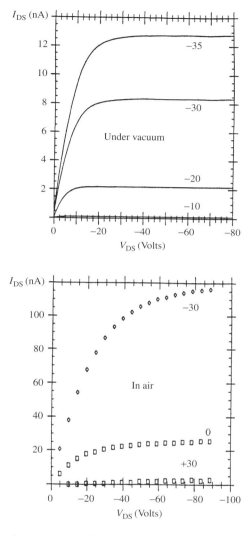

Figure 6.63 Curves I_{DS} versus V_{DS} for different source-to-grid voltages (V_{GS}) for a thin film of PcNi ($h = 300$ Å, deposition rate = 5 Å/s). The source-to-drain distance is 50 μm. The dielectric is SiO_2/Si_3N_4. (After Refs. [180] and [181])

At $V_{GS} < 0$, a third mechanism of creation of charge carriers is possible in which positive charges are generated (cation of the type PcM$^+$) in the molecular material layer.

A knowledge of the details concerning the three previous mechanisms is necessary in order to interpret the electrical characteristics of the molecular FET qualitatively. Modulation of the source-to-drain current in air occurs both for positive and negative grid voltages. Since air increases the density of PcM$^+$ in the metallophthalocyanine layer, a negative gate voltage will further increase this concentration or neutralize the ionized impurities I_A^-:

(a) $\text{PcM}//\text{gate} \xrightarrow{E} \text{PcM}^+//\text{gate}(-)$

(b) $\text{PcM}^+, I_A^-//\text{gate} \xrightarrow{E} \text{PcM}^+, I_A//\text{gate}(-)$

Path (a) leads to a generation of charge carriers whereas path (b) suppresses the Coulomb interaction between the oxidized PcM and the ionized impurity.

For positive gate voltages, no charge carriers can be created until the impurity levels are filled:

(c) $\text{PcM}, I_A//\text{gate} \xrightarrow{E} \text{PcM}, I_A^-//\text{gate}(+)$

The PcNi thickness dependence on the FET mobilities and on the threshold voltage has also been studied. Between 100 and 7000 Å, no drastic differences have been noticed [142]. The conduction channel is therefore probably related to the SiO$_2$/PcNi interface.

When the samples are exposed to air, two kinetically different processes take place:

(a) Quasi-instantaneously, the amplification ratio is lowered, the threshold voltage is increased and the FET mobility is lowered.
(b) After a long time in air (several weeks or months), the absolute value of the threshold voltage decreases to -2 V (instead of -10 V) and the mobility is approximately twice the initial mobility.

Some first conclusions can be drawn:

(a) There is no diode effect between the source or the drain electrode and the molecular material.
(b) The intrinsic conductivity of PcNi is very small. This is the cause of the very high resistance at $V_{GS} = 0$ in the absence of air.
(c) Charge carriers may be generated by applying a gate voltage.
(d) Doping with air (O$_2$ or related species) increases significantly the conductivity at $V_{GS} = 0$.
(e) PcNi (or PcZn) alone can lead to well-behaved field-effect transistors.

It was then necessary to know the FET characteristics of Pc$_2$Lu thin films. A device consisting of SiO$_2$ (3000 Å)/Si$_3$N$_4$ (2500 Å)/Pc$_2$La (100–400 Å) (La = Lu or Tm) has been made [143]. The FET characteristics are shown in Figure 6.64. The wafers (without Pc$_2$Lu) were heated *in vacuo* at 150 °C for 24 h for degassing. The measurements were made without breaking the vacuum [143].

Pc$_2$Lu and Pc$_2$Tm both lead to semiconducting molecular layers with conductivities of 5×10^{-5} and 2.5×10^{-4} Ω^{-1} cm^{-1}, respectively. In the last case, it

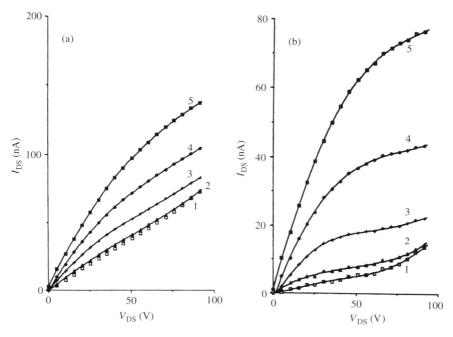

Figure 6.64 Intensity of the drain current (I_{DS}) as a function of the drain voltage (V_{DS}) for various grid voltages (V_{GS}) for thin films of Pc$_2$Lu (200 Å thick) on Si$_3$N$_4$ (measurements under vacuum) (1, $V_{GS} = 0$ V; 2, $V_{GS} = 10$ V; 3, $V_{GS} = 30$ V; 4, $V_{GS} = 50$ V; 5, $V_{GS} = 70$ V): (a) experimental curves, (b) the bulk conductivity has been abstracted [143]

has been suspected that the conduction is not purely intrinsic [143]. The drain-to-source current is more than 50 times larger than for PcNi under vacuum and of the same order of magnitude compared to air-doped PcNi. Bulk current corresponds for Pc$_2$Lu to a conductivity $\sigma = 2.5 \times 10^{-5}$ Ω^{-1} cm^{-1}. The calculation is carried out by assuming that the diameter of the channel is negligible compared to the overall thickness of the molecular material. The bulk current flowing relatively far from the interface is not modulated by the grid voltage. Modulation is effective with $V_{GS} > 0$: this means that an n-type conductivity must occur in the molecular material [143].

The stability of the device under vacuum (10^{-7} torr) is excellent, with no change in the FET characteristics for several days. However, annealing at 150 °C *in vacuo* affords both n- and p-type behaviors.

Instead of Si$_3$N$_4$, SiO$_2$ treated with C$_{18}$H$_{37}$SiCl$_3$ can be used. Whereas SiO$_2$ recovered with Pc$_2$Lu and Pc$_2$Tm does not show any FET characteristics, surface-treated SiO$_2$ yields an n-type FET under vacuum and a p-type one in air [143]. The main parameters that can be deduced from the I_{DS}/V_{DS} curves are shown in Table 6.16.

Table 6.16 FET characteristics for thin films of Pc$_2$Lu and Pc$_2$Tm under various conditions [143]

	Conditions[a]	Carrier	σ_b^b (Ω^{-1} cm^{-1})	V_T^c (V)	μ^d (cm^2/V s)	g_m^e (nS)	A^f
Pc$_2$Lu	1	n	1.5×10^{-5}	8	2×10^{-4}	2.2	50
	2	n, p[g]	4×10^{-4}	1.5	2×10^{-3}	10	100
	3	p	8×10^{-4}	1	3×10^{-3}	14	100
Pc$_2$Tm	1	n	7×10^{-5}	10	3×10^{-4}	2.8	40
	2	n	1.2×10^{-4}	10	1.4×10^{-3}	16	80
	3	p	3.7×10^{-4}	4	5×10^{-3}	20	40
	4	p	1.2×10^{-3}	4	1.5×10^{-2}	25	20

[a]Conditions: 1, after fabrication in vacuum; 2, after treatment at 150 °C in vacuum; 3, after 2 and exposure to air; 4, after 2 and 3 and annealing in air at 150 °C (2 days).
[b]Bulk conductivity.
[c]Threshold voltage.
[d]Mobility of charge carriers.
[e]Transconductance.
[f]$A = g_m r_d$, with r_d the drain resistance (the bulk conductivity has been abstracted).
[g]The results indicated pertain to the p-type behavior.

It is worth pointing out that fairly high mobilities ($\mu_{FET} = 1.5 \times 10^{-2}$ cm^2/V s) have been obtained with a Pc$_2$Tm thin film annealed at 150 °C in vacuum and at the same temperature in air for two days.

The conclusions of this subsection are therefore:

(a) Intrinsic molecular semiconductors (Pc$_2$Lu, Pc$_2$Tm) can be used for fabricating field-effect transistors.
(b) A large bulk current, due to the fairly high conductivity of the molecular material, is present, but cannot be modulated by a grid voltage.
(c) The source-to-drain current at $V_{GS} = 0$ could be reduced if diodes could be formed at the interfaces between the source, the drain and the metallophthalocyanine layer.

One can now study the transistor made in 1987, which consisted of n − Si/SiO$_2$/PcZn/Pc$_2$Lu [129, 136, 137]. In the most studied device [136] the thickness of PcZn and Pc$_2$Lu were 2000 and 1000 Å, respectively. The metallic electrodes were deposited upon the metallophthalocyanine layers. The FET characteristics obtained are remarkably identical to those expected for this type of device [136] (Figure 6.65). The molecular FET plots are comparable to those obtained with amorphous silicon thin films used to make transistors for a liquid crystal display [182] (Figure 6.66).

From the previous general equation established for modelling the FET, if one transforms $V_{DS} \rightarrow -V_{DS}$ and $V_{GS} \rightarrow V_{GS} - V_{DS}$, one should find $I_{DS} \rightarrow -I_{DS}$. This is indeed the case for the molecular FET characteristics for $V_{GS} < 0$ (this part is shown in Figure 6.66) as well as for $V_{GS} > 0$. However, a nonnegligible bulk current (not modulated by the gate voltage) can be observed. The thermal

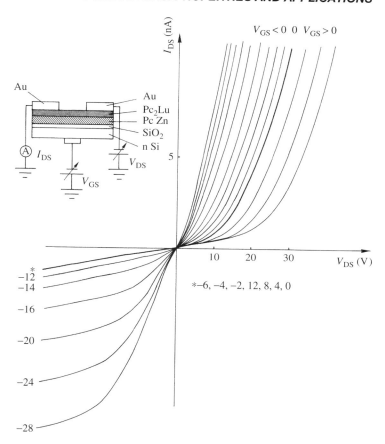

Figure 6.65 Intensity–current curves (I_{DS} versus V_{DS}) as a function of the applied gate voltage (V_{GS}). For $V_{GS} > 0$: 4, 8, 12 V; $V_{GS} < 0$: −2, −4, −6, −12, −14, −16, −20, −24, −28 V [136]

activation energy of the current is the one expected for pure Pc$_2$Lu (0.5 eV/(2kT)) [129]. The classical treatments of the data lead to the FET parameters shown in Table 6.17 together with a comparison with an amorphous silicon based device.

Because of the low mobilities of carriers, the turn on time of the molecular FET is approximately 10^3 larger than the one found for amorphous silicon devices. Taking into account the previous results concerning field-effect transistors with single layers of PcNi or Pc$_2$Lu, the following conclusions can be drawn:

(a) In the system SiO$_2$/PcZn/Pc$_2$Lu, the current which is modulated by the gate voltage arises at the SiO$_2$/PcZn interface.
(b) Bulk current flowing (relatively) far from the interface goes mainly through the Pc$_2$Lu layer. This bulk current is not modulated by the gate voltage.

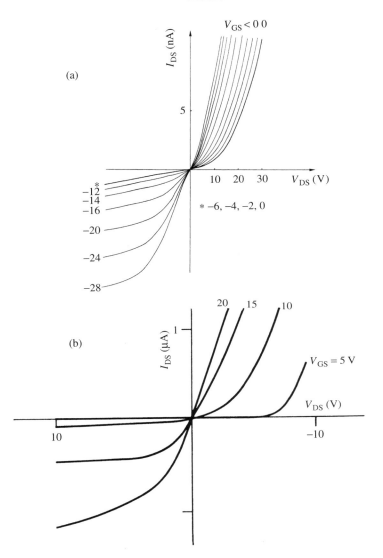

Figure 6.66 Source-to-drain current as a function of a source-to-drain voltage for: (a) $SiO_2/PcZn/Pc_2Lu$ device for $V_{GS} < 0$ [136]. (b) Amorphous Si-H FET [182]. ($V_T = 5\text{--}10\,\text{V}$, $\mu = 0.3\,\text{cm}^2/\text{V s}$)

(c) The electrical properties of the Pc_2Lu layer are close to the ones previously determined for thin film deposited on glass slides.

Subsequently, very similar studies have been published [134] and patented [184]. The system was $SiO_2/PcNi/Pc_2Sc$: the results obtained confirmed the

Table 6.17 Comparison of the characteristics of various insulated gate field-effect transistors as a function of the semiconductor used (a-SiH, hydrogenated amorphous silicon). μ is the mobility of charge carriers, calculated from the transconductance equation, V_T the threshold voltage, g_m the transconductance, A the amplification factor and T the turn on time: $T = L^2/(\mu V_{DS})$, calculated with $L = 50\,\mu m$, $V_{DS} = 10\,V$ [136]

	Pc$_2$Lu	a-SiH
μ (cm^2/V s)	10^{-4}	0.1–2
V_T (V)	-2	$-(1-5)$
g_m (Ω^{-1})	0.5×10^{-9}	50×10^{-9}
A	15	65
T	25 ms	$1 \rightarrow 25\,\mu s$
Ref.	[136]	[121, 124, 183]

previous ones. The degree of mixing occurring at the PcNi/Pc$_2$Sc interface was studied by secondary ion mass spectroscopy (SIMS) as a function of the bisphthalocyanine thickness (300 and 1000 Å).

It has been recognized, from the very beginning of the studies on FETs, that kinetic effects with long response times occur when the measurements are carried out in a conventional way with d.c. voltages. A decrease of I_{DS} of approximately 10% from one experiment to a successive one has been observed. The initial value seems to be restored after a certain time [169].

In the previous cases, a source-to-drain voltage varying linearly with time (about 10 V/s) was used to determine the FET characteristics. In this way reproducible results are obtained and drift effects are negligible. Measurements have also been carried with a ramp voltage lasting 500 ms for scanning approximately 100 V (Figure 6.67).

The measurements shown are made under vacuum and a conventional treatment of the data leads to $\mu = 6.8 \times 10^{-4}\,cm^2/V\,s$, $V_T = -24.7\,V$. The FET mobility is thermally activated and reaches $0.02\,cm^2/V\,s$ at $100\,°C$ [169].

Step voltages have also been applied on the previous molecular FET. To avoid too important an influence of the capacitance term, another electrical circuit has been adopted (Figure 6.68).

When the step voltage is applied, at very short times ($t < 3$ ms), a current associated with the RC electrical parameters of the circuit is observed. It takes a longer time (time delay τ) to observe the appearance of a source-to-drain current (Figure 6.69).

The time delay τ observed when a step voltage is applied can be interpreted qualitatively. When a source-to-drain voltage is suddenly applied, distribution of charges can be established more rapidly in the metallic gate electrode than in the poorly conducting molecular material. In the metallic gate electrode, the charge distribution is such that the potential at any point in the bulk is constant (Figure 6.70a).

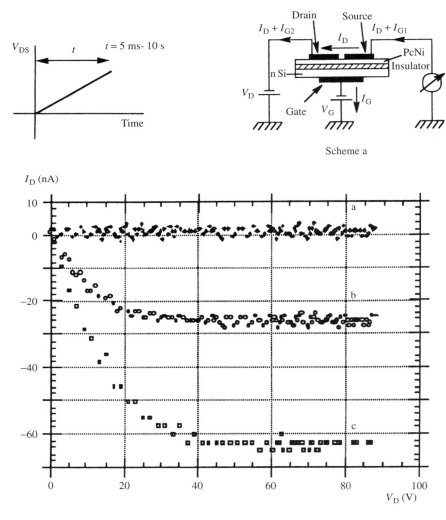

Figure 6.67 PcNi drain current as a function of the drain voltage $V_D(<0)$ for various gate voltages $V_G(<0)$. Drain voltage is a ramp voltage of 500 ms duration. Curve a: $V_G = 0$ V; curve b: $V_G = -50$ V; curve c: $V_G = -60$ V. The measurements are made without breaking the vacuum at any stage (scheme a) for the electrical circuit

At longer times, the charges migrate in the molecular material via a relatively slow hopping process driven by the field generated by the source and drain electrodes. Simultaneously, the countercharge in the gate electrode immediately follows the movement of the molecular charges. Because of those concomitant migrations, no polarization current is observed until the charges reach the counterelectrode. The time needed for the charge to go from one electrode to the other

MOLECULAR SEMICONDUCTORS: PROPERTIES AND APPLICATIONS

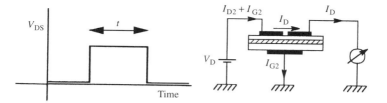

Figure 6.68 Electrical circuit used to measure transient currents generated by step voltages (scheme b) [169]

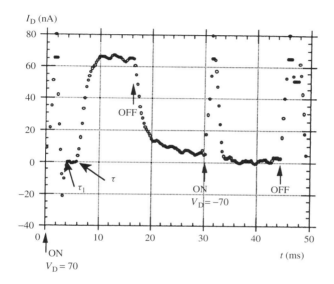

Figure 6.69 Drain current as a function of time for different drain voltages: $V_D = 70$ V and $V_D = -70$ V. The duration of the pulses is 15 ms; τ_1 = time constant associated with the electrical circuit; τ = delay observed in the appearance of the current (electrical scheme b) shown in Figure 6.68

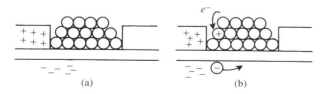

Figure 6.70 Schematic representation of the source-to-drain current when a step voltage is applied: (a) the step voltage is just applied ($t = 0$); (b) after a few ms concomitant migration of charges in the molecular material and in the gate metallic electrode occurs

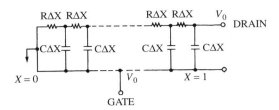

Figure 6.71 Electrical circuit used to establish a model for field-effect transistors. For a thorough development see Ref. [185]

is proportional to the transit time:

$$\frac{L^2}{\mu(V_{GS} - V_T)}$$

where

$L = $ interelectrode spacing
$\mu = $ mobility of charge carriers
$V_T = $ threshold voltage

A quantitative treatment has been previously proposed in which the device is modelled as shown in Figure 6.71. The delay time τ depends on the source-to-drain voltages. At high values of V_{DS}, it can become of the same order of magnitude as the time constant of the RC electrical circuit (τ_1) (Figure 6.72).

The characteristics of the devices are stable for months under vacuum. The time delay is thermally activated and it becomes too short to be measured if the

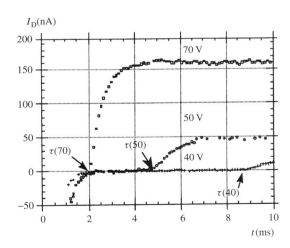

Figure 6.72 Drain current as a function of time for different V_{DS}. Pulse duration is 15 ms [169]

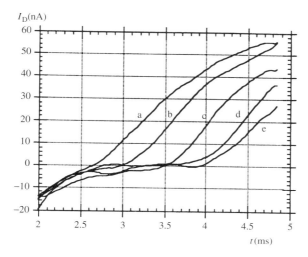

Figure 6.73 Drain current as a function of time for $V_{DS} = 70$ V (scheme b, Figure 6.68). The time elapsed between two pulses has been varied: (a) 15 ms; (b) 30 ms; (c) 100 ms; (d) 500 ms; (e) 1 s. Duration of pulse is 15 ms [169]

temperature is higher than 50 °C [169]. The time delay importantly depends on the time elapsed between two applications of the step voltage (Figure 6.73).

This experiment indicates that some deep traps are present within the material. The corresponding thermal detrapping is slow at room temperature but increases, as expected, with temperature. The description of an analytical model for molecular transistors [152] indeed indicates that 'the threshold voltage corresponds to the filling of traps and is a surface equivalent of the trap-filled limit voltage in space-charge-limited current model'.

Further step voltage measurements with PcNi [171] showed that μ and V_T are not constant as a function of drain-to-source and gate-to-source voltages. It can be demonstrated [185] that the product $I_{DS}\tau$ is given by

$$0.38 LW C_i(V_{DS} - V_T) \qquad (5.56)$$

where W = electrode width. The curve $I_{DS}\tau$ as a function of V_{DS} is indeed a straight line below 50 V (Figure 6.74).

Moreover, by extrapolation one obtains $I_{DS}\tau = 0$ when $V_{DS} = 0$; this indicates that, at least under those conditions, $V_T \sim 0$. This finding differs from the previous determinations using the ramp voltage method. Kinetic effects related to charge carrier trapping–detrapping probably play a role in this apparent discrepancy.

The corresponding mobilities can, in turn, be calculated by assuming $V_T = 0$ (Figure 6.75). Below 50 V, the mobility follows a $V_{DS}^{3.6}$ law, the variation being in the range 10^{-7}–3×10^{-4} cm^2/V s [171]. In the electrical scheme b presently

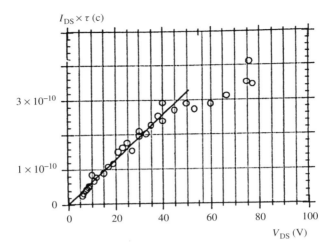

Figure 6.74 Dependence of the product $I_{DS}\tau$ on the drain voltage V_{DS} [171]

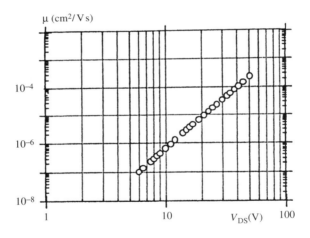

Figure 6.75 Dependence of the mobility μ on the drain voltage V_{DS} [171]

used, $V_{DS} = V_{DG}$: the variation of the mobility may therefore be related to a gate voltage which induces filling of traps [152].

Other experiments have been reported using PcCu [176] or PcM (M = Zn, Ni, Fe, Sn, Pt, H$_2$) [178]. The FET characteristics, and particularly the mobility of charge carriers (as measured from $\sqrt{I_{DS}}$ versus $V_{DS} = V_{DG}$ curves), have been shown to be highly dependent upon the temperature of the substrate on which the metallophthalocyanine is deposited. The maximum mobility ($\mu = 0.02\,\text{cm}^2/\text{V}\,\text{s}$) is found for a substrate temperature around 125 °C [176] (Figure 6.76).

MOLECULAR SEMICONDUCTORS: PROPERTIES AND APPLICATIONS

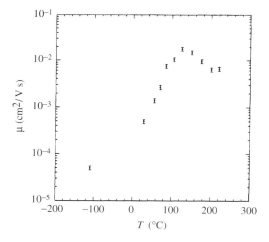

Figure 6.76 Field effect mobility of PcCu at different substrate deposition temperatures device: n-Si/SiO$_2$/PcCu (channel lengths of 12 and 25 µm). Rate of deposition is 4–5 Å/s, pressure 2.0×10^{-6} torr (Reproduced by permission of the American Institute of Physics from Ref. [176])

In the same time, the I_{ON}/I_{OFF} ratio reaches 4×10^5 [176]. It is also reported that no significant difference in the FET characteristics occurs after a long time storage in air, with the exception of a small I_{OFF} increase.

X-ray diffraction demonstrated that high temperatures of substrates favour an increase in the sizes of the crystals. The molecular units are probably edge-on relative to the surface (α-form) with the b axis parallel to the substrate (Figure 6.77). By varying the nature of the metal ion, significant differences in the FET mobilities are observed under similar conditions of deposition [178] (Table 6.18).

The results described so far demonstrate that molecular material based field-effect transistors are original from a fundamental point of view: they totally differ from silicon-based devices both for the charge transport mechanisms (hopping model against band theory) and for the physicochemical mechanisms that afford

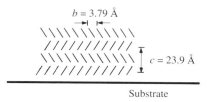

Figure 6.77 Schematic representation of the orientation of the molecular units relative to the substrate (Reproduced by permission of the American Institute of Physics from Ref. [176])

Table 6.18 FET mobilities for a substrate temperature maintained at 125 °C during deposition [178]

PcM	μ (cm^2/V s)
PcCu	2.0×10^{-2}
PcH$_2$	2.6×10^{-3}
PcZn	2.4×10^{-3}
PcNi	3.0×10^{-5}
PcPt	1.5×10^{-4}

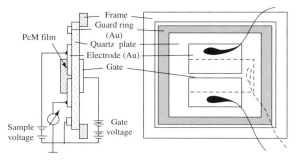

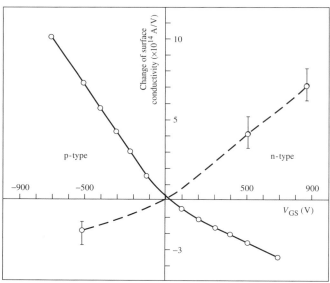

Figure 6.78 Experimental set-up used for measuring the FET characteristics of PcCu thin films: n-type, PcCu doped by deposition of lithium; p-type, O$_2$ (or another oxidant) doped PcCu. (After Ref. [130]). Reproduced with permission

the modulation of the source-to-drain current with a gate voltage. The molecular FET also shows fairly good mobilities, high I_{ON}/I_{OFF} ratios and respectable stabilities in air. When rapid switching times are not required, the molecular FET could be used in practical applications. Its cost and ease of deposition of the electroactive molecular material is in its favour when compared to thin layers of amorphous silicon, for instance. The fabrication of gas sensors is probably a realistic future industrial application of the molecular FET [192]. The effect of air on the FET characteristics has already been described in previous sections. In 1987, Laurs and Heiland described two phenomena:

(a) the effect of O_2, I_2, Br_2 on the surface conductivity of PcM layers,
(b) the fabrication of FET in which lithium is deposited on PcM [130] (Figure 6.78).

It was reported, with no detail, that 'the field effect... of p- and n-type films increased by several orders of magnitude with the dark conductivity'. Mobilities derived from field-effect measurements were in the range 10^{-9}–10^{-5} cm^2/V s [130].

Metallophthalocyanine may also be used as a gate in the FET. This type of device has been used to detect NO_2 at concentrations of the order of 1 ppb with a PcPb thin layer [186]. A similar device using Langmuir–Blodgett thin films of porphyrin derivatives has been used to detect NO_2, NH_3, CO, H_2S with various sensitivities [187].

6.11 REFERENCES

1. A. T. Vartanyan, *Zhur. Fiz. Khim.*, **22**, 769 (1948).
2. D. D. Eley, *Nature*, **162**, 819 (1948).
3. J. Simon and J.-J. André, *Molecular Semiconductors*, Springer-Verlag, Weinheim (1985).
4. E. A. Silinsh, M. Bouvet and J. Simon, *Molec. Mater.*, **5**, 1 (1995).
5. R. Hoffmann, *Angew. Chem. Int. ed.*, **26**, 846 (1987).
6. L. Salem, *The Molecular Orbital Theory of Conjugated Systems*, W. A. Benjamin, New York (1966).
7. A. Pacault, *Eléments de thermodynamique statistique*, Masson et cie, Paris (1963).
8. M. Bouvet, E. A. Silinsh and J. Simon, *Molec. Mater.*, **5**, 255 (1995).
9. S. L. Altmann, *Band Theory of Solids. An Introduction from the Point of View of Symmetry*, Clarendon Press, Oxford (1991).
10. C. Iung and E. Canadell, *Description orbitalaire de la structure électronique des solides*, Ediscience, Paris (1997).
11. C. Kittel, *Physique de l'état solide*, (trad. M. Poumellec, R. Mégy and C. Dupas), Dunod, Paris (1957).
12. N. W. Ashcroft and N. D. Mermin, *Solid State Physics*, Holt, Rinehart & Winston, New York (1976).
13. E. A. Silinsh and V. Capek, *Organic Molecular Crystals. Interaction, Localization and Transport Phenomena*, AIP Press, New York (1994); E. A. Silinsh, *Organic Molecular Crystals*, Series in Solid State Science, Vol. 16, Springer-Verlag, Berlin (1980).

14. J.-J. André, K. Holczer, P. Petit, M.-T. Riou, C. Clarisse, R. Even, M. Fourmigué and J. Simon, *Chem. Phys. Lett.*, **115**, 463 (1985).
15. P. Turek, P. Petit, J.-J. André, J. Simon, R. Even, B. Boudjema, G. Guillaud and M. Maitrot, *J. Am. Chem. Soc.*, **109**, 5119 (1987).
16. J. Simon, M. Bouvet and P. Bassoul, *Encyclopedia of Advanced Materials* (eds. D. Bloor et al.), Pergamon, Oxford (1994).
17. I. S. Kirin, P. N. Moskalev and Yu A. Makashev, *Russ. J. Inorg. Chem.*, **10**, 1065 (1965).
18. I. S. Kirin, P. N. Moskalev and Yu A. Makashev, *Russ. J. Inorg. Chem.*, **12**, 369 (1967).
19. A. T. Chang and J. C. Marchon, *Inorg. Chim. Acta.*, **53**, 241 (1981).
20. H. Sugimoto, T. Higashi and M. Mori, *Chem. Lett.*, 1167 (1983).
21. N. B. Subbotin, L. G. Tomilava, E. V. Chernikh, N. A. Kostromina and E. A. Lukyanets, *Zh. Obshch. Khim.*, **56**, 232 (1985).
22. P. N. Moskalev and I. S. Kirin, *Russ. J. Phys. Chem.*, **46**, 1019 (1972).
23. F. Castaneda, C. Piechocki, V. Plichon and J. Vaxivière, *Electrochim. Acta*, **31**, 131 (1986).
24. G. C. S. Collins and D. J. Schiffrin, *J. Electroanal. Chem.*, **139**, 335 (1982).
25. G. C. S. Collins and D. J. Schiffrin, *J. Electrochem. Soc.*, **132**, 1835 (1985).
26. M. L'Her, Y. Cozien and J. Courtot-Coupez, *C. R. Acad. Sci. Paris*, **302**, 9 (1986).
27. H. Konami, M. Hatano, N. Kobayashi and T. Osa, *Chem. Phys. Lett.*, **165**, 397 (1990).
28. M. Bouvet, P. Bassoul and J. Simon, *Molec. Cryst. Liquid Cryst.*, **252**, 31 (1994).
29. A. N. Darovskikh, A. K. Tsytsenko, O. V. Frank-Kamenetskaya, V. S. Fundamenskii and P. N. Moskalev, *Sov. Phys. Crystallogr.*, **29**, 273 (1984).
30. A. de Cian, M. Moussavi, J. Fischer and R. Weiss, *Inorg. Chem.*, **24**, 3162 (1985).
31. A. N. Darovskikh, O. V. Frank-Kamenetskaya, V. S. Fundamenskii and A. M. Golubev, *Sov. Phys. Crystallogr.*, **31**, 165 (1986).
32. S. A. Song, M. O' Connor, D. A. Barber and J. Silver, *J. Cryst. Growth*, **88**, 477 (1988).
33. L. E. Lyons, *Aust. J. Chem.*, **33**, 1717 (1980).
34. P. Bassoul, M. Bouvet and J. Simon, *Synth. Met.*, **61**, 133 (1993).
35. R. Madru, G. Guillaud, M. Al Sadoun, M. Maitrot and J. P. Schunck, *Chem. Phys. Lett.*, **168**, 41 (1990).
36. A. J. Parker, *Chem. Rev.*, **69**, 1 (1969).
37. A. Klimkans, E. A. Silinsh, M. Bouvet and J. Simon, to be published.
38. R. Even, J. Simon, D. Markovitsi, *Chem. Phys. Lett.*, **156**, 609 (1989).
39. E. Orti, J. L. Brédas and C. Clarisse, *J. Chem. Phys.*, **92**, 1228 (1990).
40. P. Petit, *Synth. Met.*, **46**, 147 (1992).
41. P. Petit, K. Holczer and J.-J. André, *J. Physique*, **48**, 1363 (1987).
42. P. Petit and J.-J. André, *J. Phys. Fr.*, **49**, 2059 (1988).
43. H. Homborg and W. Kalz, *Z. Naturforsch.*, **B33**, 1067 (1978).
44. M. A. Petit, T. Thami and R. Even, *J. Chem. Soc. Chem. Commun.*, 1059 (1989).
45. P. Turek, J.-J. André, A. Giraudeau and J. Simon, *Chem. Phys. Lett.*, **134**, 471 (1987).
46. H. Sugimoto, M. Mori, H. Masuda and T. Taga, *J. Chem. Soc. Chem. Commun.*, 962 (1986).
47. P. Turek, M. Moussavi, P. Petit and J.-J. André, *Synth. Met.*, **29**, F65 (1989).
48. P. Turek, J.-J. André and J. Simon, *Solid State Commun.*, **63**, 741 (1987).
49. P. Turek, M. Moussavi, J.-J. André and G. Fillion, *J. Physique*, **C8**, 835 (1988).
50. J.-J. André, in *Nanostructures Based on Molecular Materials* (eds. W. Göpel and Ch. Ziegler), VCH, Weinheim (1992).
51. G. H. Heilmeier, G. Warfield and S. E. Harrison, *J. Appl. Phys.*, **34**, 2278 (1963).

52. C. B. Duke and L. B. Schein, *Physics Today*, 42 February (1980).
53. N. Karl, in *Defect Control in Semiconductors* (ed. K. Sumino), p. 1725, Elsevier Scientific and North-Holland, Amsterdam (1990).
54. W. Warta, R. Stehle and N. Karl, *Appl. Phys.*, **A36**, 163 (1985).
55. W. Warta and N. Karl, *Phys. Rev.*, **B32**, 1172 (1985).
56. P. Devaux and G. Delacote, *Chem. Phys. Lett.*, **2**, 337 (1968).
57. I. Chen, *J. Chem. Phys.*, **51**, 3241 (1969).
58. M. Sukigara and R. C. Nelson, *Molec. Phys.*, **17**, 387 (1969).
59. P. Devaux and G. Delacote, *J. Chem. Phys.*, **52**, 4922 (1970).
60. S. C. Mathur, J. Singh and D. C. Singh, *J. Phys. Solid State Phys.*, **4**, 3122 (1971).
61. J. Singh, *Phys. Stat. Solidi*, **B82**, 263 (1977).
62. E. Canadell and S. Alvarez, *Inorg. Chem.*, **23**, 573 (1984).
63. E. Orti and J. L. Brédas, *Synth. Metals*, **29**, F115 (1989).
64. T. Hahn (ed.), *International Tables for Crystallography*, Kluwer Academic, Dordrecht (1989).
65. C. Piechocki, J. Simon, A. Skoulios, D. Guillon and P. Weber, *J. Am. Chem. Soc.*, **104**, 5245 (1982).
66. S. Chandrasekhar, B. K. Sadashiva and K. A. Suresh, *Pramana*, **9**, 471 (1977).
67. C. Piechocki, J. Simon, J.-J. André, D. Guillon, P. Petit, A. Skoulios and P. Weber, *Chem. Phys. Lett.*, **122**, 124 (1985).
68. Z. Belarbi, C. Sirlin and J. Simon, J.-J. André, *J. Phys. Chem.*, **93**, 8105 (1989).
69. Z. Belarbi, *J. Phys. Chem.*, **94**, 7334 (1990).
70. Z. Belarbi, G. Guillaud, M. Maitrot, J. Huck, J. Simon and F. Tournilhac, *Rev. Phys. Appl.*, **23**, 143 (1988).
71. D. Kearns and M. Calvin, *J. Chem. Phys.*, **29**, 950 (1958).
72. H. Meier, *Umschau. Wiss. Tech.*, **66**, 438 (1966); *CA*, **63**, 14202g (1966).
73. J. Simon, F. Tournilhac and J.-J. André, *J. Appl. Phys.*, **62**, 3304 (1987).
74. F. A. Haak and J. P. Nolta, *J. Chem. Phys.* **38**, 2648 (1963).
75. M. Martin, J.-J. André and J. Simon, *J. Appl. Phys.*, **54**, 2792 (1983).
76. H. B. Michaelson, *J. Appl. Phys.*, **48**, 4730 (1977).
77. K. Hall, J. S. Bonham and L. E. Lyons, *Aust. J. Chem.*, **31**, 1661 (1978).
78. C. Tantzscher and C. Hamann, *Phys. Status Solidi*, **A26**, 443 (1974).
79. C. Hamann and H. Wagner, *Kristall und Technik*, **6**, 307 (1971).
80. B. Boudjema, G. Guillaud, M. Gamoudi, M. Maitrot, J.-J. André, M. Martin and J. Simon, *J. Appl. Phys.*, **56**, 2323 (1984).
81. D. M. Chapin, C. S. Fuller and G. L. Person, *J. Appl. Phys.*, **25**, 676 (1954).
82. J. L. Stone, *Solar Energy Mater and Solar Cells*, **34**, 41 (1994).
83. R. Dalven, *Introduction to Applied Solid State Physics*, Plenum Press, New York (1980).
84. H. J. Hovel, *Semiconductors and Semimetals*, Vol. 11, *Solar Cells*, Academic Press, New York (1975).
85. G. Tollin, D. R. Kearns and M. Calvin, *J. Chem. Phys.*, **32**, 1013 (1960).
86. F. I. Vilesov, A. A. Zagrubskii and D. F. Garbuzov, *Fiz. Tverd. Tela*, **5**, 2000 (1963); *Sov. Phys. Solid. State*, **5**, 1460 (1964).
87. M. Pope, *J. Chem. Phys.* **36**, 2810 (1962).
88. C. W. Tang, *Research Disclosure*, **162**, 71 (1977).
89. M. I. Fedorov, E. P. Zinov'eva, V. A. Shorin and L. I. Nutrikhina, *Izv. Vyssh. Uch. Zav. Fiz.*, **20**, 159 (1977); *CA*, **86**, 149-485 v.
90. G. A. Stepanova, L. S. Volkova, M. A. Gainullina and F. F. Yumakulova, *Zh. Fiz. Khim.*, **51**, 1771 (1977).
91. S. Ross and J. P. Oliver, *On Physical Absorption*, Interscience Publishers (1964); cited in M. Passard, Thèse de doctorat, Université, B. Pascal, Clermont Ferrand (1995).

92. R. P. Feynman, R. B. Leighton and M. Sands, *Le cours de physique* (ed. R. P. Feynman), Mécanique 2, InterEditions (French translation), Paris (1979).
93. M. Passard, Thèse de doctorat, Clermont Ferrand (1995).
94. R. P. Linstead and J. M. Robertson, *J. Chem. Soc.*, 1195 and 1636 (1936).
95. J. M. Robertson, R. P. Linstead and C. E. Dent, *Nature*, **135**, 506 (1935).
96. J. M. Robertson, *J. Chem. Soc.*, 615 (1935).
97. J. M. Robertson and I. J. Woodward, *J. Chem. Soc.*, 219 (1937); 36 (1940).
98. E. Suito and N. Uyeda, *J. Phys. Chem.*, **84**, 3223 (1980).
99. E. Suito and N. Uyeda, *Kolloid Z. u. Z. Polym.*, **7**, 193 (1963).
100. J. H. Sharp and M. Lardon, *J. Phys. Chem.*, **72**, 3230 (1968).
101. G. Susich, *Anal. Chem.*, **22**, 425 (1950).
102. F. R. Tarantino, D. H. Stubbs, T. F. Cooke and L. A. Melsheimer, *Am. Ink Mater.*, **29**, 35 and 425 (1950).
103. A. A. Ebert Jr and H. B. Gottlieb, *J. Am. Chem. Soc.*, **74**, 2806 (1952).
104. F. W. Karasek and J. C. Decius, *J. Am. Chem. Soc.*, **74**, 4716 (1952).
105. M. Shigemitsu, *Bull. Chem. Soc. Japan*, **32**, 607 (1959).
106. T. Kobayashi, Y. Fujiyoshi, F. Iwatsu and N. Uyeda, *Acta. Cryst.*, **A37**, 692 (1981).
107. J. D. Wright, *Prog. in Surf. Sci.*, **31**, 1 (1989).
108. H. Mockert, D. Schmeisser and W. Göpel, *Sensors and Actuators*, **19**, 159 (1989).
109. C. Maleysson, D. Bouché-Pillon, O. Thomas, J. P. Blanc, S. Dogo, J. P. Germain, M. Passard and A. Pauly, *Thin Solid Films*, **239**, 161 (1994).
110. K. Ukei, *Acta Cryst.*, **B29**, 2290 (1973).
111. M. K. Engel, *Rep. Kawamura Inst. Chem. Res.*, 11–54 (1997).
112. B. Bott and T. A. Jones, *Sensors and Actuators*, **5**, 43 (1984).
113. A. Wilson and J. D. Wright, *Molec. Cryst. Liquid. Cryst.*, **211**, 321 (1992).
114. A. Wilson, G. P. Rigby, J. D. Wright, S. C. Thorpe, T. Terui and Y. Maruyama *J. Mater. Chem.*, **2**, 303 (1992).
115. P. Roisin, J. D. Wright, R. J. M. Nolte, O. E. Sielcken and S. C. Thorpe, *J. Mater. Chem.*, **2**, 131 (1992).
116. A. Schütze, U. Weber, J. Zacheja, D. Kohl, W. Mokwa, M. Rospert and J. Werno, *Sensors and Actuators*, **A37–38**, 751 (1993).
117. A. Schütze, N. Pieper and J. Zacheja, *Sensors and Actuators*, **B23**, 215 (1995).
118. M. Trometer, R. Even, J. Simon, A. Dubon, J. Y. Laval, J. P. Germain, C. Maleysson, A. Pauly and H. Robert, *Sensors and Actuators*, **B8**, 129 (1992).
119. M. Passard, J. P. Blanc and C. Maleysson, *Thin Solid Films*, **271**, 8 (1995).
120. P. Bassoul, T. Toupance and J. Simon, *Sensors and Actuators*, **B26–27**, 150 (1995).
121. S. M. Sze, *Physics of Semiconductor Devices*, John Wiley & Sons, New York (1969).
122. M. Le Contellec, F. Morin, J. Richard, J. L. Favennec, M. Bonnel and B. Vinouze, in IEE Colloquium (April 1991).
123. D. E. Carlson and C. R. Wronski, in *Topics in Applied Physics*, Vol. 36, *Amorphous Semiconductors* (ed. M. H. Brodsky), Springer-Verlag, Berlin (1979).
124. M. J. Thompson, *J. Vac. Sci. Technol.*, **B2**, 827 (1984).
125. M. L. Petrova and L. D. Rozenshtein, *Soviet Phys. Solid State*, **12**, 756 (1970).
126. M L. Petrova and P. N. Zanadvorov, *Soviet Phys. Solid State*, **14**, 1581 (1972).
127. P. N. Zanadvorov and M. L. Petrova, *Soviet Phys. Solid State*, **21**, 1423 (1979).
128. F. Ebisawa, T. Kurokawa and S. Nara, *J. Appl. Phys.*, **54**, 3255 (1983).
129. R. Madru, G. Guillaud, M. Al Sadoun, M. Maitrot, C. Clarisse, M. Le Contellec, J.-J. André and J. Simon, *Chem. Phys. Lett.*, **142**, 103 (1987).
130. H. Laurs, G. Heiland, *Thin Solid Films*, **149**, 129 (1987).
131. H.Koezuka, A. Tsumura and T. Ando, *Synth. Metals*, **18**, 699 (1987).
132. N. Oyama, F. Yoshimura, T. Ohsaka, H. Koezuka and T. Ando, *Jpn J. Appl. Phys.*, **27**, 488 (1988).
133. A. Tsumura, H. Koezuka and T. Ando, *Synth. Metals*, **25**, 11 (1988).

134. C. Clarisse, M.-T. Riou, M. Gauneau and M. Le Contellec, *Electronic Lett.*, **24**, 11 (1988).
135. J. H. Burroughes, C. A. Jones and R. H. Friend, *Nature*, **335**, 137 (1988).
136. R. Madru, G. Guillaud, M. Al Sadoun, M. Maitrot, J.-J. André, J. Simon and R. Even, *Chem. Phys. Lett.*, **145** 343 (1988).
137. R. Madru, G. Guillaud, M. Al Sadoun, M. Maitrot, J.-J. André, J. Simon and R. Even, *C. R. Acad. Sci.*, **306**, 1427 (1988).
138. P. Petit, Ph. Turek, J.-J. André, R. Even, J. Simon, R. Madru, M. Al Sadoun, G. Guillaud and M. Maitrot, *Synth. Metals*, **29**, F59 (1989).
139. H. Koezuka and A. Tsumura, *Synth. Metals*, **28**, C753 (1989).
140. G. Horowitz, D. Fichou, X. Peng, Z. Xu and F. Garnier, *Solid State Commun.*, **72**, 381 (1989).
141. H. Koezuka and A. Tsumura, *Synth. Metals*, **28**, C753 (1989).
142. G. Guillaud, R. Madru, M. Al Sadoun and M. Maitrot, *J. Appl. Phys.*, **66**, 4554 (1989).
143. G. Guillaud, M. Al Sadoun, M. Maitrot, J. Simon and M. Bouvet, *Chem. Phys. Lett.*, **167**, 503 (1990).
144. J. Paloheimo, P. Kuivalainen, H. Stubb, E. Vuorimaa and P. Yli-Lahti, *Appl. Phys. Lett.*, **56**, 1157 (1990).
145. G. Horowitz, X. Peng, D. Fichou and F. Garnier, *J. Appl. Phys.*, **67**, 528 (1990).
146. X. Peng, G. Horowitz and D. Fichou, F. Garnier, *Appl. Phys. Lett.*, **57**, 2013 (1990).
147. F. Garnier, G. Horowitz, X. Peng and D. Fichou, *Advanced Mater*, **2**, 592 (1990).
148. M. Willander, A. Assadi and C. Svensson, *Synth. Metals*, **55**, 4099 (1993).
149. A. Assadi, M. Willander, C. Svensson and J. Hellberg, *Synt. Metals*, **58**, 187 (1993).
150. F. Garnier, X. Z. Peng, G. Horowitz and D. Fichou, *Molec. Engng*, **1**, 131 (1991).
151. F. Garnier, G. Horowitz, X. Z. Peng and D. Fichou, *Synth. Metals*, **45**, 163 (1991).
152. G. Horowitz and P. Delannoy, *J. Appl. Phys.*, **70**, 469 (1991).
153. G. Horowitz, X. Z. Peng, D. Fichou and F. Garnier, *J. Molec. Electron.*, **7**, 85 (1991).
154. G. Horowitz, D. Fichou, X. Z. Peng and F. Garnier, *Synt. Metals*, **41**, 1127 (1991).
155. G. Horowitz and P. Delannoy, *J. Chim. Phys., Phys. - Chim. Biol.*, **89**, 1037 (1992).
156. G. Horowitz, X. Z. Peng, D. Fichou and F. Garnier, Synth. Metals **51**, 419 (1992).
157. X. Z. Peng, G. Horowitz and F. Garnier, *J. Chim Phys., Phys. - Chim. Biol.*, **89**, 1085 (1992).
158. F. Garnier, A. Yassar, G. Horowitz and F. Deloffre, *Molec. Cryst. Liquid. Cryst.*, **230**, 81 (1993).
159. F. Garnier, A. Yassar, R. Hajlaoui, G. Horowitz, F. Deloffre, B. Servet, S. Ries and P. Alnot, *J. Am. Chem. Soc.*, **115**, 8716 (1993).
160. G. Horowitz, F. Deloffre, F. Garnier, R. Hajlaoui M. Hmyene and A. Yassar, *Synth. Metals*, **54**, 435 (1993).
161. B. Servet, S. Ries, M. Trotel, P. Alnot, G. Horowitz and F. Garnier, *Advanced Mater.*, **5**, 461 (1993).
162. A. Assadi, C. Svensson, M. Willander and O. Inganas, *Appl. Phys. lett.*, **53**, 195 (1988).
163. J. Paloheimo, E. Punkka, H. Stubb and P. Kuivalainen, in Proceeding of NATO Advanced Study Institute, Spetses, Greece (1989).
164. H. Akimichi, K. Waragai, S. Hotta, H. Kano and H. Sasaki, *Appl. Phys. Lett.*, **58**, 1500 (1991).
165. Y. Ohmori, K. Muro, M. Onoda and K. Yoshino, *Jpn J. Appl. Phys.* **31**, L646 (1992).
166. K. Nishimura, K. Kumagai, R. Nakamura and K. Kobashi, *J. Appl. Phys.*, **76**, 8142 (1994).
167. C. Pearson, A. J. Moore, J. E. Gibson, M. R. Bryce and M. C. Petty, *Thin Solid Films*, **244**, 932 (1994).

168. P. Hesto, L. Aguilhon, G. Tremblay, J. P. Bourgoin, M. Vandevyver and A. Barraud *Thin Solid Films*, **242**, 7 (1994).
169. G. Guillaud and J. Simon, *Chem. Phys. lett.*, **219**, 123 (1994).
170. F. Garnier, R. Hajlaoui, A. Yassar and P. Srivastrava, *Science*, **265**, 1684 (1994).
171. G. Guillaud, R. Ben Chaabane, C. Jouve and M. Gamoudi, *Thin Solid Films*, **258**, 279 (1995).
172. R. C. Haddon, *J. Am. Chem. Soc.*, **118** 3041 (1996).
173. G. Horowitz, F. Garnier, A. Yassar, R. Hajlaoui and F. Kouki, *Advanced Mater.*, **8**, 52 (1996).
174. A. Dodabalapur, L. Torsi and H. E. Katz, *Science*, **268**, 270 (1995).
175. G. Horowitz, *Advanced Mater.*, **8**, 177 (1996).
176. Z. Bao, A. J. Lovinger and A. Dodabalapur, *Appl. Phys. Lett.*, **69**, 3066 (1996).
177. R. C. Haddon, A. S. Perel, R. C. Morris, T. T. M. Palstra, A. F. Hebard and R. M. Fleming, *Appl. Phys. Lett.*, **67**, 121 (1995).
178. Z. Bao, A. J. Lovinger and A. Dodabalapur, *Advanced Mater.*, **9** 42 (1997).
179. A. R. Brown, A. Pomp, C. M. Hart and D. M. de Leeuw, *Science*, **270**, 972 (1995).
180. G. Guillaud, R. Ben Chaabane and M. Gamoudi, *L'Onde électrique*, **74**, 14 (1994).
181. G. Guillaud, J. Simon and J. P. Germain, *Coord. Chem. Rev.*, **178–180**, 1433 (1998).
182. F. Morin and M. Le Contellec, *Displays*, January 1983, p. 3.
183. G. W. Neudeck and A. K. Malhotra, *Solid State Electron.*, **19**, 721 (1976).
184. C. Clarisse, M. Le Contellec and M. -T. Riou, French Patent **87**, 15 490 (1987).
185. J. R. Burns, *RCA Rev*, **68**, 15 (1969).
186. P. M. Burr, P. D. Jeffery, J. D. Benjamin and M. J. Uren, *Thin Solid Films*, **151**, L111 (1987).
187. Liangyan Sun, Chanzhi Gu, Ke Wen, Xizhang Chao, Tiejin Li, Guoyuan Hu and Jiyun Sun, *Thin Solid Films*, **210/211**, 486 (1992).
188. W. P. Zhang, K. H. Kuo, Y. F. Hou and J. Z. Ni, *J. Solid State Chem.*, **74**, 239 (1988); **75**, 373 (1988).
189. D. Lexa and M. Reix, *J. Chimie Physique* **71**, 24 (1974).
190. C. M. Guzy, J. B. Raynor, L. P. Stodulski and M. C. R. Symons, *J. Chem. Soc. A*, 997 (1969).
191. M. Bouvet Thèse de doctorat, ESPCI, Paris (1992).
192. G. Guillaud, J. Simon, PCT/FR 99/01395 priority: FR 98/07383 (June 12, 1998).

7 Molecular Dielectrics

7.1 Cause, Effect, Cooperativity and Nonlinearity	298
7.1.1 Local Field	298
7.1.2 Nonlinear Effects: Experimental Observations	299
7.1.3 Mechanism Associated with Nonlinear Effects	300
7.2 Ferroelectricity	301
7.2.1 Introduction	301
7.2.2 History	302
7.2.3 An Inorganic Example: KH_2PO_4	303
7.2.4 Molecular Examples: Chiral Smectic C Phases	304
7.3 Pyroelectricity	308
7.3.1 Introduction	308
7.3.2 History	308
7.3.3 A Molecular Crystal: 1-Bromo-4-cyanobenzene	309
7.3.4 Chiral Smectic C Phases	311
7.4 Piezoelectricity	312
7.4.1 Introduction	312
7.4.2 A Polymeric Example: PVF_2	314
7.5 Polarizability and Hyperpolarizability in Optics	317
7.5.1 Introduction	317
7.5.2 Refraction	318
7.5.3 Absorption Band/Molecular Polarizability Relationship	321
7.5.4 Polarizability and Refractive Index	324
7.5.5 Ordinary and Extraordinary Indexes	325
7.5.6 Second Harmonic Generation (SHG)	328
7.5.7 Hyperpolarizability of a Molecular Unit: Experimental Results and Calculations	332
7.5.8 Frequency Doubling and Phase Matching	337
7.5.9 SHG in Single Crystals	338
7.6 References	340

*D'r Hàns im Schnokeloch
het àlles, wàs er will:
Un wàs er het, diss will er nitt,
Un wàs er will, diss het er nitt.
D'r Hàns im Schnokeloch
het àlles, wàs er will.*

*'Jean du Trou-aux-Moustiques
a tout ce qu'il veut:
ce qu'il a, il n'en veut pas,
ce qu'il veut, il ne l'a pas.
Jean du Trou-aux-Moustiques
a tout ce qu'il veut'.*

Traditional Alsatian Song
(R. Matzen, *Anthologie des expressions d' Alsace*, Rivages, Paris, 1989)

7.1 CAUSE, EFFECT, COOPERATIVITY AND NONLINEARITY

7.1.1 Local Field

When molecular units are gathered to make molecular assemblies, materials or condensed phases, the physical properties of these are not simply the sum of the effect induced on a single molecular unit by some cause. One of the simplest way to realize this is to consider the notion of *local field* ([1]; see also Section I of Ref. [2]). Let us examine, for example, the effect of an external electrical field on a condensed phase made originally of neutral (but polarizable) molecular units. An induced dipole is going to be created on every molecular unit (Figure 7.1).

Within the bulk, far from the surface and far from an arbitrary cavity in which the field is going to be determined, the induced positive and negative charges annihilate each other. However, the charges arising at the external surface of the sample do not (depolarization field E_1). In the same way, the induced charges at the surface of an arbitrary inner cavity also induce a field (E_2). One must finally consider the field generated by the units contained within the cavity (E_3) to obtain the overall local field, E_{loc}:

$$E_{loc} = E_0 + E_1 + E_2 + E_3 \qquad (7.1)$$

The contribution of E_1, E_2, E_3 is never negligible, even for nonpolar molecular units. It is generally assumed that only E_3 depends on the crystalline structure of the condensed phase.

In the further sections, we will assume the notion of local field to have a more extensive meaning. It will be the field at a given point within a condensed phase

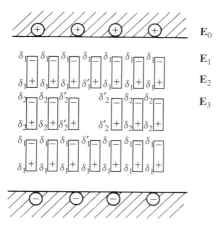

Figure 7.1 The various contributions to the local field ($\mathbf{E}_{loc}$): $\mathbf{E}_{loc} = \mathbf{E}_0 + \mathbf{E}_1 + \mathbf{E}_2 + \mathbf{E}_3$, where $\mathbf{E}_0$ = external applied field, $\mathbf{E}_1$ = depolarization field created by the charges on the external surface of the sample, $\mathbf{E}_2$ = field generated by the surface of the cavity considered within the sample (Lorentz's field), $\mathbf{E}_3$ = field generated by the units contained within the cavity (not represented)

resulting from the external cause and any type of internal (within the condensed phase) response.

7.1.2 Nonlinear Effects: Experimental Observations

Ferroelectric or ferromagnetic molecular materials can show nonlinear effects. Above the Curie temperature the interactions between the dipoles (when they can be defined) are less than the thermal energy kT and a paraelectric or paramagnetic phase is observed. At sufficiently low temperatures ($T < T_{curie}$), ordering occurs due to dipole–dipole interactions. In some cases, 'domains' of homogeneous orientation can form and the magnitude of the external field can vary the size of some domains relative to others (Figure 7.2).

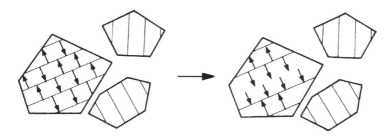

Figure 7.2 The transition between the paraelectric and the ferroelectric states of crystals

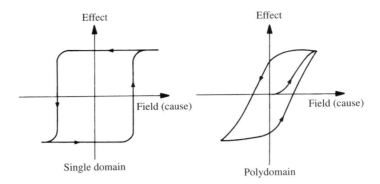

Figure 7.3 Hysteresis effects observed for single-domain or polydomain materials

By measuring the macroscopic physical properties of the material under the influence of the field, a hysteresis effect is observed. Two stable states can be reached and there is a nonlinear transition between these two states (Figure 7.3).

The hysteresis phenomenon can be qualitatively explained in a simple way in the case, for example, of ferroelectricity. The external field extends the size of some domains at the expense of others. In this way a polar state can be reached since the population of one domain is larger than the other. When the field is removed, because the transformation of one type of domain to another one implies boundary migrations (and therefore energy), the polar state is maintained. It is necessary to apply a field in the reverse direction to cancel the polarization.

7.1.3 Mechanism Associated with Nonlinear Effects

A general explanation can be given to the previous observations [3, 4]. In order to have a *bistable* behaviour, it is necessary to have simultaneously two processes:

(a) nonlinear relationship between the cause and the effect,
(b) feedback action of the effect on the cause.

This can be written as

$$\text{Effect} = f(\text{cause}, \text{effect})$$

where f is a nonlinear function; f can be taken as a polynomial of the type:

$$ax^3 + bx^2 + cx + d$$

In this case, the cause/effect relationship indeed indicates that in some area (II) of the graph in Figure 7.4, three stationary states are possible for a given magnitude of the cause whereas only one state is found in the other areas (I, III). This type of curve leads to a hysteresis: when the magnitude of the *cause* increases, the *effect* must increase. This implies that at certain points (A and C

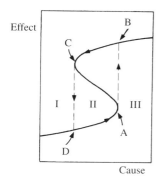

Figure 7.4 The cause/effect plot for a cubic function f

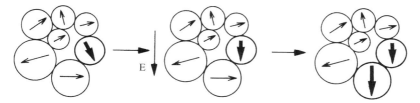

Figure 7.5 The orientation of the dipoles in one domain facilitates the orientation of a neighbouring domain

in Figure 7.4) one must jump directly from one part of the curve to the other instead of going back.

The nature of the feedback effect can be visualized. The field tends to increase the size of some of the domains. These domains will in turn facilitate the creation of identical domains because of the local field they generate: external and internal fields cooperate to favour some of the domains (Figure 7.5).

7.2 FERROELECTRICITY

7.2.1 Introduction

In some molecular single crystals, the molecular electrical dipole moments can be packed in such a way that polar materials can be obtained (see, for example, Refs. [5] and [6]). However, this is not sufficient to find ferroelectric behavior. It is also necessary to be able to switch between two different stable states. In conventional polar van der Waals crystals, this is not possible due to the large volume of the elementary polar units. It will be shown that such a process is, however, possible in liquid crystalline phases. It will also be seen further on that ferroelectricity is in general due to small movements of ions under the influence of the electrical field. These can be found in mixed organic/inorganic materials like 'sel polychreste' [7].

In terms of symmetry requirement, ferroelectric materials are among the easiest to describe. Under the application of all symmetry elements of a given crystalline class, the polar axis must remain unchanged. This means that x and/or y and/or z must be associated with the totally symmetrical irreducible representation usually noted A. The symmetry group $C_{\infty v}$ and its subgroups (C_∞, C_{nv}, C_n, C_s, C_1) fulfil this requirement.

7.2.2 History (after Ref. [7])

In approximately 1665, Elie Seignette fabricated a salt he called 'sel polychreste' which means 'salt with various uses'. About 65 years later, the French pharmacist and chemist Simon Boulduc found out by analysis that the 'sel polychreste' was some 'tartaric derivative of soda' (Figure 7.6).

In 1880, Jacques and Pierre Curie [9] discovered the phenomenon of piezo-electricity and they noticed that the 'Rochelle salt' ('sel de Seignette') was by far more active than other substances like quartz. F. Pockels in 1893 [10] studied in detail the influence of an electrical field on the optical properties of various crystals including 'sel de Seignette'. He then discovered what is now known as the linear electro-optic effect (Pockels' effect).

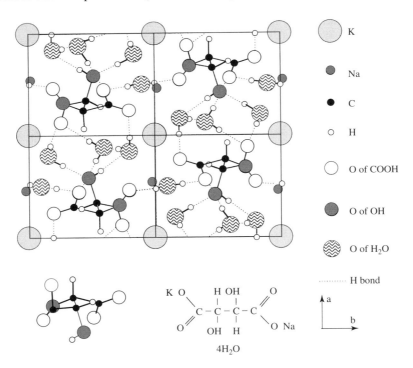

Figure 7.6 Structure of 'sel de Seignette' (also called 'sel polychreste' or 'Rochelle salt') [8]

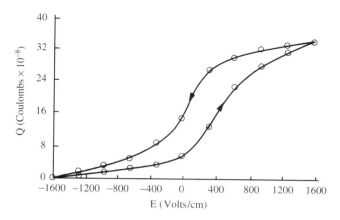

Figure 7.7 Hysteresis loop observed in 1920 by J. Valasek with Rochelle salt (sel de Seignette). (From Refs. [5] and [13])

In 1900, F. Beaulard [11] published a paper entitled 'Sur l'hystérésis diélectrique', which dealt with the study of the dielectric constant of solids as a function of the applied electrical field. He found a dielectric hysteresis analogous to ferromagnetic hysteresis loops with a material called 'diélectrine' (a mixture of paraffin and sulphur).

Joseph Valasek [12] in 1920 discovered that a hysteresis of the electrical polarization analogous to magnetic hysteresis can be observed with 'Rochelle salt' (Figure 7.7). However, works carried out until 1934 demonstrated that 'in view of the complex crystal structure of Rochelle salt, no real physical explanation for the experimental results could be provided beyond the idea that some kind of unknown electric dipoles might be oriented in an electric field or even spontaneously' [7].

In 1935, G. Busch and P. Scherrer discovered that KH_2PO_4 (often referred to as KDP for potassium dihydrogen phosphate) was indeed ferroelectric [14]. Busch and Scherrer suggested that the hydrogen bonds between adjacent oxygen atoms were responsible for ferroelectricity. J. Slater in 1941 provided the first significant molecular theory of ferroelectricity involving hydrogen bonds [15]. As an example, we will give the main results found for KH_2PO_4.

7.2.3 An Inorganic Example: KH_2PO_4

At 121 K, KH_2PO_4 undergoes a transition yielding a ferroelectric phase in which the polarization is along the c axis of a tetragonal unit cell (Figure 7.8):

$$\bar{4}2\,m(D_{2d}) \xrightarrow{121\,K} mm2(C_{2v})$$

The structure of KH_2PO_4 is made of tetrahedral $(PO_4)^{3-}$ in interaction with potassium ions. They are linked between each other via hydrogen bonds.

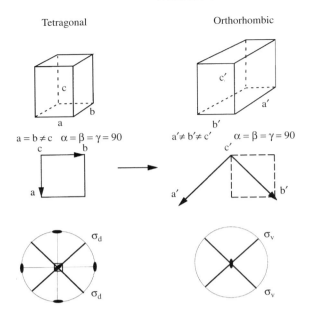

Figure 7.8 Distortion occurring at 121 K in KH_2PO_4

Neutron diffraction studies [16, 17] have shown that:

(a) At $T > 121$ K the hydrogen are disordered on positions close to oxygen atoms.
(b) At $T < 121$ K the hydrogen are ordered, forming O–H bonds with an oxygen atom of a tetrahedron and H-bonds with another one (Figure 7.9).

This temperature corresponds to the paraelectric–ferroelectric transition. The symmetry group D_{2d} is not polar (it is not a subgroup of $C_{\infty v}$) but C_{2v} is. Concomitant to the ordering of the protons, a cooperative displacement of the heavier ions occurs (Figure 7.10).

7.2.4 Molecular Examples: Chiral Smectic C Phases

The use of liquid crystalline phases (mesophases) allows the molecular reorientations necessary to obtain ferroelectricity. 'Speculations' on the possibility of obtaining ferroelectric mesophases were reported in 1973 by McMillan [18]. However, symmetry arguments were later convincingly used to demonstrate that smectic C and H liquid crystals can yield ferroelectrics using optically pure chiral mesogens. At the same time, the synthesis of the corresponding compounds was achieved, confirming previous predictions [19].

Smectic C phases have a symmetry C_{2h}. Since this group is not a subgroup of $C_{\infty v}$ it cannot yield a polar condensed phase. However, if optically pure chiral

MOLECULAR DIELECTRICS

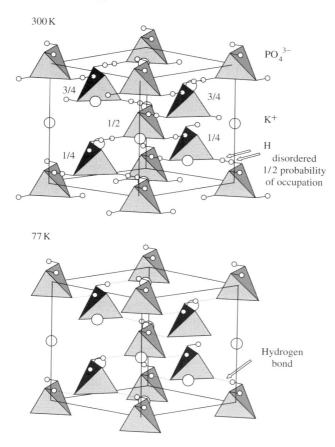

Figure 7.9 Crystalline structures of the high- and low-temperature phases of KH_2PO_4

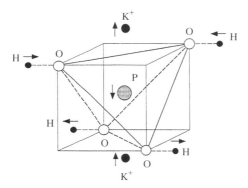

Figure 7.10 Atomic displacements occurring during the paraelectric–ferroelectric transition at 121 K in KH_2PO_4

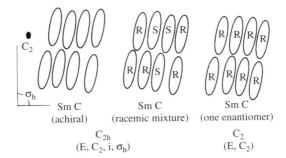

Sm C (achiral)

Sm C (racemic mixture)

Sm C (one enantiomer)

C_{2h}
(E, C_2, i, σ_h)

C_2
(E, C_2)

Figure 7.11 Symmetry of achiral and chiral smectic C phases. C_2 is a subgroup of $C_{\infty v}$ and the polar axis must be parallel to the C_2 axis

molecular units are used, the symmetry plane perpendicular to the C_2 axis is removed, leading to the symmetry C_2 (Figure 7.11).

By considering, for simplicity reasons, a (quasi) two-dimensional molecular shape, a unit must have a polar vector perpendicular to the plane of the molecular unit in order to be chiral (Figure 7.12, case c).

A smectic-type layer may be formed with the chiral two-dimensional molecular units. In this case, the dipole moments perpendicular to the plane are all oriented in the same direction. When a field is applied in order to reverse the polar direction, the molecular units must change the angle with the normal of the layer (Figure 7.13).

The Smectic C phase is in general fairly fluid, allowing molecular translations and rotations. In particular, rotation around the long axis of the molecular unit is possible. However, some concerted rotations between neighbours must be effective to avoid an average to zero of the polarity.

The molecular units yielding chiral smectic C liquid crystalline phases (see Figure 7.14) always have a chiral centre situated not too far from the central

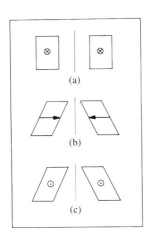

Figure 7.12 Achiral (cases a and b) and chiral (case c) (quasi) two-dimensional molecular units. (After Ref. [20])

MOLECULAR DIELECTRICS

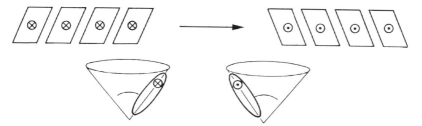

Figure 7.13 Schematic representation of the effect of an electrical field on a chiral smectic C mesophase (the electrical field is perpendicular to the plane)

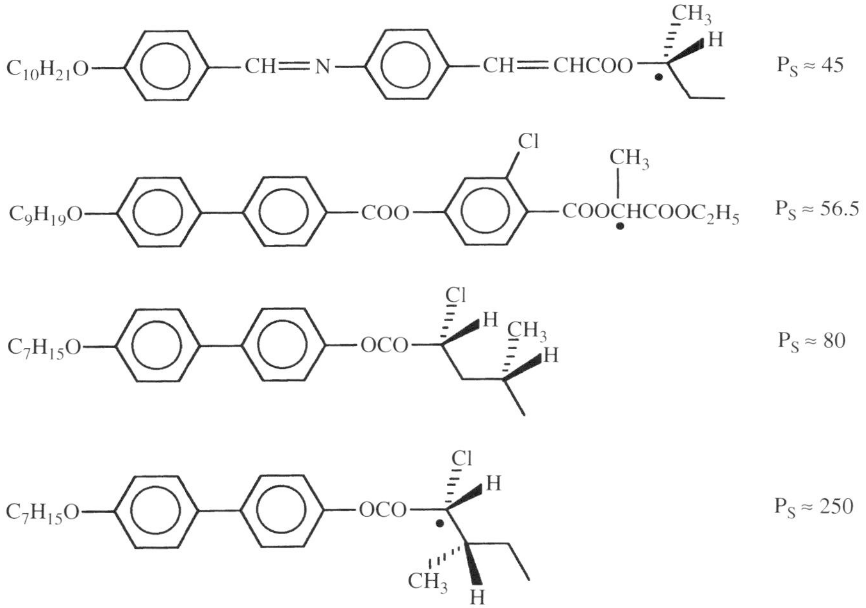

$C_{10}H_{21}O-\bigcirc-CH=N-\bigcirc-CH=CHCOO-\overset{CH_3}{\underset{H}{C}}$ $\quad P_S \approx 45$

$C_9H_{19}O-\bigcirc-\bigcirc-COO-\bigcirc(Cl)-COOCH(CH_3)COOC_2H_5$ $\quad P_S \approx 56.5$

$C_7H_{15}O-\bigcirc-\bigcirc-OCO-\overset{Cl,H}{\underset{}{C}}\overset{CH_3}{\underset{H}{}}$ $\quad P_S \approx 80$

$C_7H_{15}O-\bigcirc-\bigcirc-OCO-\overset{Cl,H}{\underset{CH_3,H}{}}$ $\quad P_S \approx 250$

Figure 7.14 Some of the molecular units used for making chiral smectic C phases and the corresponding spontaneous polarization measured (after Ref. [21]). The first derivatives synthesized can be found in Ref. [22]

rigid core [21, 22]. The chiral nature of the constitutive molecular unit and the polarity of the mesophase produce a small rotation of the average inclination angle of the molecules from one layer to another. The pitch is generally very large compared to a monolayer thickness; it represents around 10^4–10^5 layers. The spontaneous polarization must therefore be measured with thin-film cells in which the pitch is larger than the cell thickness. The electrodes — in many cases glass plates covered with indium tin oxide (ITO) — are also processed with aligning agents.

7.3 PYROELECTRICITY

7.3.1 Introduction

Pyroelectricity is related to an electrical polarization change when the temperature is varied:

$$\Delta P_i = p_i \Delta T \tag{7.2}$$

The Curie principle can be applied by considering that the cause ΔT is isotropic and remains unchanged by any operation of the symmetry group of the material. ΔT transforms as the irreducible representation A of the point symmetry group of the material. The effect is a vectorial phenomenon and the components x and/or y and/or z must be associated with the irreducible representation A. The condensed phase must therefore be polar and must belong to the $C_{\infty v}$ symmetry group or one of its subgroups. Pyroelectrics and ferroelectrics are both polar but, in the former, the polarity is not necessarily switched by an external electrical field. In most cases, constant crystal dimensions (the primary pyroelectric coefficient) are assumed. If the dimensions of the molecular material change [23], secondary piezolectricity may result from anisotropic dilatations.

7.3.2 History

Pyroelectricity has been (indirectly) reported in ancient times. The Greek philosopher Theophrastos (372–287 BC) observed a material 'hard as a real stone and with an attraction power like amber' (cited in Ref. [24]). Dutchmen in 1703 imported from Ceylon a stone called 'tourmaline' which has the property of becoming electrified by heating and then rapid cooling. This stone can attract small particles as well as small pieces of paper. They also observed that tourmaline could attract ashes from their pipe and they gave it the nickname 'Aschentrekker' (that which attracts ashes) [25] (Figure 7.15, Plate 5).

The formula of tourmaline is $(Na, Ca)(Li, Al)_3 Al_6 (OH)_4 (BO_3)_3 Si_6 O_{18}$. The structure (space group R3m, crystalline class 3m or C_{3v}) belongs to a polar symmetry group. Many other cations derived from Mn, Mg, Fe, Ti, Cr, etc., can be incorporated in the lattice and yield many different colors from yellow to blue. The structure may be described by the packing of three rings [26] (Figure 7.16, Plate 5).

The sequences ABC, CAB, BCA are crosslinked by Al^{3+} ions to form sheets. Sodium ions can be found between the sheets. The structure is polar with all the silica tetrahedra oriented in the same direction, pointing up the c axis. This is also the case for the NaO_9 polyhedra and Mg and AlO(OH) octahedra [26]. Hydrogen atoms on OH^- also point up the c axis on $(OH)_3$ ions but down for (OH) ions (see Figure 7.16).

The value of the pyroelectric coefficient of tourmaline is 4×10^{-16} C/m² K. A temperature change of 1 °C is equivalent to a field of 7.4×10^4 V/m applied to the crystal.

7.3.3 A Molecular Crystal: 1-Bromo-4-cyanobenzene

One can consider a material constituted of molecular units possessing a permanent dipole moment μ and polarizability α. In the condensed phase an internal (or local) field $\mathbf{E}_{loc}$ results from:

(a) the electrical field produced by the permanent dipoles of the neighbouring molecules:

$$\mathbf{E}_N = \frac{1}{4\pi\varepsilon_0} \sum_{N_j} \frac{q_j \mathbf{r}_j}{|\mathbf{r}_j|^3} \qquad (7.3)$$

(b) the Lorentz's field $\mathbf{E}_L$ defined in Figure 7.1:

$$\mathbf{E}_{loc} = \mathbf{E}_N + \mathbf{E}_L \qquad (7.4)$$

The total dipole moment of the molecular unit in the condensed phase is thus given by

$$\boldsymbol{\mu}_{tot} = \boldsymbol{\mu} + \alpha \mathbf{E}_{loc} \qquad (7.5)$$

P is the corresponding macroscopic polarization of the condensed phase resulting from the summation of the individual dipole moments, μ_{tot}. Two angles may be defined:

(a) the δ angle between $\boldsymbol{\mu}$ and **P**,
(b) the β angle between $\mathbf{E}_N$ and **P**.

If one neglects the size changes of the material with temperature, the pyroelectric coefficient at constant stress may be written as [24]

$$\frac{d\mathbf{P}}{dT} = \mathbf{p} = \mathbf{p}_1 + \mathbf{p}_2 + \mathbf{p}_3 \qquad (7.6)$$

where $\mathbf{p}_1$ is proportional to $\partial(\cos\delta)/\partial T$ and $\mathbf{p}_2$ to $\partial(\cos\beta)/\partial T$. These terms originate in the change of the molecular dipole moment through motions. $\mathbf{p}_3$ corresponds to $\partial\varepsilon_p/\partial T$ where ε_p is the dielectric constant of the material. Detailed studies have been reported for single crystals of 1-bromo-4-cyanobenzene [24]. The growth and purification of the crystal were carried out by the zone-melting technique. This compound crystallizes in the space group Cm (crystalline class m or C_s) (Figure 7.17). The axis of the polarization **P** is within the mirror plane.

The pyroelectric coefficient **p** of 1-bromo-4-cyanobenzene has been experimentally determined and compared to calculations (INDO-SCF) [24] (Figure 7.18). The increase of **p** at $T > 320\,K$ can be due to some molecular dipole motions changing their mean direction within the (a, c) plane.

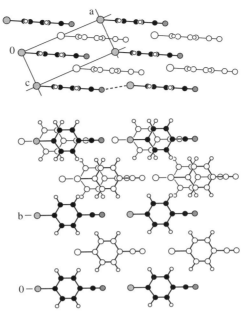

Figure 7.17 Structure of 1-bromo-4-cyanobenzene. Monoclinic; $a = 9.55$, $b = 8.58$, $c = 4.15$ Å, $\beta = 90.03°$; space group Cm, $Z = 2$ (class m, C_s). (From Ref. [27])

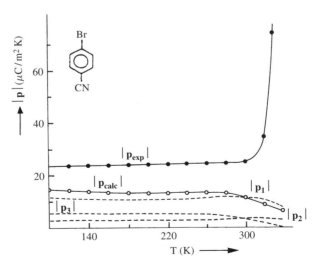

Figure 7.18 Experimental pyroelectric coefficient p_{exp} compared to calculated values see equation (7.6). (Modified from Ref. [24])

7.3.4 Chiral Smectic C Phases

As an example, we will detail the pyroelectric properties of chloro-2-propyl-p'-hexyloxybenzylidene-p'-aminocinnamate:

$$C_6H_{13}O-\langle\bigcirc\rangle-CH=N-\langle\bigcirc\rangle-CH=CH$$

with substituents Cl, CO_2, CH_3, $\overset{*}{C}H$, CH_2

This compound forms two ordered smectic phases [28] as well as conventional chiral smectic C and smectic A phases:

$$K^* \rightleftharpoons S^{1*} \rightleftharpoons S^{2*} \rightleftharpoons S_C^* \rightleftharpoons S_A \rightleftharpoons I$$

The phase S^{2*}, sometimes called S_G^*, presents a two-dimensional hexagonal lattice. The axes of the columns are tilted with respect to the lattice, as shown in Figure 7.19.

The pyroelectric coefficient $\gamma = dP_s/dT$ (P_s is spontaneous polarization) was determined by locally heating a liquid crystalline layer containing a small amount (<0.5%) of a dye by irradiation at 1.06 μm [29]. A d.c. external electric voltage is applied to unwind the helical structure. The temperature dependence of γ is shown in Figure 7.20.

The pyroelectric coefficient varies drastically at both $S_G^* \rightarrow S_C^*$ and $S_C^* \rightarrow S_A$ transitions because of the high molecular mobilities of the units in these frontier regions. For the same reason, the pyroelectric coefficient is larger in the S_C^* domain than in the S_G^* phase. On the contrary, the spontaneous polarization is higher in the S_G^* phase than in S_C^* because of the better average orientation. The S_A phase is not polar for symmetry reasons. In S_G^* and S_C^*, the polar axis is parallel to the C_2 axis of the condensed phases. By varying the laser pulse duration, information concerning the characteristic time of movement of the director tilt can be gained [29].

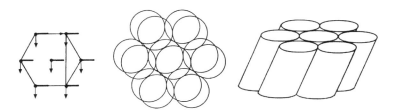

Figure 7.19 Schematic representation of the S_G^* phase of chloro-2-propyl-p-hexyloxybenzylidene-p'-aminocinnamate. The arrows indicate schematically the dipole direction

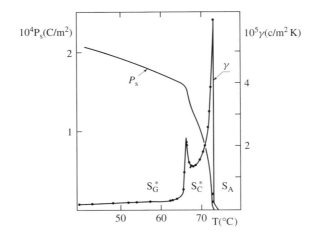

Figure 7.20 Temperature dependence of the pyroelectric coefficient γ and the spontaneous polarization P_s for chloro-2-propyl-p-hexyloxybenzylidene-p'-aminocinnamate. (From Ref. [29])

7.4 PIEZOELECTRICITY

7.4.1 Introduction

Piezoelectricity (or pressure-induced electricity) comes from the Greek 'piezo' (to press). It was coined in 1880 by Pierre and Jacques Curie by studying quartz, blende, tourmaline and related crystals [9]. They also noted that Rochelle salt was more active than quartz. Piezoelectric materials were used nearly 50 years after their discovery by Langevin, who applied the piezoelectric properties of quartz for underwater sound signaling and receiving [30].

The cause/effect relationship [31, 32] can be written in that case as

$$P_i = d_{ijk}\sigma_{jk} \tag{7.7}$$

where P_i is the i component of the polarization **P** and σ_{jk} is the stress applied on the material. The elementary stresses shown in Figure 7.21 may be found.

The piezoelectric coefficient d_{ijk} differs from zero whenever the cause (jk) and the effect (i) are associated with the same irreducible representation in the point symmetry group of the condensed phase. To examine this point, we will determine the consequences of a stress applied on an apolar condensed phase. In appropriate symmetry groups, the crystal can become polar under the influence of the stress.

The structure of KH_2PO_4 has already been described in Section 7.2.3. At room temperature the stable phase is not ferroelectric. Its symmetry group D_{2d}, although not polar, is compatible with piezoelectric properties.

The use of the symmetry group approach allows the nonzero tensor coefficients to be predicted (Table 7.1). Only the irreducible representations B_2 and

MOLECULAR DIELECTRICS

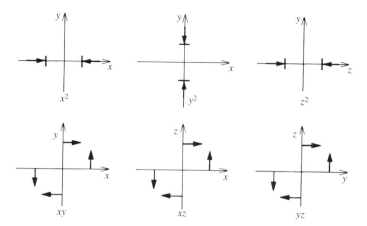

Figure 7.21 The various elementary stresses

Table 7.1 Prediction of the nonzero tensor coefficients for piezoelectricity

D_{2d}	E	$2S_4$	C_2	$2C_2'$	$2\sigma_d$	Effect	Cause
A_1	1	1	1	1	1		$x^2 + y^2$, z^2
A_2	1	1	1	−1	−1	R_z	
B_1	1	−1	1	1	−1		$x^2 - y^2$
B_2	1	−1	1	−1	1	z	xy
E	2	0	−2	0	0	(x, y), (R_x, R_y)	(xz, yz)

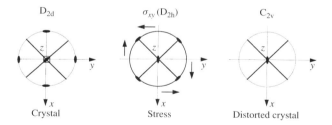

Figure 7.22 Stereographical representation of the point symmetry group D_{2d} of the crystal, of a σ_{xy} shear stress indicated by arrows (D_{2h}) and a distorted crystal (C_{2v})

E associate a vectorial effect (components x, y or z) with a stress perturbation (xy and (xz, yz)). However, as already encountered, the order of the irreducible representation E (2) does not allow direct determination of the nonzero tensor coefficients. The method taken in Section 4.3.3 can be followed. However, by coming back to the original Curie principle, another method can be used. A shear stress σ_{xy} has a symmetry D_{2h}. The binary axis is taken along Oz and the two other axes form diagonals of the Ox, Oy axes (Figure 7.22). When a σ_{xy} shear stress is applied to the crystal of symmetry D_{2d} only the C_2 axis (contained in S_4) and the two mirror planes are conserved, leading to the subgroup C_{2v}. Consequently, a polarity can only appear along the z axis:

$$P_z = d_{zxy}\sigma_{xy}$$

When σ_{xz} and σ_{yz} stresses are applied, the unchanged crystalline axes are y and x respectively. The uniaxial stresses of symmetry x^2, y^2, z^2 cannot give rise to any polarization. The tensor that describes the piezoelectric properties of KH_2PO_4 therefore possesses three terms [33]:

$$\begin{pmatrix} \cdot & \cdot & \cdot & d_{xyz} & \cdot & \cdot \\ \cdot & \cdot & \cdot & \cdot & d_{yxz} & \cdot \\ \cdot & \cdot & \cdot & \cdot & \cdot & d_{zxy} \end{pmatrix}$$

7.4.2 A Polymeric Example: PVF₂

Poly(vinylidene fluoride), abbreviated to PVF₂, whose chemical formula is $(CH_2-CF_2)_n$, exhibits high piezoelectric activities [34]. As in the previous cases, an attempt will be made to relate the molecular (microscopic) properties of this material to some macroscopic physical property.

The van der Waals radius of fluorine (1.35 Å) is slightly larger than that of hydrogen (1.20 Å) and the C–F bond is polar (1.9D) due to the difference of electronegativity between C (2.55) and F (3.98). For the chain configuration, the following notation is used:

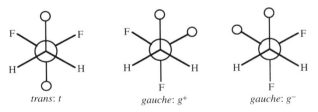

where o is the continuation of the polymeric chain. 95 to 97% of the polymeric material show a regular $(-CH_2-CF_2-)$ chain without head-to-head or tail-to-tail units. PVF₂ adopts (at least) three regular conformations of comparable energies: all-*trans*, tg^+tg^- and $tttg^+tttg^-$. The first two are the most common ones [34] (Figure 7.23). The tg^+tg^- conformation yields components of the dipole moment both parallel and perpendicular to the chain axis, while the all-*trans* conformation has dipoles essentially normal to the molecular axis [34].

Chain packing is also important for ensuring polar properties of the polymeric material. The most common phase, called α, is obtained by melt-solidification: this phase shows a cancellation of the dipolar moment (Figure 7.24). A polar analog of the α-phase, called δ, can be obtained by applying an electrical field (Figure 7.25). The most polar phase of PVF₂ is the β-phase, in which the polymeric chains are in an all-*trans* conformation (Figure 7.26).

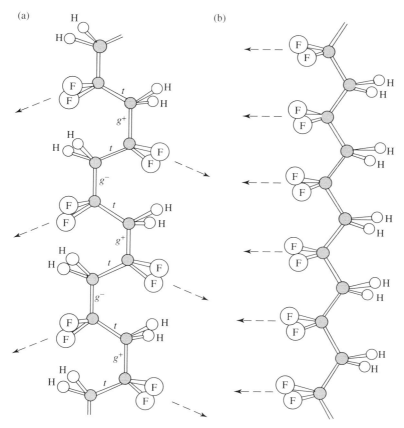

Figure 7.23 Schematic representation of the two most common crystalline chain conformations in PVF$_2$: (a) tg^+tg^- and (b) all-*trans*. The arrows indicate projections of the $-CF_2$ dipole directions on planes defined by the carbone backbone. (After Ref. [34])

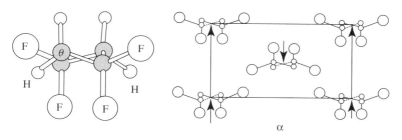

Figure 7.24 One molecule of PVF$_2$ in the tg^+tg^- conformation and schematic representation of the α-phase. (Redrawn from Refs. [30] and [34] with permission from Elsevier Science)

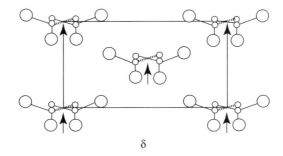

Figure 7.25 Schematic representation of the δ-phase of PVF$_2$. (Reprinted from Ref. [30] by permission of Elsevier Science)

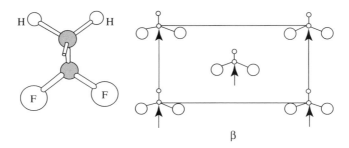

Figure 7.26 The β-phase of PVF$_2$. The chains are in an all-*trans* conformation. (Reprinted from Refs. [30] and [34] by permission of Elsevier Science)

Table 7.2 Piezoelectric coefficients of PVF$_2$ and, for comparison, of a few mineral compounds. For PVF$_2$ the first subscript indicates the electrical poling direction and the second one the draw direction

	Piezoelectric coefficient (pC/N)
PVF$_2$ (β-phase)	$d_{31} = 20$–30 $d_{32} = 2$–3 $d_{33} = -30$
PVF$_2$ (δ-phase)	$d_{31} = 10$–17 $d_{32} = 2$–3 $d_{33} = 10$–25
Quartz	$d_{11} = 2$
Rochelle salt	$d_{11} = 150$
Tourmaline	$d_{11} = 2$

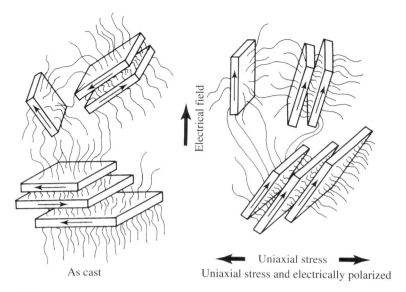

Figure 7.27 Schematic representation of an as-cast polymer and the same material oriented by stretching (the draw direction is indicated) and by application of an electrical field (poling). (Reprinted with permission from Ref. [30] by permission of Elsevier Science)

The most common technique for obtaining macroscopically polar films of PVF$_2$ consists in (a) stretching the films with a mechanical extension and (b) the application of an electrical field to orient the dipole moments (poling) (Figure 7.27). The application of a high poling field has been reported to produce concomitantly dipole alignment and transition to the β-phase [30]. The δ-phase may also be obtained. The piezoelectric coefficients are usually noted d_{31}, d_{32} and d_{33}, the first subscript defining the electrical field direction and the second one the mechanical stress with the following notation: 1, 2, 3, x, y, z for the first subscript and xx, yy and zz for the second one (see Appendix 8) (Table 7.2).

7.5 POLARIZABILITY AND HYPERPOLARIZABILITY IN OPTICS

7.5.1 Introduction

In the following paragraphs only a qualitative description of the phenomena will be given. A demonstration of the corresponding equations has been the object of many books, or even of series of books, and is far beyond the scope of this manuscript.

An electromagnetic wave can be considered, for the properties we deal with here, as an oscillating electrical field propagating in the medium:

$$E = E_0 \cos[2\pi\nu(t - \varphi)] \tag{7.8}$$

When the wave is observed at a distance r, the time necessary to reach this point is given by $r/c = \varphi$, the phase difference between the origin and the point considered. The time dependence of the physical phenomena must now be considered.

7.5.2 Refraction

W. Snell van Royen (1580–1626) first discovered the law of refraction of light, studied latter on by R. Descartes (1596–1650). A light ray is deflected when passing from a medium of refractive index n_1 to another one of refractive index n_2 as follows:

$$n_1 \sin\theta_1 = n_2 \sin\theta_2$$

where θ_1, θ_2 are the angles between the light beam and the normal of the plane. A molecular interpretation of this effect may be given on qualitative grounds (Figure 7.28).

An electromagnetic wave may be considered to arise from an oscillating dipole moment:

$$\mu = \mu_0 \cos 2\pi\nu t \tag{7.9}$$

where

$$\mu = qd$$

and

q = charge constituting the dipole
d = intercharge distance

The magnitude of the electrical field thus generated varies as slowly as $1/r$, where r is the distance between the source and the point considered. As a consequence, a galaxy situated at a 5 billion light year distance is detectable with a radiotelescope [35].

The wavelength of visible photons is comprised of between 3000 and 7000 Å while typical molecular units have dimensions of the order of 10 Å. Each molecular unit therefore experiences an electrical field constant in space but periodically varying with time. In the present case the incident photon is considered not to be absorbed but it only polarizes the molecular units (Figure 7.28).

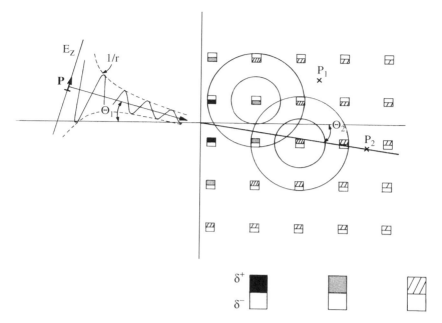

Figure 7.28 Molecular description of the phenomenon of refraction (largely modified from Ref. [35]). (Note that the representation does not respect the relative sizes of the constituents)

The field in a point P_i within the condensed phase is a result of the incident electromagnetic wave and the one resulting from each polarized molecular unit. These units possess an induced dipole moment and oscillate at the same frequency as the source (forced vibration regime) but with a difference of phase $\varphi = r/c$ which depends on the distance r from the excitating source. The combination of the electrical fields arising from the source and from the various polarized molecular units yields interferences. If the molecular units can be considered as point dipoles, their size being negligible compared to the wavelength of the excitating light, the interferences will be destructive in all points of the condensed phase (P_1) except in a direction that corresponds to the refracted light beam (P_2).

The molecular units have their own frequency of oscillation which corresponds to the absorption of light:

$$E_{\text{abs}} = h\nu = hc/\lambda \tag{7.10}$$

Numerically that is

$$E(\text{kcal/einstein}) = \frac{28\,560}{\lambda(\text{nm})}$$

where 1 einstein = $\mathcal{N}$ quanta.

The first transition — unless symmetrically forbidden — corresponds to the HOMO/LUMO energy difference. The interaction of an electromagnetic wave with a molecular unit can therefore be separated into two processes:

(a) When the energy of the photon is smaller than the HOMO/LUMO difference, the molecular unit is simply polarized: it is the domain of transparency of the material (out-of-resonance process).
(b) The photon is absorbed when the light energy and the HOMO/LUMO difference (or other couples of molecular orbitals) are equal (resonance process).

The interaction between the light and a molecular unit may be modelled by postulating that the electrical force:[†]

$$F_e = qE = -eE_0 \cos 2\pi \nu t \quad (7.11)$$

is counterbalanced by a force proportional to the displacement Δx of the charge:

$$F = -k_c \, \Delta x \quad (7.12)$$

Newton's law gives

$$m \frac{\partial^2 \Delta x}{\partial t^2} = -k_c \, \Delta x - eE_0 \cos 2\pi \nu t$$

In the forced regime,

$$\Delta x(t) = A \cos 2\pi \nu t$$

Then

$$-A(2\pi \nu)^2 \cos 2\pi \nu t + \frac{k_c}{m} A \cos 2\pi \nu t = -\frac{e}{m} E_0 \cos 2\pi \nu t$$

$$A = -\frac{e}{m} \frac{E_0}{k_c/m - \omega^2} \quad \text{with } \omega = 2\pi \nu$$

In the case where there is no external force, the equation corresponding to the harmonic oscillator is

$$m = -\frac{\partial^2 \Delta x}{\partial t^2} = -k_c \, \Delta x$$

with a solution of the form

$$\Delta x = \cos 2\pi \nu_0 t \qquad \omega_0 = 2\pi \nu_0$$

It can be readily found that

$$\omega_0^2 = k_c/m$$

[†] The positive charges are displaced in the direction of the electric field vector, the negative charges in the opposite direction.

It follows that

$$\Delta x(t) = -\frac{eE_0}{m(\omega_0^2 - \omega^2)} \cos \omega t \qquad (7.13)$$

The demonstration has been taken from Ref. [36].

In molecular terms, the energy $h\nu_0$ corresponds to one of the transitions possible between the ground and excited electronic states. The distance $\Delta x(t)$ may be understood as a displacement of the negative cloud away from positive charges of the nuclei. It is therefore proportional to an induced dipole:

$$\mu_{ind} = -e\Delta x$$

and therefore

$$\boxed{\mu_{ind} = \frac{e^2 E_0}{m(\omega_0^2 - \omega^2)} \cos \omega t} \qquad (7.14)$$

At a macroscopic scale, the induced polarization of the condensed phase $\mathbf{P}_{ind}$ is the vectorial summation of the individual induced molecular dipole moments:

$$\mathbf{P}_{ind} = \sum_i \mu_{ind}^i \qquad (7.15)$$

7.5.3 Absorption Band/Molecular Polarizability Relationship

It has been seen in the previous section that the absorption of a photon by a molecular unit occurs at the resonance between the excitating radiation and the electronic excitation of the molecular unit: the material is nontransparent and there is an energy transfer from the radiation to the material. Out-of-resonance (transparency domain), the molecular unit is merely polarized and no energy transfer occurs between the electromagnetic radiation and the molecular unit.

The polarizability coefficient is

$$P_i = \alpha_{ij} E_j \qquad \text{with } \alpha = \varepsilon_0 \chi$$

where χ is the dielectric susceptibility, can be derived from the characteristics of the absorption bands; in general one considers only the closest band to the incident electromagnetic radiation. The polarization effect will tend to more or less extensively distort the electronic cloud towards a distribution existing in the excited state; this is exemplified in Figure 7.29. Depending on the field and on the energy of the excitating radiation, the electronic cloud is more or less distorted towards the one in the excited state.

Absorption properties are usually expressed as a function of the transition moment $\mathbf{R}^{nm}$, which is a vectorial quantity having magnitude and direction. It is given by

$$\mathbf{R}^{nm} = \int \Psi_n^* \mu \Psi_m \, d\tau \qquad (7.16)$$

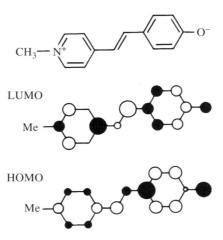

Figure 7.29 The electron density in the HOMO and LUMO orbitals of a betaine (CNDO/INDO calculations). (After Ref. [37])

when a transition between the electronic states m and n is considered. $\boldsymbol{\mu}$ is the electric dipole moment operator and is also a vector:

$$\boldsymbol{\mu} = \sum_i q_i \mathbf{r}_i \qquad (7.17)$$

where q_i, $\mathbf{r}_i$ are the charge and position of the particle i.

The transition moment may be regarded as an oscillating dipole caused by a transition between the levels m and n. It interacts with the electric field component of the photon to polarize the molecular unit or, at the resonance, to absorb it. Experimentally, the absorption spectrum of the molecular unit in solution is expressed as a function of the absorbance A at a given frequency:

$$A = \log(I_0/I) = \varepsilon c l$$

where
I_0 = incident radiation intensity
I = emerging radiation intensity
ε = molar absorption coefficient
c = concentration
l = length of the cell

The magnitude of the transition moment may be calculated as [31, 38]

$$\int_{\nu_2}^{\nu_1} \varepsilon(\nu)\,d\nu = K_1 \bar{\nu}_{nm} |\mathbf{R}^{nm}|^2 \qquad (7.18)$$

$$K_1 = \text{constant} \qquad (7.19)$$

where

$\bar{v}_{nm}$ = average wavenumber of the transition ($v = 1/\lambda$, with λ = wavelength)

The maximum value of the molar absorption coefficient, ε_{max}, is often used as an approximate measure of the total absorption intensity given by the integral of the curve, but it does not take into account the width of the absorption. The symmetry selection rules for the absorption of a photon may be derived from the symmetry properties of the orbitals involved and of the dipole moment operator μ. The vector μ has components μ_x, μ_y, μ_z along the cartesian axes. The transition moment may be decomposed into three contributions

$$R_x^{nm} = \int \Psi_n^* \mu_x \Psi_m \, dx$$
$$R_y^{nm} = \int \Psi_n^* \mu_y \Psi_m \, dy \qquad (7.20)$$
$$R_z^{nm} = \int \Psi_n^* \mu_z \Psi_m \, dz$$

The overall transition movement is given by

$$|\mathbf{R}^{nm}|^2 = (R_x^{nm})^2 + (R_y^{nm})^2 + (R_z^{nm})^2 \qquad (7.21)$$

The transition nm is allowed as long as the symmetry of the quantity to be integrated is totally symmetric, i.e. it must possess the symmetry of the molecular unit:

$$\Gamma(\Psi_m) \times \Gamma(\mu) \times \Gamma(\Psi_n^*) = A$$

where

$\Gamma(\Psi)$ = irreducible representation of the molecular orbitals in the symmetry group of the molecular unit
A = totally symmetric representation

This can be decomposed into:

$$\Gamma(\Psi_m) \times \Gamma(x) \times \Gamma(\Psi_n^*) = A$$
$$\text{and/or} \quad \Gamma(\Psi_m) \times \Gamma(y) \times \Gamma(\Psi_n^*) = A$$
$$\text{and/or} \quad \Gamma(\Psi_m) \times \Gamma(z) \times \Gamma(\Psi_n^*) = A$$

since μ_x, μ_y and μ_z transform as x, y, z, respectively.

In the case of degenerate states, the product gives a sum of irreducible representations. The transition is allowed if A is contained in this sum. The components μ_x, μ_y, μ_z transform as x, y and z, whose corresponding irreducible representations may be found in the character table of the symmetry point group of the molecular unit. This approach only concerns a pair of orbitals and the overall polarizability should be obtained by a summation over all possible transitions.

Time-dependent perturbation theory permits us to find the expression of the polarizability coefficients α_{ij} as a function of the transition moment (derived from Refs. [35] and [37]):

$$\alpha_{ii} \sim \frac{\omega_0 |\mathbf{R}_i^{nm}|^2}{\omega_0^2 - \omega^2} \qquad i = x, y, z$$

In consequence, when the excitating radiation is close in energy to the absorption band ($\omega_0 \sim \omega$), the polarizability coefficient increases until absorption occurs when $\omega_0 = \omega$. This is also reflected in the refractive index since it is related to the polarizability of the constitutive molecular units. The dispersion effect is the change of the refractive index due to the term ($\omega_0^2 - \omega^2$) when the incident radiation wavelength is varied.

Near the absorption band ($\omega \sim \omega_0$), the molecular transition moments absorb and reemit a non negligible quantity of energy; an additional term must be taken into account which characterizes the emitting state. Emission is out-of-phase relative to the excitating radiation and depends on the lifetime τ_{ex} of the excited state [39–41]:

$$\alpha(\omega_0) \sim \omega_0 |\mathbf{R}^{nm}|^2 \tau_{ex}^2 \qquad (7.22)$$

7.5.4 Polarizability and Refractive Index

The time needed for an electrical dipole to reorient in an oscillating electric field is in the range 10^{-11} s^{-1}. Ion migration through organic media starts to be important above 10^{-2} s. Consequently, in the rest of the section only the electronic polarization (α_e) will be considered.

The Lorenz–Lorentz relationship relates the refraction, the refractive index and the polarizability α [42, 43] (see Section 5.3).

$$R = \frac{n^2 - 1}{n^2 + 2} \frac{M_w}{d} = \frac{4}{3}\pi\alpha$$

where

R = refraction (cm^3)
n = refractive index
M_w = molar mass
d = density
α = polarizability

The numerical values for ions [44, 45] or bonds [46–50] have been given in Chapter 5. This equation is valid only for isotropic media.

In the case where absorption of the molecular units is considered, it has been seen that an out-of-phase phenomenon takes place which can be written as $i\Gamma\omega$

MOLECULAR DIELECTRICS

(Γ is the damping coefficient). Additionally, one can consider several oscillators whose natural resonance occurs at ω_k. It can then be demonstrated [35] that the refractive index, n, is given by

$$n = 1 + \frac{e^2}{2\varepsilon_0 m} \sum_k \frac{N_k}{\omega_k^2 - \omega^2 + i\Gamma\omega} \qquad (7.23)$$

where N_k is the number of electrons per unit of volume. In this model, we found a relationship between a macroscopic parameter — the refractive index — and the elementary oscillators associated with the absorption bands of a molecular unit (Figure 7.30).

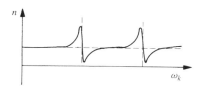

Figure 7.30 Refractive index as a function of ω, where $\omega_1, \omega_2, \ldots, \omega_i, \ldots$ correspond to the various absorption bands represented as oscillators. (After Ref. [35])

7.5.5 Ordinary and Extraordinary Indexes

It has been previously seen that the polarization tensor may be defined by only three coefficients, α_{xx}, α_{yy}, α_{zz}, whenever an appropriate choice of axes is taken. A material may be defined in the same way by three refractive indices, n_1, n_2, n_3, and the corresponding ellipsoïd is given by (Figure 7.31)

$$(x/n_1)^2 + (y/n_2)^2 + (z/n_3)^2 = 1 \qquad (7.24)$$

Three cases may be distinguished:

(a) $n_1 = n_2 = n_3$: isotropic medium. The corresponding condensed phase can be of symmetry K_h; the cubic crystalline classes (T_h, O_h, T, O, T_d), which have at least two C_n axes ($n > 2$), also yield isotropic optical properties.

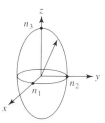

Figure 7.31 Ellipsoid of the refractive indexes

(b) $n_1 = n_2 \neq n_3$: unaxial medium. There is only one C_n axis ($n > 2$); the tetragonal, trigonal and hexagonal systems belong to this category. The z direction is called the *optical axis*.

(c) $n_1 \neq n_2 \neq n_3$: biaxial medium. There is no C_n axis ($n > 2$): this is the case for triclinic, monoclinic orthorhombic systems.

When an electromagnetic wave goes through an interface separating two media of refractive indices n_1 and n_2, the rates of propagation are given respectively by c/n_1 and c/n_2. Since all the oscillations must occur at the same frequency (forced regime domain), this means that the spacings between maxima of the wave must be the same at the surface delimitating the two media. The only way to fulfil this requirement is to have a wave propagating in another direction than the incident one when crossing the interface (Figure 7.32).

In a uniaxial medium, two different refractive indexes, and therefore two propagation velocities, must be considered. The two different refractive indexes are called the ordinary and extraordinary indices. This leads to the effect of double refraction [51] (Figure 7.33).

The double refraction in uniaxial media may be explained at the price of some approximations. It will be considered that the directions of the external applied field E and the macrospic induced dipole moment P_{ind} are approximately the same (although the medium is anisotropic). If the medium is considered to be quasi-isotropic, it can be written as

$$n^2 = 1 + \alpha$$

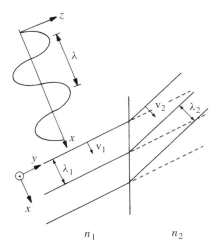

Figure 7.32 Relationship that can be found between the direction of refraction and the velocity change of light propagation. Two different scales for λ. (Modified from Ref. [35])

MOLECULAR DIELECTRICS

Figure 7.33 Double refraction observed with a crystal of spath, the trigonal form of $CaCO_3$. (Reproduced by permission of Hachette Livre from Ref. [51])

The ellipsoid of indexes can be, within previous limitations, represented by

$$\frac{x^2}{1+\alpha_{xx}} + \frac{y^2}{1+\alpha_{yy}} + \frac{z^2}{1+\alpha_{zz}} = 1$$

The polarizabilities (and the corresponding refractive indexes) are different in the x (or y) directions and in the z direction.

For a uniaxial medium (such as the trigonal form of $CaCO_3$), the ellipsoid (Figure 7.34) allows the geometrical determination of the refractive indexes of the ordinary (n_o) and extraordinary (n_e) beams. It can be assumed that the overall ellipsoid is a sum of a sphere of radius n_o and a deviation from this spherical shape. By turning back to the tensors of polarizability, this can be written as:

$$\begin{pmatrix} P_x \\ P_y \\ P_z \end{pmatrix} = \begin{pmatrix} \alpha_{xx} & & 0 \\ & \alpha_{yy} & \\ 0 & & \alpha_{zz} \end{pmatrix} \begin{pmatrix} E_x \\ E_y \\ E_z \end{pmatrix}$$

$$\alpha_{xx} = \alpha_{yy} = \alpha_0$$

$$\begin{pmatrix} P_x \\ P_y \\ P_z \end{pmatrix} = \begin{pmatrix} \alpha_0 & & 0 \\ & \alpha_0 & \\ 0 & & \alpha_0 \end{pmatrix} \begin{pmatrix} E_x \\ E_y \\ E_z \end{pmatrix} + \begin{pmatrix} 0 & & \\ & 0 & \\ & & \alpha_{zz} - \alpha_0 \end{pmatrix} \begin{pmatrix} E_x \\ E_y \\ E_z \end{pmatrix}$$

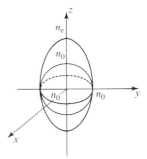

Figure 7.34 Geometrical representation of the ordinary and extraordinary indexes

The first term produces a polarization of constant magnitude whatever the direction of the electrical field. This term will give rise to the ordinary refractive index, n_0, which is also constant in any direction within the material. For the second term, which reflects the deviation from the sphere, the magnitude depends on E_z (projection of the electrical vector $\mathbf{E}$ on O_z). It is therefore not independent of the direction of the vector $\mathbf{E}$. The inequality $\alpha_{xx} = \alpha_{yy} \neq \alpha_{zz}$ imposes the presence of a second refractive index, n_e (extraordinary refractive index). A far more detailed description may be found in [52].

7.5.6 Second Harmonic Generation (SHG)

Franken and coworkers in 1961 detected ultraviolet light ($\lambda = 347$ nm) at twice the frequency of a ruby laser beam ($\lambda = 694$ nm) (cited in Ref. [53]). The use of laser allowed high electrical fields (10^3–10^6 V/cm) to be obtained that are necessary to obtain a non-negligible nonlinear term in the polarization of an unit. This field can be compared with the one arising from a solar photon (10 V/cm) or the one linking electrons to nuclei (approximately 10^9 V/cm).

From a symmetrical point of view, SHG may be treated in the same way as piezoelectricity (see the previous section and Section 4.3). The hyperpolarizability of a molecular unit may be written as

$$P_i^{2\omega} = \sum_{i,j} \beta_{ijk} E_j E_k \qquad (7.25)$$

where β_{ijk} is a third-rank tensor with 27 components which can be reduced to 18 when $E_i E_j = E_j E_i$. The magnitude of β can be tentatively estimated if the sigma, π and charge transfer contributions can be separated [54–56]:

$$\beta_{\text{TOT}} = \beta_\sigma + \beta_\pi + \beta_{\text{CT}} \qquad (7.26)$$

For singly-substituted benzenes, the contribution of σ and π electrons is comparable with the exception of nitro derivatives in which charge transfer between the benzene ring and the substituent occurs. A more systematic estimation may be made by considering that the hyperpolarizability term β is related to the dissymmetry of the electron cloud brought by the substituent. The magnitude of this dissymmetry may be estimated from mesomeric moments [57–59]. Mesomeric moments (μ_M) are defined by the difference between the dipole moment of alkyl and aryl derivatives:

$$\mu_M = \mu_{RX} - \mu_{ArX} \qquad (7.27)$$

where μ_M reflects the magnitude of the interaction between the substituent X and the π aromatic cloud. The most efficient substituents are $-\text{NMe}_2(1.66)$, $-\text{NH}_2(1.02)$ and $-\text{OMe}(0.96)$ for electron donors and $-\text{NO}_2(-0.76)$, $-\text{SO}_2\text{Me}(-0.6)$, $-\text{CCl}_3(-0.5)$ and $-\text{CN}(-0.45)$ for electron acceptors [60]. The extent of the aromatic core also influences the ground-state dipole moment and correlatively the mesomeric moment of the molecular unit (Table 7.3) [60].

Biphenyl (4.36D) and stilbene (4.56D) nitro derivatives demonstrate significantly higher dipole moments than nitrobenzene (4.01D). The isomers of positions of naphthalene also yield very different electrical moments: (see Table 7.3).

A relationship between the mesomeric moment and the hyperpolarizability coefficient β may be calculated when the extent of interaction between the substituent and the aromatic cloud is not too large [58, 61, 62]. Substitution is considered to be equivalent to an internal electric field E_0 which induces a dipole μ_M on the benzene ring:

$$\mu_M = \alpha E_0$$

If an external field E is superimposed, the overall field $E_T = E + E_0$ is effective on the molecular unit [55, 58, 59]. The polarization becomes

$$\alpha(E + E_0) + \gamma(E + E_0)^3 + \cdots$$

The coefficient of E^2 therefore leads to

$$\beta = 3\gamma \frac{\mu_M}{\alpha}$$

Comparison of the mesomeric moments of monosubstituted benzene derivatives and their β coefficients is shown in Table 7.4.

Halogeno derivatives show a relationship between the β value and the electronegativity of the halogen. However, it reflects a contribution from the β_σ term since the π term is approximately constant in the series. In these cases the absolute values of the σ and π contributions are of the same order of magnitude. The correlation with the mesomeric moment is satisfactory. For nitro-derivatives the σ contribution becomes negligible [55, 63]. This observation seems to be general: substituents that interact with the benzene ring through inductive effects lead to comparable σ and π contributions. The β_σ term is negligible when charge transfer

Table 7.3 Dipole moments of various nitro and amino derivatives

CH_3-X
X = NO_2, μ = 3.10D
X = NH_2, μ = 1.46D

(phenyl)-X
X = NO_2, μ = 4.01D
X = NH_2, μ = 1.53D

(biphenyl)-X
X = NO_2, μ = 4.36D
X = NH_2, μ = 1.74D

(naphthyl-1)-X
X = NO_2, μ = 3.98D
X = NH_2, μ = 1.49D

(naphthyl-2)-X
X = NO_2, μ = 4.36D
X = NH_2, μ = 1.77D

(phenyl)-CH=CH-(phenyl)-X
X = NO_2, μ = 4.56D
X = NH_2, μ = 2.07D

Table 7.4 Correlation between the hyperpolarizability coefficients β, the mesomeric moment μ_M, the electronegativity of the substituent and the maximum wavelength of the absorption band λ_{max} for mono-substituted benzene derivatives. (After Refs. [55] and [63])

 X	β^a	β_σ^b	β_π^c	μ_M (D)	El(X)^d
F	0.5 ± 0.2	0.06	0.4	0.41	4.1
Cl	0.2 ± 0.2	-0.1	0.3	0.41	2.83
Br	0 ± 0.2	-0.43	0.4	0.43	2.74
I	-0.6 ± 0.3	-1.1	0.5	0.5	2.21
Me	0.2			0.35	
NO$_2$	-2.4	Negligible	-2.4^e	-0.76	
NH$_2$	0.9–1.2			1.02	
NMe$_2$	1.5–1.7			1.66	

^{a}Projection of β on the direction of the dipole moment of the molecule (in 10^{-30} esu).
^{b}Measurement on saturated compounds.
$^c\beta\pi = \beta - \beta\sigma$ (projection).
^{d}Electronegativity of the group X.
^{e}From Ref. [63].

can occur. The β values determined in solution are affected by large uncertainties due to the method of measurement and important solvent effects [64].

π-Electron acentricity provided by substitution may also be estimated from optical absorption spectra [59], from ^{13}C NMR shifts [65] or from other physicochemical properties [66]. The most general method is, however, to use Hammett's coefficients which relate the effect of a substituent on the rate, k, of a given reaction [67, 68]:

$$\log k/k_0 = \rho\sigma \qquad (7.28)$$

where

ρ = constant for a given type of reaction
σ = constant associated with the substituent

Inductive (σ_I) and resonance (σ_R) effects are usually separated. CNDO calculations on substituted benzenes show that σ_R is directly related to the amount of charge transfer occurring between the aromatic ring and the substituent [69].

In the case where electron donor and electron acceptor groups simultaneously substitute the benzene ring, the previous methods of estimating β no longer hold. The charge transfer contribution, β_{CT}, is usually larger than the π and σ terms. The interaction moment, μ_{int}, may be used to estimate the amount of charge transfer [60]. μ_{int} is the difference between the experimental value of the electrical dipole moment, μ_{exp}, and the one calculated by postulating a vectorial

MOLECULAR DIELECTRICS

additivity of the contributions of the various substituents:

$$\mu_{int} = \mu_{exp} - \mu_{calc} \tag{7.29}$$

Alternatively, the amount of charge transfer may be determined from the wavelength of the charge transfer band in the optical spectra [70]. Values obtained with *para*-disubstituted benzenes are shown in Table 7.5 [54, 71, 72].

The β value of *p*-nitroaniline is 45×10^{-30} esu, 10 and 20 times larger than for nitrobenzene or aniline, respectively [64]. Correspondingly, the charge transfer band in *p*-nitroaniline is at about 378 nm, while it occurs at around 250 nm for the singly-substituted derivatives. The maximum of absorption of the charge transfer band is not sufficient to estimate the efficiency of the second harmonic generation since the molecular hyperpolarizability coefficient β also depends on the difference between the excited-state and ground-state dipole moments ($\Delta\mu$). Perturbation theory can permit the hyperpolarizability coefficients to be related to the transition moment [73]. In the case where only two levels are considered:

$$\beta_{ijk} = \frac{1}{2\hbar^2}\left[\frac{\Delta\mu_i R_j R_k}{\omega_0^2 - \omega^2} + R_i(R_j\,\Delta\mu_k + R_k\,\Delta\mu_j)\frac{\omega_0^2 + 2\omega^2}{(\omega_0^2 - 4\omega^2)(\omega_0^2 - \omega^2)}\right] \tag{7.30}$$

Table 7.5 Hyperpolarizability coefficients (β) of various disubstituted benzenes as a function of the interaction moment (μ_{int}) and the wavelength of the charge transfer optical band (λ_{max}). (After Refs. [54, 71, 72])

D	A	μ_{exp} (D)	μ_{int} (D)	λ_{max} (nm)	β^a (10^{-30} esu)
NMe$_2$	NO$_2$	6.93	1.48	418	51
	COMe	5.05	1.20		
	CF$_3$	4.62	0.61		
	CN	5.90	0.41	297	14
	Cl	3.29	0.2		
NH$_2$	SO$_2$CF$_3$	6.88	1.48		
	NO$_2$	6.33	1.19	378	45
	COMe	4.43	0.67	310	23
	Cl	2.99	0.15		
OH	NO$_2$	5.07	0.82	313	
	SO$_2$Me	5.32	0.71		
	CN	4.95	0.63	283	
	Cl	2.27	0.06		

[a] Values obtained at 1.89 μm from Ref. [71] (electric field induced second harmonic generation method).

For a one-dimensional charge transfer, the β is noted β_{CT} and numerically:

$$\beta_{CT} = 103 \times 10^{-30} \Delta\mu f_{osc} F \tag{7.31}$$

where

β is in esu
$\Delta\mu$ is in debye
$f_{osc} = 4.6 \times 10^{-9} \varepsilon_{max} \Delta_{1/2}$
$\Delta_{1/2}$ = half-height width (cm^{-1}) of the absorption band
ε_{max} = maximum extinction coefficient of the CT band (l/mole.cm)

with

$$F = \frac{E_0}{[E_0^2 - (2E)^2](E_0^2 - E^2)} \tag{7.32}$$

where

E_0 = energy of the absorption band (eV) (1 eV = 1240/λ (nm))
$E, 2E$ = energies of the incident radiation and its second harmonic (eV)

Note that eV = $\mathcal{N}$ electrons in a potential of one volt. β_{CT} therefore depends on the energy of the incident radiation. A term independent of the wavelength of the incident electromagnetic wave, β^0, may, however, be found [74]:

$$\beta = \beta^0 \frac{\lambda^4}{(\lambda^2 - 4\lambda_0^2)(\lambda^2 - \lambda_0^2)} \tag{7.33}$$

where

λ_0 = absorption band (nm)
λ = incident radiation (nm)

The difference in molecular dipole moments between the excited state and the ground state ($\Delta\mu$) can be determined from the shift of the absorption band when the polarity of the solvent is varied (solvatochromy). Abundant literature is available on this subject [75–80].

7.5.7 Hyperpolarizability of a Molecular Unit: Experimental Results and Calculations

Molecular units dissolved in solutions yield a medium of symmetry K_h (achiral units) or K (chiral units) which cannot lead to second harmonic generation (see Chapter 4). By applying an electrical field to the solution, a small amount of the polar molecular units will orient following Langevin's law:

$$\tau_{or} = \frac{\mu_g E_{loc}}{3kT} \tag{7.34}$$

MOLECULAR DIELECTRICS

where

μ_g = ground-state dipole moment of the molecular unit
E_{loc} = local field

For usual electrical fields (10^3–10^5 V/cm) and ground-state dipole moments (3–5D), between 0.01 and 5% of the molecular units are noncentrosymmetrically oriented. The symmetry of the medium is then $C_{\infty v}$, which is appropriate for second harmonic generation.

The type of cell used for electric field induced second harmonic generation (EFISHG) is shown in Figure 7.35.

The overall field applied on a molecular unit is given by (neglecting local field contributions)

$$E_{TOT} = E_\omega + E_{ext} \tag{7.35}$$

where

$E_\omega = E_0 \cos \omega t$
E_{ext} = external electrical field

The polarity is then, with μ_g parallel to the applied field,

$$\mu = \mu_g + \alpha E_{TOT} + \beta E_{TOT}^2 + \gamma E_{TOT}^3 + \cdots$$
$$= (\mu_g + \alpha E_{ext} + \beta E_{ext}^2 + \gamma E_{ext}^3)$$
$$+ (\alpha + 2E_{ext}\beta + 3E_{ext}^2\gamma)E_\omega$$
$$+ (\beta + 3E_{ext}\gamma)E_\omega^2 + \gamma E_\omega^3 + \cdots$$

The coefficient of E_ω^2 can be written as

$$\beta' = \beta + 3\gamma E_{ext}$$

With this method of measurement only the vectorial part of the tensor is determined:

$$\beta_z = \beta_{zzz} + \tfrac{1}{3}(\beta_{zxx} + 2\beta_{xzx} + \beta_{zyy} + 2\beta_{yzy})$$

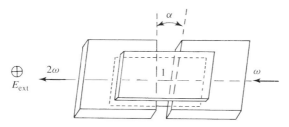

Figure 7.35 Cell used for EFISHG measurements. (Reproduced from Ref. [81])

where z is the direction of the permanent dipole moment of the molecule. Kleinman's symmetry [71, 91] relations $\beta_{ijk} = \beta_{kij} = \beta_{jik}$ allow this to be simplified to

$$\beta_z = \beta_{zzz} + \beta_{zxx} + \beta_{zyy} \tag{7.36}$$

As an example, a molecular unit that can act as a ligand towards transition metal ions will be studied in some detail. This compound derives from the reaction of glyoxal with benzene derivatives substituted with the electron donor groups $-OCH_3$ or $-N(CH_3)_2$ and a phenylhydrazine derivative. This compound, depending on the solvent, can exist in the *cis* and *trans* conformations (Figure 7.36).

Experimental determination of the permanent dipole moments in $CHCl_3$ (predominantly the *cis* form) and in dioxane (predominantly the *trans* form) have been compared with theoretical calculations (MNDO) (Table 7.6). The experimental values of the molecular dipole moments have been calculated taking into account local field effects due to the solvent [60, 83].

Knowing the permanent dipole moments, it is possible to determine the hyperpolarizability coefficients of the molecular units in solution. A schematic representation of the molecular arrangements relative to the electrical and electromagnetic fields during the EFISHG measurements is shown in Figure 7.37. The incident radiation at ω goes through the partially oriented molecular units, creating a polarization at 2ω:

$$P^{2\omega} = \Gamma^{2\omega} E_\omega E_\omega E_{\text{ext}}$$

where $\Gamma^{2\omega}$ is the macroscopic susceptibility of the solution.

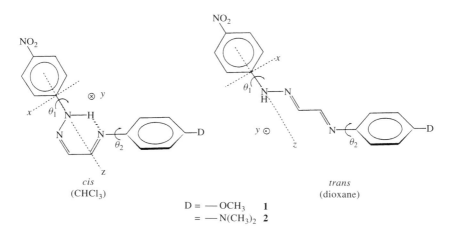

Figure 7.36 The *cis* and *trans* conformations of the ligand studied (after Ref. [82]). The conformations of the ligands have been optimized using the AM1 method [83]

MOLECULAR DIELECTRICS

Table 7.6 Calculated (MNDO) and experimental values of the permanent dipole moment of the *cis* and *trans* forms of compounds **1** and **2** of Figure 7.36. The axes (x, y, z) are as shown in Figure 7.36. (After Ref. [83])

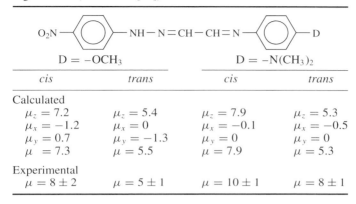

	cis	trans	cis	trans
Calculated	$\mu_z = 7.2$	$\mu_z = 5.4$	$\mu_z = 7.9$	$\mu_z = 5.3$
	$\mu_x = -1.2$	$\mu_x = 0$	$\mu_x = -0.1$	$\mu_x = -0.5$
	$\mu_y = 0.7$	$\mu_y = -1.3$	$\mu_y = 0$	$\mu_y = 0$
	$\mu = 7.3$	$\mu = 5.5$	$\mu = 7.9$	$\mu = 5.3$
Experimental	$\mu = 8 \pm 2$	$\mu = 5 \pm 1$	$\mu = 10 \pm 1$	$\mu = 8 \pm 1$

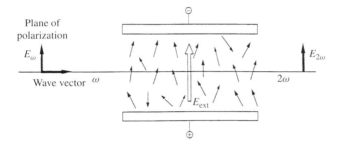

Figure 7.37 Schematic representation of the orientations of the electrical and electromagnetic fields relative to the molecular dipole moment

For a polar molecular unit, it is necessary to consider [55, 58] the scalar product $\mu_g \beta$:

$$\gamma^0 \approx \frac{\mu_g \beta}{5kT} \tag{7.37}$$

where γ^0 is the third-order hyperpolarizability coefficient and β is the vectorial part of the hyperpolarizability tensor.

If z is the direction of the permanent dipole moment of the molecular unit, this expression simplifies to

$$\mu_g \beta = \mu_z \beta_z \quad \text{with } \beta_z = \beta_{zzz} + \beta_{zxx} + \beta_{zyy} \tag{7.38}$$

The β_z^0 values (see equation (7.33)) of the compounds **1** and **2** have been determined for both the *trans* and *cis* conformations (Table 7.7).

Table 7.7 EFISHG values of the hyperpolarizability coefficients β_z^0 (in 10^{-30} esu: electrostatic units in the system CGS) determined at the wavelengths indicated (1.9 or 1.06 μm). (After Ref. [83])

| | NO$_2$–Ar–D | | | |
| | –OCH$_3$ | | –N(CH$_3$)$_2$ | |
	λ_0 (nm)	β_z^0	λ_0 (nm)	β_z^0
	—⟨O⟩—NH–N≡N—⟨O⟩—			
cis	414[a]	8[c] ± 6	459[a]	30[c] ± 10
trans	401[b]	25[c] ± 15	434[b]	35[c] ± 15
	—⟨O⟩—			
	304[b]	4.4[c]	382[b]	18[c]
	—⟨O⟩—CH=CH—⟨O⟩—			
	379[a]	35[d]	436[a]	90[d]

[a] In CHCl$_3$.
[b] In dioxane.
[c] At 1.9 μm.
[d] At 1.06 μm.

The –N(CH$_3$)$_2$ group leads to higher β values than –OCH$_3$. The compounds **1** and **2** do not allow a full conjugation between the electron donor and electron acceptor groups. The results, however, show that it does provide a better β value than a single benzene ring. Calculations have been carried out to determine all the hyperpolarizability coefficients of the tensor [83] (Table 7.8).

It has previously been shown that the main contribution to the permanent dipole moment is along the Z direction; therefore the β_z^0 determined by EFISHG (Table 7.7) must be compared with the sum $\beta_{zzz} + \beta_{zxx} + \beta_{zyy}$. The last two terms are negligible and, taking into account the uncertainties, there is good agreement between the experimental and calculated values.

The term β_{zzz} must correspond to a charge transfer between one nitrogen heteroatom of the hydrazine moiety to the nitro group. The nature of the electron donor (–N(CH$_3$)$_2$ or –OCH$_3$) has a weak influence on this charge transfer.

The term β_{xxx} is negative in the *cis* form; it indicates a displacement of electrons from the lateral groups to the central core. The electron donor groups do influence this transfer and are involved in this process.

The terms β_{xzz} and β_{xxx} both depend strongly on the nature of the donor group –D and the corresponding excitation is mainly centered on this part of the molecule (see Ref. [83]).

MOLECULAR DIELECTRICS

Table 7.8 Calculated values of the hyperpolarizability coefficient of compounds **1** and **2** (calculations carried out by M. Barzoukas, see Ref. [83]) (The axes are those shown in Figure 7.36)

	cis form		trans form	
	−OMe	−NMe$_2$	−OMe	−NMe$_2$
β_{zzz}	18.0	28.1	22.4	34.6
β_{xxx}	−2.6	−10.4	−0.7	5.8
β_{yyy}	0.2	0	0.3	0
β_{zxx}	0	8.2	−2.5	5.5
β_{zyy}	0	−0.6	−0.2	−0.5
β_{xzz}	−2.2	−11.2	1.1	11.2
β_{xyy}	0	0.5	−0.1	−0.5
β_{yzz}	0.3	0	−0.7	−0.6
β_{yxx}	0.3	0.1	0	0

7.5.8 Frequency Doubling and Phase Matching

Under the influence of a plane wave of frequency ν, each molecular unit constituting the material becomes an oscillating dipole. The molecular units situated in a single plane perpendicular to the propagation direction are in-phase. Their individual induced dipole may be written as

$$\mu_{\text{ind}} = \alpha E + \beta E^2 + \gamma E^3 + \cdots$$

since

$$E^2 = E_0^2 \cos^2 \omega t$$
$$= \frac{E_0^2}{2}(1 + \cos 2\omega t)$$

A non-zero β coefficient therefore yields to an apparent fusion of two low-energy photons (frequency ν) to a single one of double energy (frequency 2ν).

However, the refractive index, because of the dispersion term $(\omega_0^2 - \omega^2)$, depends on the frequency. The two waves at ω and 2ω do not propagate at the same velocity. As a consequence, destructive interferences (Figure 7.38) may occur. In order to have constructive interferences, it is necessary to have (phase

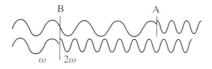

Figure 7.38 Second harmonic generation may occur in A or B. The 2ω waves can then form destructive interferences depending on the distance between A and B

matching condition)
$$n^\omega = n^{2\omega} \tag{7.39}$$

These conditions may be reached for uniaxial and biaxial crystals. In uniaxial media it has been shown that two different refractive indexes (ordinary n_o and extraordinary n_e) coexist. Phase matching can be obtained [92] when

$$n_o^{2\omega} = n_e^\omega \tag{7.40}$$

7.5.9 SHG in Single Crystals

In conventional SHG devices, single crystals of inorganic materials (KH_2PO_4 and related salts) are used [84]. Such salts can be grown from water solutions to give very large single crystals. The growth of single crystals of molecular materials is by no means easy and most of the limitations to the use of organic compounds in commercial devices arise from the difficulty of obtaining large single crystals of high optical quality [85–87]. The efficiency of second harmonic generation may be tested on powders [88] on qualitative grounds before any effort towards crystal growth is made.

A large variety of organic molecules have been tested for SHG. Polarizable cores were mostly based on aromatic derivatives: benzene and related compounds, pyridine, stilbene, biphenyl. Donors such as alkylamino-, ether and thioethers were the most frequently used; nitro-, amido-, halide, cyano and esters are frequently encountered for acceptor groups. The selection of the most suitable crystals may be achieved from examination of their space groups. Besides the unavoidable noncentrosymmetry of the crystal class, crystal symmetries may also be selected from their ability to give rise to optical phase matching; in the case of a quasi one-dimensional charge transfer such as in *p*-nitroaniline, the crystal point groups 1, 2, m and mm2 are the most favorable in this respect [2]. The materials most thoroughly studied are at the present time: *N*-(4-nitrophenyl)-(L)-prolinol (NPP), *N*-(4-nitrophenyl)-*N*-methylamino-acetonitrile (NPAN), methyl-(2,4-dinitrophenyl)-aminopropanoate (MAP), 3-methyl-4-nitropyridine-*N*-oxide (POM) and related compounds [71] (Figures 7.39 and 7.40). Organic materials are, in the most favorable cases, 100 times more efficient than the inorganic

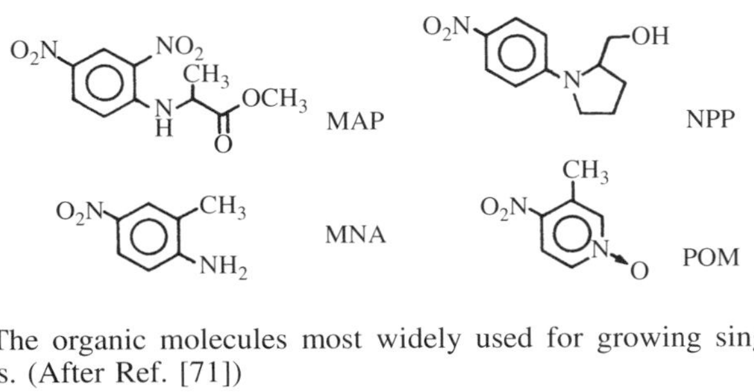

Figure 7.39 The organic molecules most widely used for growing single crystals for nonlinear optics. (After Ref. [71])

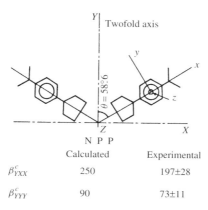

Figure 7.40 Comparative values of the crystalline nonlinear coefficients (β^c) of NPP (expressed in units of 10^{-9} esu) where x, y, z is molecular reference frame and X, Y, Z is the crystal reference frame. Determinations are at 1.17 eV. Calculated values are from CNDOVSB. (After Ref. [90].

standard lithium niobate [89]. The crystal structure of NPP is monoclinic (space group P2$_1$) with two molecules per unit cell [90].

The macroscopic polarization (reference frame XYZ) is obtained by summing all the contributions arising from the individual molecular unit in which each possesses its own reference frame (x_i, y_i, z_i). Only the relative orientations of the molecular unit must be considered: a representation in the crystal class (2 or C_2) is sufficient to find the contribution of the various molecules within the unit cell ($Z = 2$) to the overall macroscopic polarization. For NPP, it will be considered that only β_{xxx} is not negligible. First, the macroscopic components of the incident field E_X, E_Y, E_Z must be expressed at the level of the molecular units and in their reference frame (by separating E_x and E_y contributions):

$$E_x = E_X[\cos(\eta/2 - \theta)] = E_X \sin\theta$$
$$E_x = E_Y \cos\theta$$

Then

$$(\mu_x^{2\omega})_1 = \beta_{xxx}(E_Y \cos\theta + E_X \sin\theta)^2$$
$$(\mu_x^{2\omega})_2 = \beta_{xxx}(E_Y \cos\theta - E_X \sin\theta)^2$$

for the two molecules 1 (x_1, y_1, z_1) and 2 (x_2, y_2, z_2). The overall macroscopic polarization is the sum of all individual contributions expressed in the crystal frame (X, Y, Z). The molecular polarizations $(\mu_x^{2\omega})_1$ and $(\mu_x^{2\omega})_2$ must be expressed in the reference frame of the material (X, Y, Z)—

along OX: $(\mu_x^{2\omega})_1 \sin\theta - (\mu_x^{2\omega})_2 \sin\theta = P_X$

along OY: $(\mu_x^{2\omega})_1 \cos\theta + (\mu_x^{2\omega})_2 \cos\theta = P_Y$

$$P_X = 4\beta_{xxx}(\cos\theta \sin^2\theta)E_X E_Y$$

therefore

$$\beta^c_{XXY} = 2\beta_{xxx}\cos\theta\sin^2\theta \quad \text{(by considering } 2E_X E_Y\text{)}$$
$$P_Y = (2\cos^3\theta E_Y^2 + 2\cos\theta\sin^2\theta E_X^2)\beta_{xxx}$$

therefore

$$\beta^c_{YYY} = 2\cos^3\theta\beta_{xxx}$$
$$\beta^c_{YXX} = 2\cos\theta\sin^2\theta\beta_{xxx}$$
$$= \beta^c_{XXY}$$

The induced dipole moment has been calculated for one unit cell ($Z = 2$). The polarity at a macroscopic level is obtained by multiplying by the number of unit cells, for one mole, it is the Avogadro number (corrections due to local field factors should also be taken into account). Experimental and calculated values of β^c are in good agreement (Figure 7.40).

The stability under laser irradiation (damage threshold) seems to be comparable for organic and inorganic materials [87]: 2.3×10^4 MW/cm^2 for KH$_2$PO$_4$, 1.2×10^2 for LiNbO$_3$, 2×10^3 MW/cm^2 for POM and 3×10^3 MW/cm^2 for MAP (at $1.064\,\mu$m).

7.6 REFERENCES

1. Modified from C. Kittel, *Introduction à la physique de l'état solide* (trad. E.-L. Huguenin and R. Papoular), Dunod, Paris (1958).
2. J. Zyss and J. L. Oudar, *Phys. Rev. A*, **26**, 2028 (1982).
3. D. Gourier, personal communication.
4. L. Binet, Thèse de doctorat, Université P. & M. Curie, Paris (1995).
5. L. Y. Chiang, A. F. Garito, D. J. Sandman (eds.) *MRS Symposium Proceedings of Electrical, Optical and Magnetic Properties of Organic Solid State Materials*, Vol. 247, Pittsburg (1992).
6. D. Y. Curtin and I. C. Paul, *Chem. Rev.*, **81**, 525 (1981).
7. G. Busch, *Condensed Matter News*, **1**, 20 (1991).
8. B. C. Frazer, M. McKeown and R. Pepinsky, *Phys. Rev.*, **94**, 1435 (1954).
9. J. Curie and P. Curie, *Compt. Rend. Acad. Sci. Paris*, **91**, 294 and 383 (1880).
10. F. Pockels, *Abh. der Königl. Ges. der Wissenschaften zu Göttingen*, **39**, 1 (1893).
11. F. Beaulard, *J. de Phys. (Paris)*, **9**, 422 (1900).
12. J. Valasek, *Phys. Rev.*, **15**, 475 and 537 (1920).
13. J. Valasek, *Ferroelectrics*, **2**, 239 (1971).
14. G. Busch and P. Scherrer, *Naturwiss.*, **23**, 737 (1935).
15. J. C. Slater, *J. Chem. Phys.*, **9**, 16 (1941).
16. W. Känzing, *Ferroelectrics and Antiferroelectrics Solid State Physics*, Academic Press, New York Vol. 4 (1957) and references therein.
17. W. Cochran, *Adv. in Phys.*, **2**, 387 (1961).
18. W. L. McMillan, *Phys. Rev.*, **A8**, 1921 (1973).
19. R. B. Meyer, L. Liébert, L. Strzelecki and P. Keller, *J. de Physique-lettres (Paris)*, **36**, L-69 (1975).
20. C. Escher, *Kontakte (Darmstadt)*, **2**, 3 (1986).

21. J. W. Goodby, *Ferroelectric Liquid Crystals Ferroelectricity and Related Phenomena*, Vol. 7, Gordon & Breach, Philadelphia (1991).
22. P. Keller, L. Liébert and L. Strzelecki, *J. de Physique Colloque (Paris)*, **C3**, 3 (1976).
23. R. G. Kepler, *Ann. Rev. Phys. Chem.*, **29**, 497 (1978).
24. A. Weiss and S. Fleck, *Ber. Bunsenges Phys. Chem.*, **91**, 913 (1987).
25. W. Schumann, *Guide des pierres précieuses, pierres fines et pierres ornementales* (trad. F. Georges-Catroux and J. E. Dietrich; ed. D. Perret) Delachaux et Niestlé S. A., Neuchâtel (1992).
26. H. W. Jaffe, *Crystal Chemistry and Refractivity*, Cambridge University Press, Cambridge (1988).
27. D. Britton, J. Konnert and S. Lam, *Cryst. Struct. Commun.*, **6**, 45 (1977).
28. J. Doucet, P. Keller, A.-M. Levelut and P. Porquet, *J. de Phys. (Paris)*, **39**, 548 (1978).
29. L. M. Blinov, V. A. Baikalov, M. I. Barnik, L. A. Beresnev, E. P. Pozhidayev and S. V. Yablonski, *Liquid Cryst.*, **2**, 121 (1987).
30. P. E. Dunn and S. H. Carr, *MRS Bull.*, 22 (February 1989).
31. J. Simon, P. Bassoul and S. Norvez, *New J. Chem.*, **13**, 13 (1989).
32. J. Simon, *C. R. Acad. Sci. Paris*, **324** (IIb), 47 (1997).
33. J. F. Nye, *Propriétés physiques des cristaux*, Dunod, Paris (1961).
34. A. J. Lovinger, *Science*, **220**, 1115 (1983).
35. R. Feynman, R. B. Leighton and M. Sands, *Le cours de physique de Feynman*, Inter-Editions, Paris (1979).
36. P. Le Barny, personal communication.
37. J. Catalan, E. Mena, W. Meutermans and J. Elguero, *J. Phys. Chem.*, **96**, 3615 (1992).
38. J. M. Hollas, *High Resolution Spectroscopy*, Butterworths, London (1995).
39. J. G. Calvert and J. N. Pitts Jr, *Photochemistry*, John Wiley & Sons, New York (1967).
40. A. Hinchliffe and R. W. Munn, *Molecular Electromagnetism*, John Wiley & Sons, New York (1985).
41. J. R. F. S. Crawford, *Ondes Berkeley: Cours de physique* (trad. P. Lena), Vol. 3, A. Colin, Paris (1972).
42. L. Lorenz, *Wied. Ann. Phys.*, **11**, 70 (1880).
43. H. A. Lorentz, *Wied. Ann. Phys.*, **9**, 641 (1880).
44. A. Heydweiller, *Phys. Z.*, **26**, 526 (1925).
45. A. Dalgarno, *Adv. Phys.*, **11**, 281 (1962).
46. A. I. Vogel, W. T. Cresswell, G. H. Jeffery and J. Leicester, *J. Chem. Soc.*, 514 (1952).
47. R. G. Gillis, *Rev. Pure Appl. Chem.*, **10**, 21 (1960).
48. R. J. W. Le Fèvre, in *Advances in Physical Organic Chemistry* (ed. V. Gold) Academic Press, London (1965).
49. Sheng-Nien Wang, *J. Chem. Phys.*, **7**, 1012 (1939).
50. P. L. Davies, *Trans. Faraday Soc.*, **47**, 789 (1952).
51. M. Françon, *L'Optique moderne et ses développements depuis l'apparition du laser*, Hachette-CNRS, Paris (1986).
52. H. Curien, *Cours d'optique cristalline*, Association corporative des étudiants en Sciences, Paris (1960).
53. N. Bloembergen, *Nonlinear Optics*, W. A. Benjamin, New York (1965).
54. V. J. Dogherty, D. Pugh and J. O. Morley, *J. Chem. Soc. Faraday Trans. II*, **81**, 1179 (1985); *J. Chem. Soc. Perkin Trans. II*, 1351; 1357; 1361 (1987).
55. B. F. Levine and G. G. Bethea, *Appl. Phys. Lett.*, **24**, 455 (1974); *J. Chem. Phys.*, **63**, 2666 (1975).
56. C. W. Dirk, R. J. Twieg and G. Wagniere, *J. Am. Chem. Soc.*, **108**, 5387 (1986).
57. L. E. Sutton, *Proc. Roy. Soc. Lond.*, **133**, 668 (1931).

58. J. L. Oudar, *J. Chem. Phys.*, **67**, 446 (1977).
59. B. F. Levine, *J. Chem. Phys.*, **63**, 115 (1975).
60. V. I. Minkin, O. A. Osipov and Yu. A. Zhdanov, *Dipole Moments in Organic Chemistry*, Plenum Press, New York (1970).
61. J. L. Oudar and H. Le Person, *Opt. Commun.*, **15**, 258 (1975).
62. D. S. Chemla, J. L. Oudar and J. Jerphagnon, *Phys. Rev.*, **B12**, 4534 (1975).
63. G. Hauchecorne, F. Kervhervé and G. Mayer, *J. Phys. (Paris)*, **32**, 47 (1971).
64. B. F. Levine and C. G. Bethea, *J. Chem. Phys.*, **69**, 5240 (1978).
65. J. Gasteiger and H. Salber, *Angew. Chem. Int. Ed. Engl.*, **24**, 687 (1985).
66. A. R. Katrizky and R. D. Topsom, *Angew. Chem. Int. Ed. Engl.*, **9**, 87 (1970).
67. R. W. Taft and I. C. Lewis, *J. Am. Chem. Soc.*, **81**, 5343 (1959).
68. L. P. Hammett, *Physical Organic Chemistry*, McGraw-Hill, New York (1940).
69. R. T. C. Brownlee and R. W. Taft, *J. Am. Chem. Soc.*, **90**, 6537 (1968).
70. M. E. Peover, *Trans. Faraday Soc.*, **58**, 1956 and 2370 (1962); **60**, 479 (1964); **61**, 1516 (1965).
71. D. S. Chemla and J. Zyss (ed.), *Nonlinear Optical Properties of Organic Molecules and Crystals*, Academic Press, Orlando, Florida (1987).
72. A. Dulcic and C. Sauteret, *J. Chem. Phys.*, **69**, 3453 (1978).
73. J. L. Oudar and J. Zyss, *Phys. Rev.*, **A26**, 2016 (1982).
74. F. Tournilhac, Thèse de doctorat, ESPCI-Paris (1989).
75. C. A. G. O. Varma and E. J. J. Groenen, *Rec. Trav. Chim. Pays-Bas*, **91**, 295 (1972).
76. T. Abe and I. Iweibo, *Bull. Chem. Soc. Jpn*, **58**, 3415 (1985).
77. P. Suppan and C. Tsiamis, *Spectrochim. Acta*, **36A**, 971 (1980).
78. H. Labhart, *Experientia*, **22**, 65 (1966).
79. N. G. Bakhshiev, M. I. Knyazhanskii, V. I. Minkin, O. A. Osipov and G. V. Saidov, *Russian Chem. Rev.*, **38**, 740 (1969).
80. W. Liptay, *Angew. Chem. Int. Ed. Engl.*, **8**, 177 (1969).
81. D. J. Williams, *Angew. Chem. Int. Ed. Engl.*, **23**, 690 (1984).
82. T. Thami, P. Bassoul, M. A. Petit, J. Simon, A. Fort, M. Barzoukas and A. Villaeys, *J. Am. Chem. Soc.*, **114**, 915 (1992).
83. T. Thami, Thèse de doctorat, ESPCI-Paris (1992). The calculations mentioned in the dissertation were carried out by M. Barzoukas.
84. Chuang-tian Chen and Guang-zhao Liu, *Ann. Rev. Mater. Sci.*, **16**, 203 (1986).
85. M. C. Etter, D. A. Jahn, B. S. Donahue, R. B. Johnson and C. Ojala, *J. Cryst. Growth*, **76**, 645 (1986).
86. H. Tabei, T. Kurihara and T. Kaino, *Appl. Phys. Lett.*, **50**, 1855 (1987).
87. P. V. Vidakovic, M. Coquillay and F. Salin, *J. Opt. Soc. Am. B*, **4**, 998 (1987) and references therein.
88. S. K. Kurtz and T. T. Perry, *J. Appl. Phys.*, **39**, 3798 (1968).
89. J. Zyss and I. Ledoux, *L'écho des recherches*, **127**, 19 (1987).
90. M. Barzoukas, D. Josse, P. Fremaux, J. Zyss, J.-F. Nicoud and J. O. Morley, *Opt. Soc. Am. B*, **4**, 977 (1987).
91. D. A. Kleinman, *Phys. Rev.*, **126**, 1977 (1962).
92. J. E. Midwinter and J. Warner, Br. *J. Appl. Phys.*, **16**, 1135 (1965).

8 Industrial Applications of Molecular Materials

8.1 Introduction 344
8.2 Soaps 344
  8.2.1 Introduction 344
  8.2.2 The Main Amphiphilic Molecules 346
  8.2.3 Structural Characteristics of Soap-Based Molecular Materials 351
    8.2.3.1 Ordered Lipophilic Parts 352
    8.2.3.2 Disordered Lipophilic Parts and Ordered Polar Heads 353
    8.2.3.3 Disordered Hydrophilic and Hydrophobic Parts 355
  8.2.4 Conclusions 358
8.3 Organic Pigments and Dyes (colorants) 359
  8.3.1 The Notion of Color 360
  8.3.2 Chemical Nature of Colorants 362
  8.3.3 Physical Characterization of Pigments 367
  8.3.4 Coatings 368
    8.3.4.1 Paints 369
    8.3.4.2 Inks 376
8.4 Photoconductors and Photocopying Machines 385
  8.4.1 Introduction 385
  8.4.2 Photoconduction in Molecular Materials 386
  8.4.3 Electrophotographic Processes 388
  8.4.4 Xerographic Photoreceptors 391
  8.4.5 Charge Transport Layer 393
    8.4.5.1 Polymer Binders 393
    8.4.5.2 Molecular Units Used in Charge Transport Layers 393
    8.4.5.3 Physicochemical Properties 394
  8.4.6 Charge Generation Layer 394
    8.4.6.1 Azo Pigments 396
    8.4.6.2 Dye Polymer 396

		8.4.6.3 Charge Transfer Complexes	399
		8.4.6.4 Perylene Pigments	401
		8.4.6.5 Squaraine Pigments	401
		8.4.6.6 Phthalocyanine Pigments	402
	8.4.7	Comparative Xerographic Performances	403
8.5	Liquid Crystal Displays		405
	8.5.1	Introduction	405
	8.5.2	Dynamic Scattering Displays	406
	8.5.3	Twisted Nematic Liquid Crystal Displays	407
	8.5.4	Dichroic Liquid Crystal Displays	413
	8.5.5	Ferroelectric Liquid Crystal Displays	415
		8.5.5.1 Orientation of S_C^* Mesophases by Surfaces	415
		8.5.5.2 Chronology of the Ferroelectric Liquid Crystal Display Technological Advances	421
8.6	References		424

> *Jede Anziehung ist wechselseitig*
> J. W. Goethe (1749–1832)
> *Die Wahlverwandtschaften*

8.1 INTRODUCTION

Industrial applications may be qualified in different ways: present (in the sense of real), future and possible. In many cases no clear difference is made between the three categories. Technologies associated with a world production of thousands of tons are sometimes associated with emerging discoveries. In the following, only widespread, well-established, industrial applications will be mentioned. They are obviously arbitrarily chosen and many others could have been cited.

8.2 SOAPS

8.2.1 Introduction

Soap is a typical functional molecular material (Figure 8.1) since:

(a) It consists of molecular units.
(b) Its condensed phase structure is governed by the amphiphilic nature of the molecule.
(c) It is not used for its mechanical properties and ensures another function.

Soap has been fabricated since antiquity from greases treated with a base like *natron*, a mineral of hydrous sodium carbonate often crystallized with other salts. The name 'natron', with variants, was found in ancient times in Spanish, Arab, Greek and even Semitic or Hittite languages (2000–1200 BC). The greases contain 92–98% of glycerides which can be saponified to give the corresponding salt of the soap and glycerin. This latter was characterized by Chevreul in 1823.

INDUSTRIAL APPLICATIONS OF MOLECULAR MATERIALS

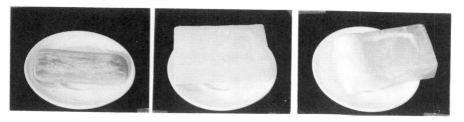

Figure 8.1 The various states of hydration of soap. At each soap/water ratio a large number of different lyotropic mesophases are formed

There are more than 100 different fatty acids in vegetals and several hundreds in animals. They contain between 4 and 22 carbon atoms; they can be linear, branched, saturated, unsaturated with an odd or even number of carbon atoms.

Industrially, soap is fabricated from various sources: beef fat, groundnut, palm, coconut, olive, etc. The choice of the constituents is linked to economical considerations and to the characteristics required: detergency, foaming properties, softening, dampening, etc. The conventional process ('Marseillais') is shown schematically in Figure 8.2 [1].

The world production of soap in 1994 was approximately [2]:

(a) 6–7 Mtons of ordinary soap made from fat of various origins (Mtons = millions of tons),
(b) 2.8 Mtons of anionic surface agents,

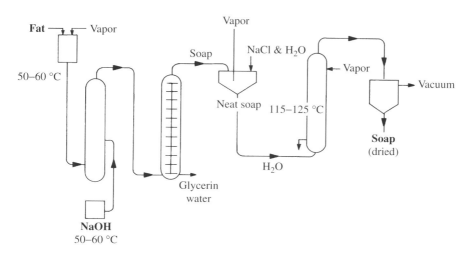

Figure 8.2 Schematic industrial process used to produce soap from fat and coconut oil. (Redrawn from Ref. [1])

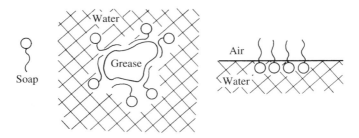

Figure 8.3 Schematic representation of the interaction of amphiphilic molecules with a greasy particle and structure of a Langmuir–Blodgett thin film

(c) 2.9 Mtons of nonionic surface agents,

(d) 0.43 Mtons of cationic surface agents,

(e) 0.07 Mtons of amphoteric surface agents.

The raw materials can be derived from natural substances as well as from the petroleum industry. Soaps have in common the ability to possess both hydrophilic *and* hydrophobic moieties: they are consequently called 'amphiphiles' (from the Greek *amphi*, meaning two sides, double, around and *phile*, *philos*, friend and *philein*, to love).

The 'Principe des affinités électives' is particularly well exemplified in the case of soaps. In condensed phases, the lipophilic and hydrophilic fragments form segregated domains. The hydrophobic tail of the soap will stick, for example, on the surface of a fatty particle whereas the ionized polar heads remain solvated in water (Figure 8.3).

Amphiphilic compounds, also called surfactants, decrease the surface tension of water, which is defined as

$$W = \gamma_s \Delta L \tag{8.1}$$

where

γ_s = surface tension
ΔL = length of displacement of the molecules corresponding to the work W

The surfactant effect is due to the fact that the amphiphilic molecules intercalate between surface water molecules, leading to a decrease in the mean interunit interaction.

8.2.2 The Main Amphiphilic Molecules

The triglycerides are an important source of lipids. They are called fats when solids at room temperature and oils when liquids. The triglycerides are composed of fatty acids and glycerol:

INDUSTRIAL APPLICATIONS OF MOLECULAR MATERIALS

$$\begin{array}{c} \text{—O—C(=O)—R} \\ \text{—O—C(=O)—R} \\ \text{—O—C(=O)—R} \end{array}$$

Two examples of fatty acids that can be isolated after saponification of triglycerides are shown in Figure 8.4.

The main natural sources of fats or oils are listed below:

(a) Animals
   $C_{16}-C_{18}$: beef fat
(b) Vegetals
   C_8-C_{14}: coconut, palm
   $C_{16}-C_{18}$: palm, soya bean, castor oil, colza

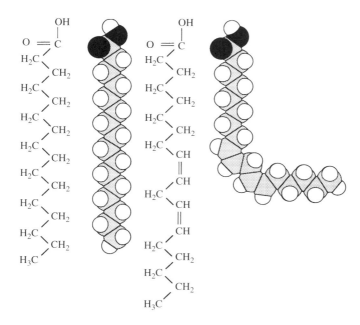

Figure 8.4 Molecular structures of palmitic acid and linoleic acid. (Modified from Ref. [3])

(c) Fishes
   C_{18} (unsaturated and others)

However, petrochemistry provides approximately 50% of the total of production of amphiphilic compounds [2].
   The amphiphiles also differ by the chemical nature of the polar head.

Anionic

Sulfonates:
   Alkylarene sulfonates

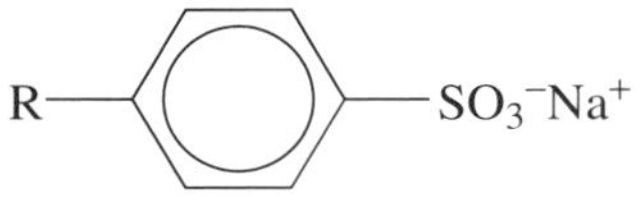

$$R-\!\!\!\bigcirc\!\!\!-SO_3^-Na^+$$

   Paraffin sulfonates

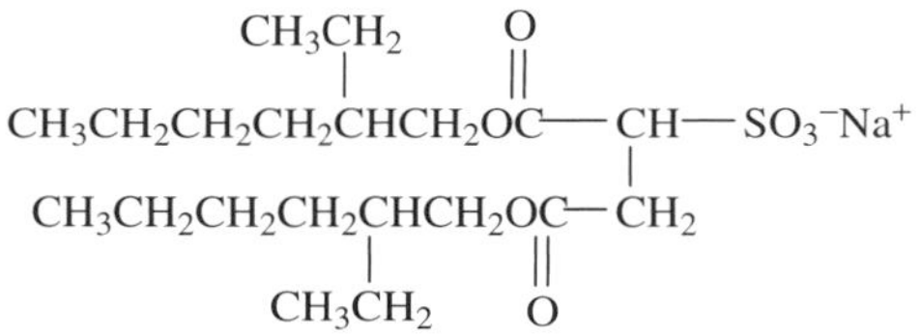

$$R-CH=CH-SO_3^-Na^+$$ (Aerosol OT structure shown)

   α-Olefine sulfonate
$$R-CH=CH-SO_3^-Na^+$$

Sulfates:
$$ROSO_3^-Na^+$$
$$RO(CH_2CH_2O)_n SO_3^-Na^+$$

Phosphates, phosphonates:

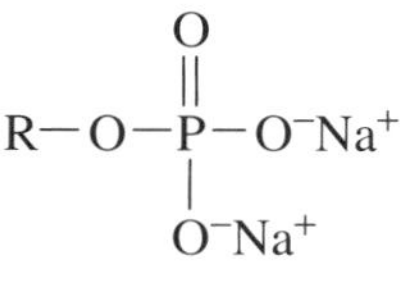

$$R-O-\underset{O^-Na^+}{\overset{O}{\underset{\|}{P}}}-O^-Na^+$$

Carboxylates:
$$R-CO_2^-M^+$$

Nonionic

Nonionic amphiphiles are very useful in producing low-foaming surfactants [4]. The hydrophilic properties are generally due to polyethyleneoxide chains.

INDUSTRIAL APPLICATIONS OF MOLECULAR MATERIALS

The addition of polypropylene oxide subunits yields more or less pronounced hydrophobic properties depending on the molar mass [4]:

$$C_nH_{2n+1}(\overset{R}{\underset{|}{C}}HCH_2O)_x(CH_2CH_2O)_yH$$

R = alkyl   BASF-Plurafac

$$HO(CH_2CH_2O)_x(\overset{CH_3}{\underset{|}{C}}HCH_2O)_y(CH_2CH_2O)_zH$$

BASF-Pluronic PE

These molecules are used as low-foaming detergents or even as antifoaming agents in the treatment of sugarbeet [4]. Amphiphiles derived from polyols (like sorbitol) or sugars (like saccharose or polyglucosides) are also produced industrially: 200 000 tons of tensioactive agents of this type are produced in the world for use as cosmetics and detergents [5] (Figure 8.5).

Cationic [6]

The products industrially used include: fatty amines, amido-amines, imidazolin, amino-esters and related derivatives. The structures of some of them are shown in Figure 8.6. These molecules are mainly used for softening and as biocides.

Zwitterionic [6]

Some naturally occurring phospholipids, like phosphatidyl ethanolamine and phosphatidyl choline, possess a zwitterionic polar head (Figure 8.7). The phospholipids constitute the major component of the biological membranes. Alkylamino acids and betaines are important classes of industrial surfactants (Figure 8.8).

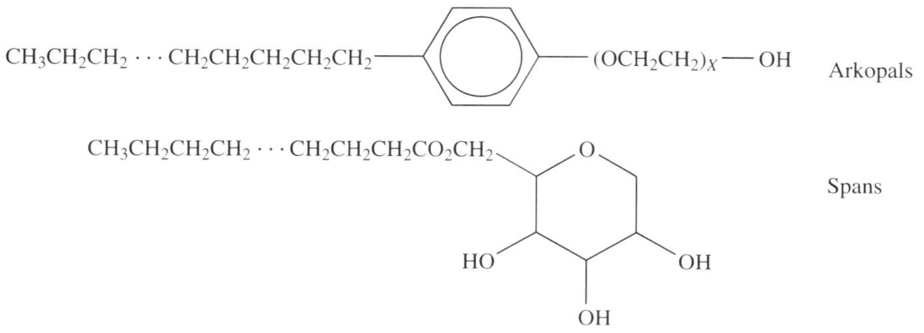

$CH_3CH_2CH_2 \cdots CH_2CH_2CH_2CH_2CH_2$—⟨phenyl⟩—$(OCH_2CH_2)_x$—OH   Arkopals

$CH_3CH_2CH_2CH_2 \cdots CH_2CH_2CH_2CO_2CH_2$—(sugar ring with HO, OH, OH)   Spans

Figure 8.5 Two commercially available neutral surfactants

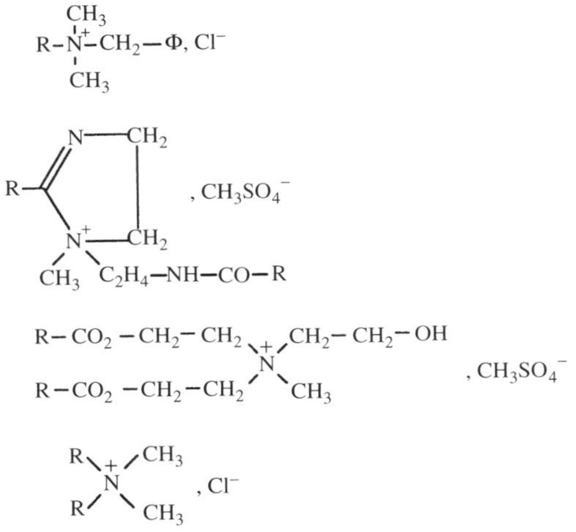

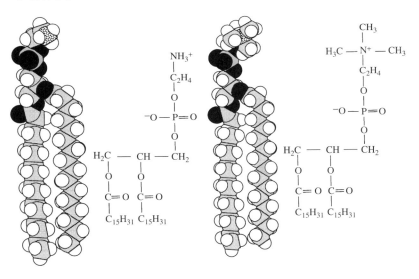

Figure 8.6 Some of the cationic amphiphiles that are used industrially, where $R = C_nH_{2n+1}$ [6]

Figure 8.7 Chemical formulae of phosphatidyl ethanolamine and phosphatidyl choline. (Drawn from Ref. [3])

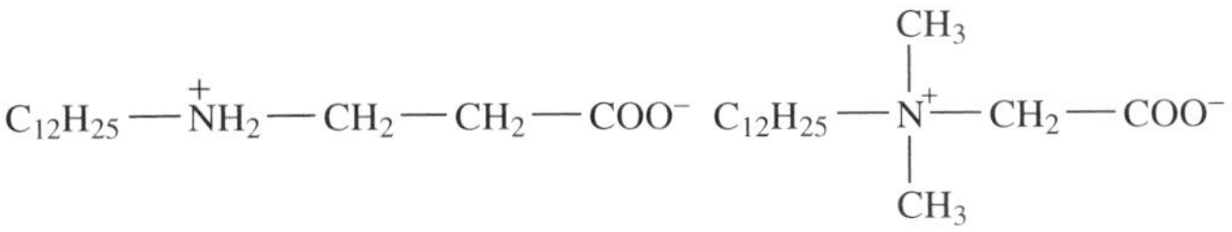

$$C_{12}H_{25}-\overset{+}{N}H_2-CH_2-CH_2-COO^- \quad C_{12}H_{25}-\underset{CH_3}{\overset{CH_3}{\overset{|}{\underset{|}{N^+}}}}-CH_2-COO^-$$

Figure 8.8 *N*-laurylaminopropionic acid and *N*-lauryldimethylbetaine [2]

8.2.3 Structural Characteristics of Soap-Based Molecular Materials

The characterization of metal ion soaps by X-ray diffraction started early in the century (see Ref. [7] and references therein). The importance of having a precise knowledge of their structures in order to understand the properties of sodium, calcium and aluminum soaps (lubricating greases) or manganese, cobalt and lead soaps (driers in the paint industry) had already been recognized at that time. The soaps mainly form lyotropic mesophases (binary mixtures of soap/water) in which both the proportion of water and the temperature can be varied.

Gibbs' law indicates that

$$v = c + 2 - \varphi \tag{8.2}$$

where

v = variance
c = number of independent constituents
φ = number of phases

When $v = 0$:

$\varphi_{max} = c + 2$ maximum number of phases (at constant pressure $\varphi_{max} = c + 1$)
$\varphi_{max} = 4$ for a binary mixture ($\varphi_{max} = 3$ at constant pressure)

The shaded regions in Figure 8.9 represent the domains in which two different phases coexist in thermodynamical equilibrium ($\varphi = 2$). For the mixture

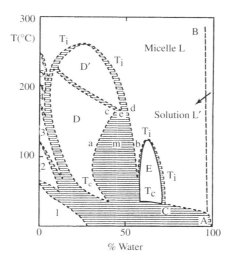

Figure 8.9 Phase diagram at constant pressure for mixtures of sodium oleate/water: 1, crystalline and gel phases; 2, 3, 4, 5, thermotropic phases (pure or highly concentrated amphiphiles); D, D', lamellar phases (neat soap); E, cylindrical phase (middle soap); L, micelle; L', molecularly dispersed solution. (Reproduced by permission of EDP Sciences from Ref. [8])

represented in Figure 8.9 as the point m, the two corresponding phases have a composition given by the points a and b. By going from point a to b, the composition of the two phases does not change: only the relative proportion varies. The point where two curves intercept (point e in Figure 8.9) is such that $\varphi = 3$. The composition of the three phases is given by points e, c and d.

The curve indicated by T_c defines the temperature at which any crystalline order and any organization of the lipophilic moieties of the surfactant molecules disappears. The phase 'gel' is situated below T_c. The curves indicated by T_i reflect the disappearance of all long-range order. It corresponds to the formation, at high temperatures, of micellar solutions [8].

The various organized phases derived from soaps may be classified depending on the degree of order of the hydrophilic and hydrophobic moieties.

8.2.3.1 Ordered Lipophilic Parts

In the *crystalline* state, the aliphatic tails are fully elongated and tilted (around 25°) relative to the normal of the layer (Figure 8.10a). The detailed structure is very rarely known [8] and it is, in many cases, only characterized by the thickness of the lamellae and the *molecular area* (area occupied by a chain in the plane of the layers) which is around 20 Å^2. The crystals are observed at temperatures lower than 100 °C. The cation associated with the anionic polar head forms a tighly bound ion pair [9].

The gel phase is observed in the presence of water. It is formed of lamellae separated by a sheet of water molecules up to 100 Å thick (depending on the concentration of the amphiphile). The soap molecules are on the average perpendicular to the surface layer and form a single layer (Figure 8.10b).

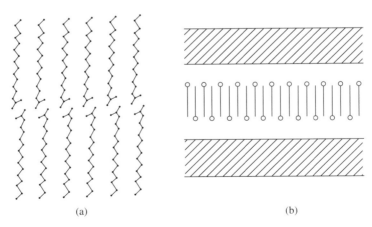

Figure 8.10 Schematic representations of (a) the crystalline state and (b) gel phase of a soap. The amphiphiles (dashed domains) are separated by water layers. (Reproduced by permission of EDP Sciences from Ref. [8])

INDUSTRIAL APPLICATIONS OF MOLECULAR MATERIALS

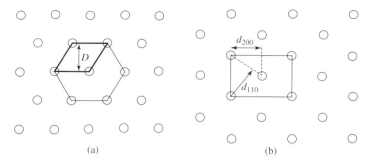

Figure 8.11 The two types of packing observed in the gel phase for usual soaps: (a) hexagonal ($D \sim 4.1$ Å); (b) orthorhombic ($d_{110} = 4.2$ Å, $d_{200} = 3.8$ Å). The values D, d_{110} and d_{200} have been determined by X-ray diffraction measurements (Reproduced by permission of EDP Sciences from Ref. [8])

In the gel phase the area per polar head is of the order of 40 Å^2 (20 Å^2 per paraffinic chains). The paraffinic chains are fully elongated and, depending on the temperature, they form either a hexagonal or a centered rectangular bidimensional lattice (Figure 8.11).

In a fully extended conformation, the length of the chain is given by [9]

$$l_{ext} = 1.265n + r_{vdw} \text{ (Me)} \qquad (8.3)$$

where

n = number of CH$_2$ units
r_{vdw} (Me) = van der Waals radius of a methyl group (2 Å)

In the gel domain, when the concentration of water is varied, the thickness of the hydrophobic part and the area per polar head remain constant.

8.2.3.2 Disordered Lipophilic Parts and Ordered Polar Heads

This type of organization is mainly found for pure amphiphilic compounds and especially for alkali or alkaline earth soaps. At high temperature, the hydrocarbon chains are disordered and their length may be expressed as

$$l_{flex} = \rho l_{ext}$$

where

l_{flex} = effective length in the disordered state
ρ = coefficient depending on the number of C atoms of the hydrocarbon chains

The coefficient ρ has been calculated for several chain lengths [10] (see Table 8.1).

The polar heads are still packed in a regular way. They can be considered as crystalline although the corresponding domain sizes are small.

Table 8.1 Characteristic lengths of the fully extended paraffinic chain (l_{ext}) in the crystalline state and the corresponding distances in the molten state (l_{flex}). (Cited in Ref. [9], from Ref. [10])

n	1.265n	ρ_{flex}	l_{flex} (Å)
4	5.06	0.90	6.3
6	7.59	0.84	8.0
8	10.12	0.80	9.7
10	12.65	0.75	11.0
12	15.18	0.72	12.4
14	17.71	0.69	13.6
16	20.24	0.66	14.7
18	22.77	0.63	15.6

Ribbon Structure

The soap molecules are arranged in quasi-infinite ribbons whose width decreases when the temperature is increased. They form double layers and the area per polar head is of the order of 20 Å² (approximately the same as for crystals). The ribbons form two-dimensional centered rectangular or oblique lattices (Figure 8.12).

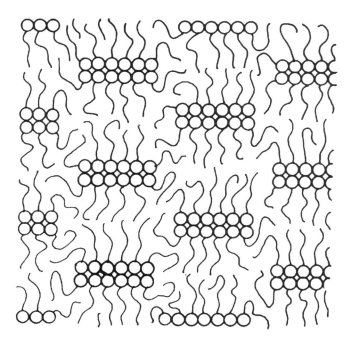

Figure 8.12 Ribbon-like structure of soaps (centered rectangular lattice). (Reproduced by permission of EDP Sciences from Ref. [8])

INDUSTRIAL APPLICATIONS OF MOLECULAR MATERIALS

Infinite Rods

This structure is well established for the high-temperature phases of alkaline earth soaps. The diameter of the rods is several angströms and they are made of the polar heads. The rods form a hexagonal lattice. The chains are in a disordered state [8].

Small-Sticks Models

The rods, contrarily to the previous case, have a limited length and they form three different structures: centred cubic, centred tetragonal and rhombohedral (Figures 8.13, 8.14 and 8.15).

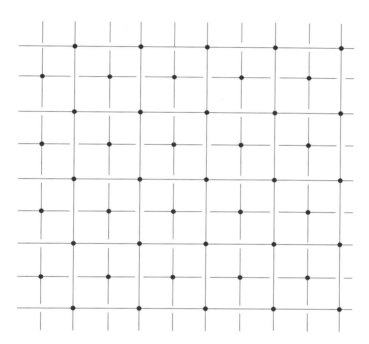

Figure 8.13 Centred tetragonal structure. The two planar networks of sticks are represented as a projection in a plane perpendicular to the C_4 axis $(0, \frac{1}{2})$. (Reproduced by permission of EDP Sciences from Ref. [8,100])

8.2.3.3 Disordered Hydrophilic and Hydrophobic Parts

Lamellar Phase (Neat Soap)

The neat phase (phase lisse) is probably the most important one from an industrial point of view. It is frequently encountered with pure soaps at high temperatures as well as in the presence of water, hydrocarbons or cosurfactants. It corresponds to the packing of lamellae, the amphiphiles being arranged in bilayers (Figure 8.16).

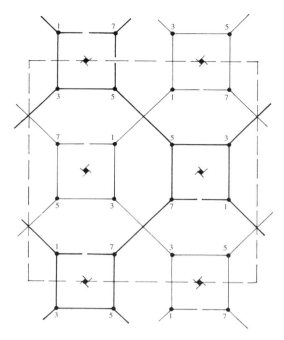

Figure 8.14 Centred cubic structure (Ia3d). There are 24 sticks per unit cell which form two interpenetrated three-dimensional lattices. The numbers (in $\frac{1}{8}$ of unit cell) indicate the heights. (Reproduced by permission of EDP Sciences from Ref. [8,101])

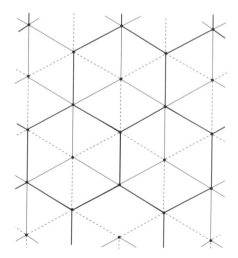

Figure 8.15 Rhombohedral structure (R3m). There are three planar honeycomb lattices of sticks represented as a projection in a plane perpendicular to the C_3 axis $(0, \frac{1}{3}, \frac{2}{3})$. (Reproduced by permission of EDP Sciences from Ref. [8,100])

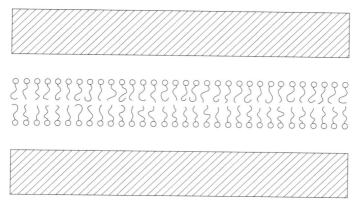

Figure 8.16 Schematic representation of a lamellar phase. (Reproduced by permission of EDP Sciences from Ref. [8])

The thickness of the soap layer is of the order of the molecular lengths and the area per polar head is approximately $40\,\text{Å}^2$.

Cylindrical Phases (Middle Soap)

The soap molecules form cylinders of (quasi) infinite length whose core is made of the hydrophobic moieties of the molecule (Figure 8.17). The cylinders are separated by a water medium. The diameter of the cylinder is of the same order

Figure 8.17 Schematic structure of middle soap (cylindrical phase). (Reproduced by permission of EDP Sciences from Ref. [8])

as the addition of the molecular length of two amphiphilic compounds. The area per polar head at the interface between the hydrophobic and hydrophilic parts is of the order of $50\,\text{Å}^2$.

Inverse Cylindrical Structure

Some amphiphiles like Aerosol OT (see Section 8.2.2) in the presence of water may lead to an inverse cylindrical structure (Figure 8.18). The core of the cylinders is made of water molecules surrounded by the polar heads of the amphiphiles. The area per polar head at the interface is of the order of $30\,\text{Å}^2$.

Miscellaneous

Many other structures have been characterized while studying the phase diagrams of soaps. Only two examples among many others will be given in this subsection (Figure 8.19).

8.2.4 Conclusions

Although soap is commonly found and routinely used, this molecular material is still not known in detail. The basic principles involved in a cleaning process are not understood with clarity:

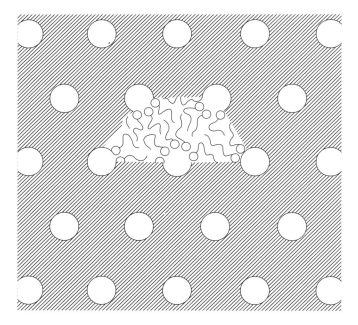

Figure 8.18 Schematic representation of the inverse cylindrical structure. (Reproduced by permission of EDP Sciences from Ref. [8])

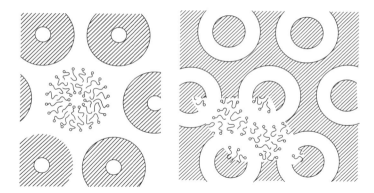

Figure 8.19 Schematic representations of the normal complex hexagonal structure (left) and of the inverse complex hexagonal structure (right). (For complementary information see also Ref. [11]). (Reproduced by permission of EDP Sciences from Ref. [8])

(a) What are the driving forces for the interaction of the hydrophobic tails with a greasy particle?
(b) What is the contribution of the structure of water?
(c) What is the effect of the polar head charge and structure on the amphiphilic properties?
(d) In the lyotropic condensed phases, what are the relative contributions of the molecular fragments on the stability of a given structure?
(e) Is it possible to predict the existence of a mesophase knowing merely the characteristics of the constituents and the corresponding physical parameters?

In other words, 'common' does not mean 'simple' and, importantly, the daily use of soaps preceded, by a very long time, the partial understanding of their behavior.

8.3 ORGANIC PIGMENTS AND DYES (COLORANTS)

Colorant is a generic name that includes dyes (molecularly dissolved in the medium) and pigments that are approximately micrometric-sized particles dispersed in various media (resins, polymers, oils, plastics, etc.).

In molecular chemistry, it is often difficult to grow large single crystals and thin films obtained by spin coating, vacuum sublimation or casting from solutions are fragile and difficult to handle. It is the reason why, in many cases, the molecular units are incorporated into a polymeric matrix for providing the required mechanical or physicochemical properties. This is the case for paints and inks.

8.3.1 The Notion of Color

The *flux* is defined as the energy received by a surface per second. The corresponding unit is the watt (W) which is independent of the wavelength. In photometry, the unit of flux is the lumen (lm) which is analogous to the watt except that it is weighted by the *luminous efficiency* of the human eye. The luminous efficiency for daylight vision as a function of the wavelength is shown in Figure 8.20.

Luminous flux (in lumen) is related to radiant flux (in watt) by the formula

$$\Phi_v = 683 \int_\lambda V_\lambda \Phi_{e\lambda} \, d\lambda \tag{8.4}$$

where

V_λ = luminous efficiency: function taking into account human eye sensitivity (for daylight, phototopic observer)

$\Phi_{e\lambda} \, d\lambda$ = radiant flux (in W) in the interval $d\lambda$

The value of 683 lumen per watt was chosen to correspond to the lumen defined in early times as the light of a standard candle. The unit of luminous intensity (I_v) is the candela (cd) which is one lumen per steradian. The *luminance* (L_v in candela per square meter) is the luminous intensity per projected area normal to the line of observation (angle θ).

The relationships which can be found between the various parameters are given below [12]:

$$E_v = \frac{\Phi_v}{a} = \frac{I_v}{d^2} \tag{8.5}$$

$$I_v = \frac{\Phi_v}{a/d^2} \tag{8.6}$$

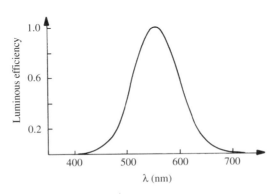

Figure 8.20 Luminous efficiency of the human eye for daylight vision. The luminous efficiency is the ratio of the fluxes at λ over the one at $\lambda_m = 555$ nm, producing the same luminous sensation. (From Williamson and Cummins, cited in Ref. [12])

INDUSTRIAL APPLICATIONS OF MOLECULAR MATERIALS

$$L_v = \frac{I_v}{a \cos \theta} \quad (8.7)$$

where

E_v = illuminance (lumen per unit area)
Φ_v = luminous flux (lumen)
L_v = luminance (lumen per steradian per square meter, or candela per square meter)
d = distance from the source with the corresponding area a
a/d^2 = solid angle

Color matching is usually carried out under conditions in which an observer compares the colors of one plate, one half illuminated by the sample color and the other half with the color mixture of three colors (Figure 8.21). A sufficiently small visual angle of 2° is generally taken in such a way that only the receptors at the centre of the retina are stimulated (conditions called CIE 1931 observer).

The primary colors R, G, B can be taken as monochromatic with red at 700 nm, green at 546.1 nm and blue at 435.8 nm. It was decided to measure the amount of green in a luminance unit of 4.5907 cd/m², blue in 0.0601 cd/m² and red 1.0000 cd/m². If R, G and B are the number of units of the primary colors needed to match the color S of the sample, then

$$S = R\,R + G\,G + B\,B \quad (8.8)$$

The relative amounts of R, G, B are chosen to match white light W:

$$W = 1\,R + 1\,G + 1\,B$$

In some cases, it is necessary to subtract one of the primary colors. Alternatively, one could add the proper primary color to the sample:

$$S = R\,R + G\,G - B\,B \quad (8.9)$$

$$S + B\,B = R\,R + G\,G \quad (8.10)$$

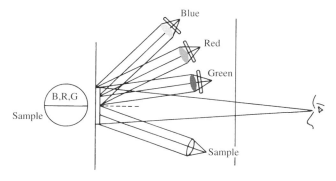

Figure 8.21 The color reconstituted from three independent sources (primaries) (additive method). One half of the plate is compared to the other half used as reference [12]

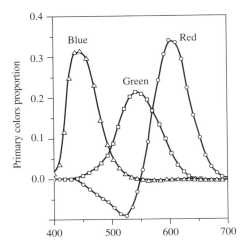

Figure 8.22 Color matching functions for a set of monochromatic primary colors R (700 nm), G (546.1 nm), B (435.8 nm). (The amount of power, in watts, is kept constant on half of the test screen)

The values of R, G and B necessary to match any given wavelength λ of the sample lead to the color matching functions (Figure 8.22).

To avoid negative amounts of primary colors three artificial primaries X, Y, Z have been proposed instead of R, G, B. X, Y, Z may be obtained from R, G, B. The chromaticity coordinates are defined as

$$x = \frac{X}{X+Y+Z}$$
$$y = \frac{Y}{X+Y+Z} \quad (8.11)$$
$$z = \frac{Z}{X+Y+Z}$$

Any color can be represented within a zone delimited by a locus (Figure 8.23, Plate 6). Since $x + y + z = 1$, only x and y are needed to specify the chromaticity.

The CIE primary colors are represented by $x = 1$, $y = 0$ for X, $x = 0$, $y = 1$ for Y and $x = 0$, $y = 0$ for Z. The point E ($x = \frac{1}{3}$, $y = \frac{1}{3}$) represents the location of the equal energy white (achromatic point).

8.3.2 Chemical Nature of Colorants

Cave paintings demonstrate that pigments, such as ocher hematite (Fe_2O_3) or brown iron ore (hydrated ferric oxide), were already used 30 000 years ago (Figure 8.24, Plate 6). Cinnabar (HgS), azurite ($Cu_3[(OH)_2/(CO_3)_2]$), malachite

INDUSTRIAL APPLICATIONS OF MOLECULAR MATERIALS

($Cu_2[(OH)_2/CO_3]$) and lapis lazuli (Na, Ca)$_8$[Al$_6$Si$_6$O$_{24}$]SO$_2$) were employed during the third millenium BC in China and Egypt [15].

The earliest synthetic pigment is probably 'Egyptian blue' (tomb of Perneh, 2650 BC) made by heating a copper silicate with sand and lime. Prussian blue, generally considered as the first inorganic pigment, was synthesized in 1704. A synthetic organic dye, picric acid was obtained in 1771 by P. Woulfe by treating indigo with nitric acid [13]. It was not, however, widely used for dyeing except for silk.

Many natural colorants can be extracted from vegetals or animals [16] (Appendix 10). The yellow dyes all come from vegetals. On the contrary, natural red colorants mainly derive from insects [16]. The structure of Tyrian purple was elucidated by Friedlander in 1909; using 12 000 molluscs (*Murex brandaris*) he obtained 1.4 g of 6,6'-dibromoindigo (Figure 8.25). The black dye logwood was known in 1500 but it became of real importance only when Chevreul discovered in 1812 that together with metallic salts it can give lakes [16].

W. H. Perkin in 1856 synthesized by accident what is considered to be the first organic dye. He was working on the synthesis of quinine, an antimalarial drug. By comparing the chemical formulae of allyltoluidine and quinine, he assumed that oxidation of the first compound could yield quinine [16]:

$$2C_{10}H_{13}N + 3[O] \longrightarrow C_{20}H_{24}N_2O_2 + H_2O$$

He thus attempted to prepare quinine by oxidation of allyltoluidine with $K_2Cr_2O_7$ in H_2SO_4. The chemical transformation is, if he had known the chemical formula, impossible (Figure 8.26).

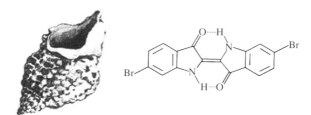

Figure 8.25 Chemical structure of 6,6'-dibromoindigo and the mollusc from which it is extracted (*Murex*). (From Ref. [17])

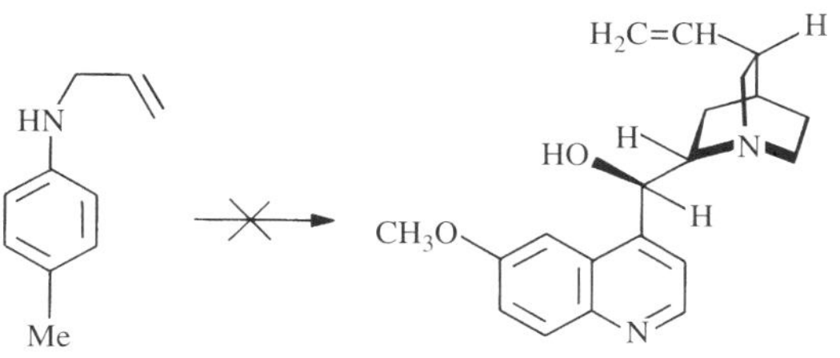

Figure 8.26 Perkin's attempt to synthesize quinine from allyltoluidine [16]

However, during his experiments, Perkin isolated purple crystals and he recognized that the corresponding compound (mauveine) might be used as a dye. This discovery was therefore based on a wrong chemical pathway, but it was also possible only because the starting material was not pure and contained aniline (Figure 8.27). Perkin's discovery of mauveine was achieved without any knowledge of the structures of the aromatic amines involved [16].

The number of synthetic organic (or molecular) colorants now exceeds 7000 [16] and the world consumption of organic pigments was, in 1990, 158 500 tons per year (powder basis) with a large proportion (about one-third) derived from phthalocyanine (Figure 8.28).

As already mentioned, the phthalocyanine molecule was synthesized early in the twentieth century. The pigments were first referred to as 'phthalocyanine blue pigments' around 1935. Partially or totally chlorinated derivatives led to

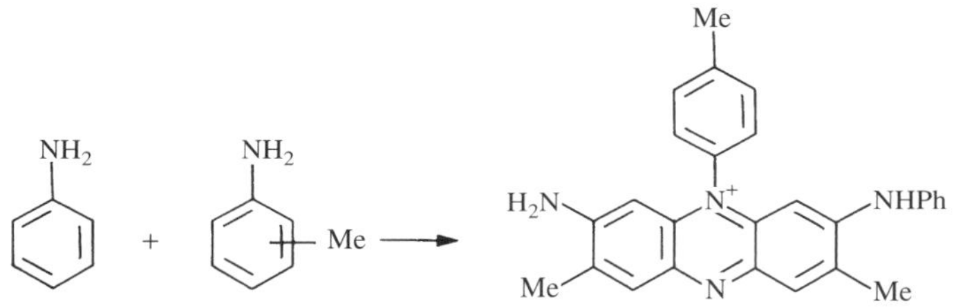

Figure 8.27 The synthesis of mauveine [16]

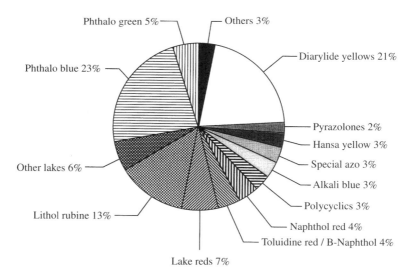

Figure 8.28 Wold consumption of organic pigments in 1990: 158 500 tons (powder basis). (The numbers shown exclude textile printing, paper, leather/fur). (After Ref. [18])

INDUSTRIAL APPLICATIONS OF MOLECULAR MATERIALS

'phthalo green pigments'. Diarylide yellows (patented in 1911) and Lithol Rubine (released on to the market in 1903) were also synthesized and used during the same period (see Appendix 10).

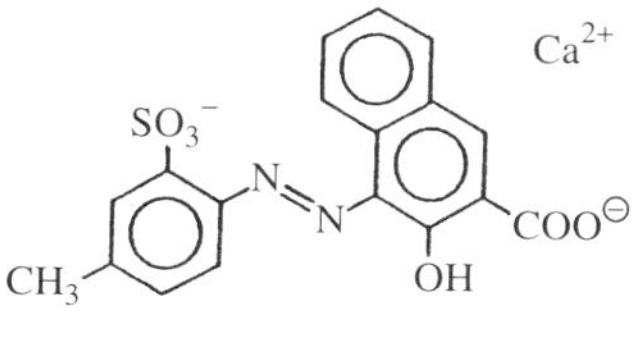

Lithol rubine

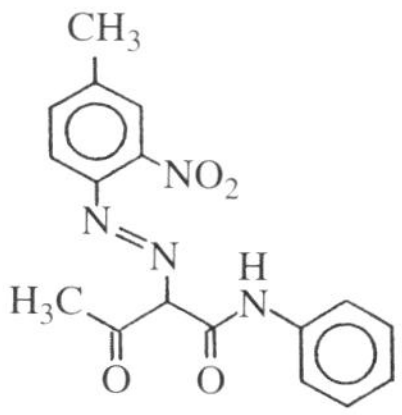

Hansa yellow

These two compounds belong to the very wide class of diazo derivatives used for making pigments such as Hansa Yellow, commercially available since 1909. In 1858, Verguin oxidized a mixture of aniline, o-toluidine and p-toluidine in nitrobenzene, obtaining a bluish-red fuchsin. This process has been industrially exploited since 1859 [15], and other derivatives have led to the class of pigments known as 'alkali blue'.

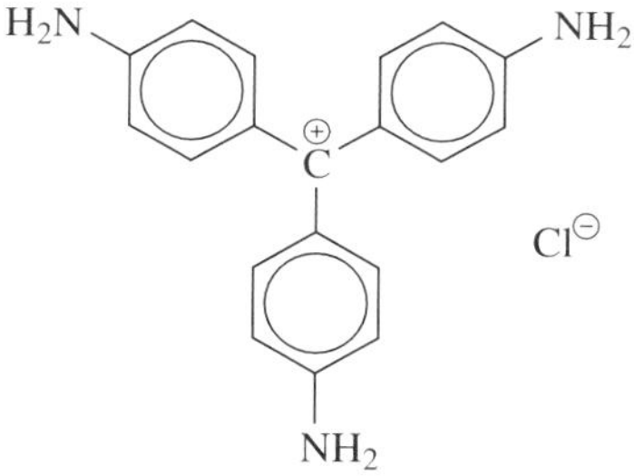

para-Fuchsin

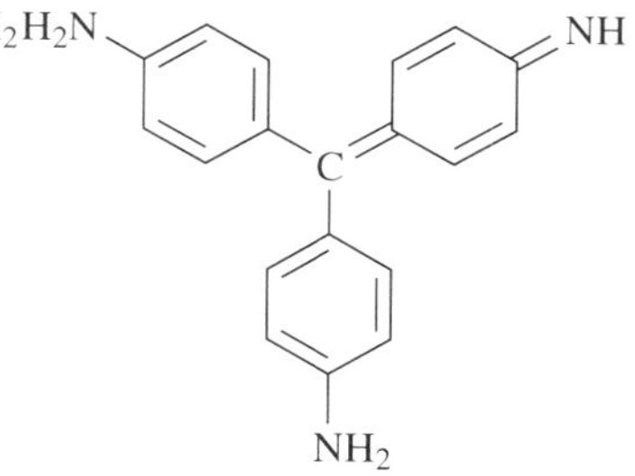

para-Rosaniline

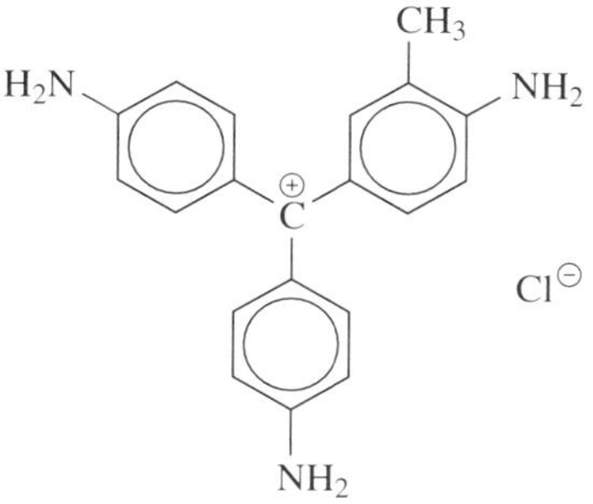

Bluish-red fuchsin

The class of pigments known as 'polycyclic pigments' is not defined on the basis of a uniform chemical structure, but because they do not derive from azo-derivatives (Table 8.2) [15].

Table 8.2 Some of the molecular units which can be used for fabricating 'polycyclic' pigments [15] (The class of pigments is indicated)

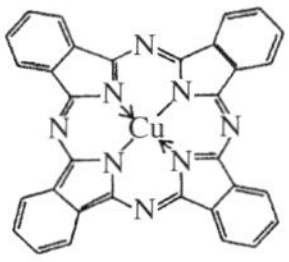

Phthalocyanine

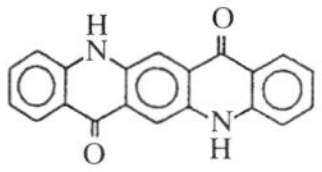

Quinacridone

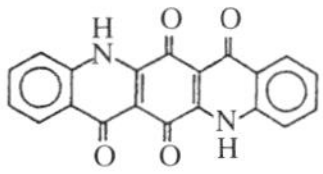

trans-Quinacridone quinone

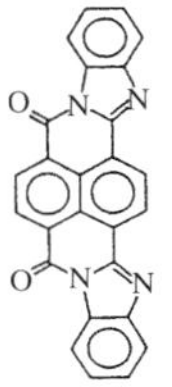

Perinone (*cis*)

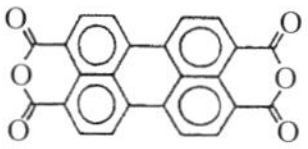

Perylene

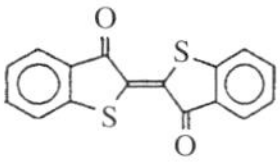

Thioindigo

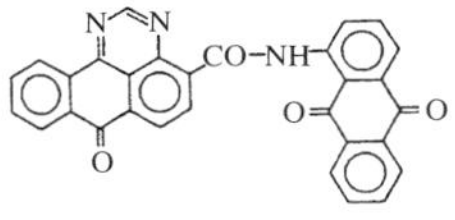

Anthrapyrimidine

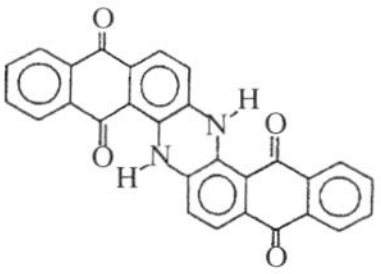

Indanthrone

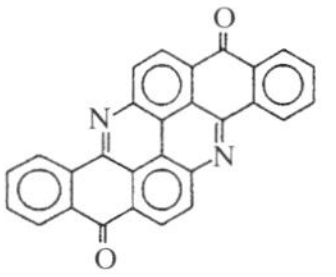

Flavanthrone

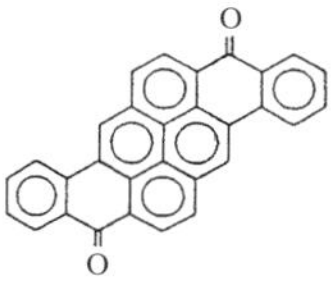

Pyranthrone

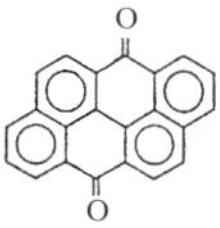

Anthanthrone

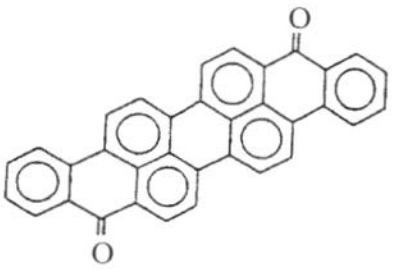

Isoviolanthrone

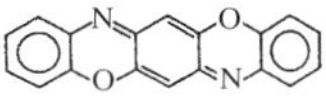

Dioxazine

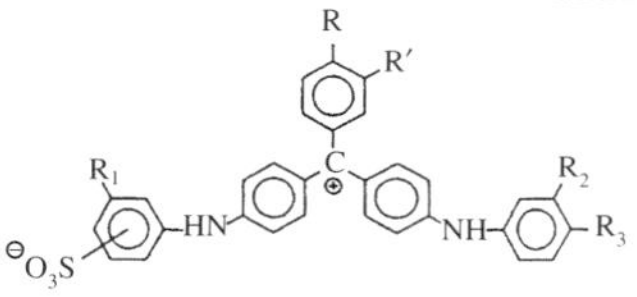

Triarylcarbonium

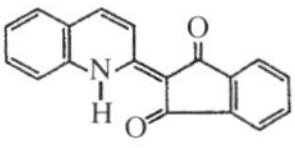

Quinophthalone

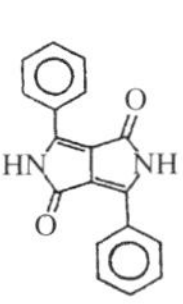

Diketopyrrolo-pyrrole

8.3.3 Physical Characterization of Pigments

Organic pigments are used for printing inks (40%), paints and dispersions (30%), incorporated into plastics (25%) and other markets (5%) [18]. Pigments do not yield molecularly dispersed molecular units (in most matrices) but are used as micrometric-sized particles. The particles are generally crystalline (at least partly) and the color properties of the pigment are mainly governed by the solid-state absorption characteristics of the molecular unit. In condensed phases, the coupling of the transition moments leads to a splitting of the absorption bands whose characteristics depend on the relative geometrical arrangement of the constitutive molecules (Figure 8.29). The number of bands appearing in the solid state depends on the number of molecules per unit cell. The effect of the coupling of transition moments is exemplified in the case of phthalocyanines (Figure 8.30).

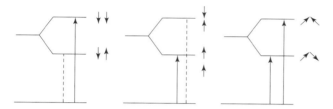

Figure 8.29 Three of the possibilities to couple transition moments (from left to right: parallel, head-to-tail, oblique): ———, allowed transition; – – –, forbidden transition. (Drawing modified from Refs. [19] and [20])

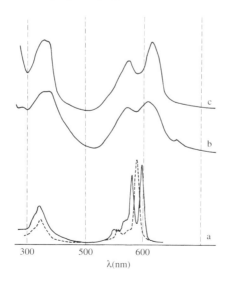

Figure 8.30 Absorption spectra of PcH_2 and PcM in solution (molecularly dispersed) (curves a) and for thin films of PcH_2 (curve b) and PcCu (curve c) deposited by vacuum sublimation. (Partly from Ref. [21], with permission)

Another important point in determining the color properties is the angle of the transition moment of the molecular unit relative to the face of the crystal on which the light is going to be reflected or absorbed. Such an example is shown in Figure 8.31 [19, 22]. Consequently, factors such as the geometry of the unit cell, the crystal lattice and the crystal shape will all influence the pigment characteristics. The properties of the particles (surface structure, crystallinity, particle size distribution) also play a major role (Figure 8.32, Plate 7).

The particles may be classified in to three categories (Figure 8.33):

(a) *Primary particles.* These are true single crystals with (eventually) some lattice disorder. Several structures may coexist. They differ by their shape (cubes, platelets, needles, bars, etc.).
(b) *Aggregates.* The aggregates are formed from primary particles grown together.
(c) *Agglomerates.* The agglomerates are groups of single crystals and/or aggregates but are not grown together.

The agglomerates can be separated during a dispersion process by the shearing forces occurring when the pigments are incorporated in a resin or in a polymeric matrix, but *not* the aggregates or the primary particles [15]. The dispersibility of a pigment is importantly determined by the nature and the density of the agglomerates.

8.3.4 Coatings

Coatings are layers of any substance spread over a surface for protection and decoration. In the following subsections an arbitrary choice has been made among the various coating processes. We will survey two of them: paints and inks.

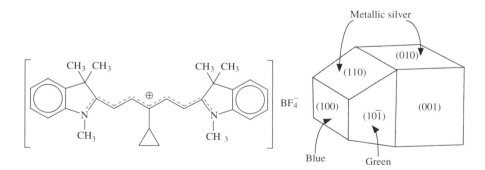

Figure 8.31 Chemical and crystalline structures of the illustrated dye. Triclinic crystal ($a = 13.19$ Å, $b = 11.04$ Å, $c = 9.87$ Å, $\alpha = 104°76'$, $\beta = 100°15'$, $\gamma = 92°58'$). The molecular axis is inclined at an angle of 9.4° relative to the {1 1 0} face. (After Refs. [19] and [22])

INDUSTRIAL APPLICATIONS OF MOLECULAR MATERIALS

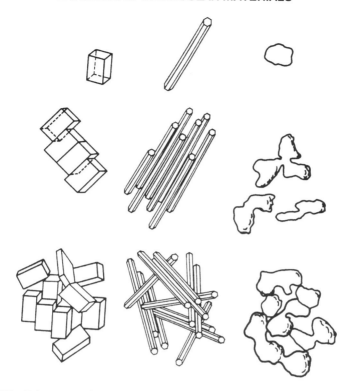

Figure 8.33 Primary particles, aggregates, and agglomerates. (Reproduced by permission of Wiley-VCH from Ref. [15])

8.3.4.1 Paints

A formulated paint is made [24–28] of pigments, solvents, additives and a 'vehicle', which is a solution or a colloidal dispersion of a polymer (binder). A distinction is made between colloids that are not influenced by gravity (and whose particle size is less than approximately 1 μm) and suspensions in which particles can decantate. The main terms encountered in this domain are the following [26]:

(a) Dispersion: finely divided solid particles in a liquid phase.

(b) Suspension: non-colloidal dispersion of a solid in a liquid.

(c) Emulsion: an immiscible liquid dispersed as droplets in another liquid.

(d) Latex: a colloidal dispersion of a polymer in a liquid (the polymer is generally prepared by emulsion polymerization).

A few examples of the main products used in industry will be given in the next subsections.

Siccative Oils

Natural siccative oils (vegetal or animal) derive from the triesters of glycerol [29]:

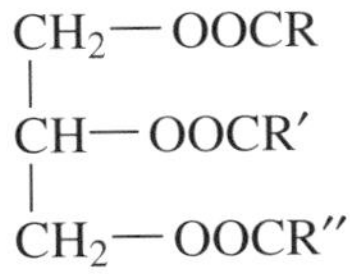

$$\begin{array}{l} CH_2-OOCR \\ | \\ CH-OOCR' \\ | \\ CH_2-OOCR'' \end{array}$$

where R, R' and R" differ by the number of C atoms, their structure and the number of double bonds.

Saturated fatty acids
  Palmitic acid $C_{15}H_{31}CO_2H$ (m.p. 62.5–63 °C)
  Stearic acid $C_{17}H_{35}CO_2H$ (m.p. 69.5–71 °C)
  Arachidic acid $C_{19}H_{39}CO_2H$ (m.p. 76–77 °C)

Unsaturated fatty acids
  Oleic acid (m.p. 13 or 16 °C depending on the crystalline form)
  The natural form is *cis*:

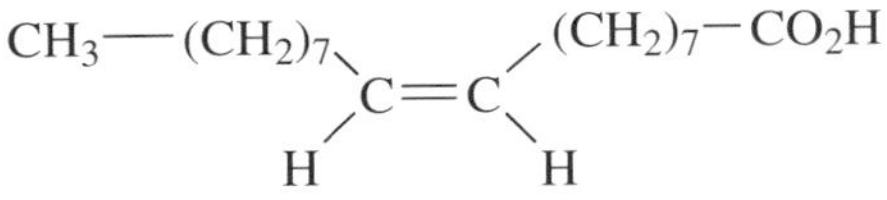

$$CH_3-(CH_2)_7\diagdown_{C=C}\diagup^{(CH_2)_7-CO_2H}_{\diagdown H}$$
$$H$$

Source: palm oil (50%)
Linoleic acid (m.p. −12 °C)

$$CH_3(CH_2)_4-CH=CH-CH_2-CH=CH-(CH_2)_7CO_2H \; (cis-cis)$$

Major constituent of many vegetal oils
Source: soya oil (54%)
Linolenic acid (m.p. 15 °C)

$$CH_3CH_2-CH=CH-CH_2-CH=CH-CH_2-CH=CH-(CH_2)_7CO_2H (cis-cis-cis)$$

Source: linseed oil (50%)
Eleostearic acid

$$CH_3-(CH_2)_3-CH=CH-CH=CH-CH=CH-(CH_2)_7-CO_2H$$

Three varieties are known.
Source: tung oil (80%)
Isanic acid (m.p. 39.5 °C)

$$CH_2=CH-(CH_2)_4-C\equiv C-C\equiv C-(CH_2)_7-CO_2H$$

Tends to polymerize and explode.

INDUSTRIAL APPLICATIONS OF MOLECULAR MATERIALS

Source: boleko oil (vegetal oils from tropical and equatorial regions)
Ricinoleic acid (m.p. 5.5 °C)

$$CH_3-(CH_2)_5-CHOH-CH_2-CH=CH-(CH_2)_7-CO_2H$$

It is not a constituent of siccative oils but it composes 90% of the fatty acids of castor oils. By dehydration under heating, it leads partly to isomers of linoleic acid.
Licanic acid (m.p. 74–75 °C or 99.5 °C)

$$CH_3-(CH_2)_3-CH=CH-CH=CH-CH=CH-(CH_2)_4-CO-(CH_2)_2-CO_2H$$

Exists in two crystalline forms

Besides triglycerides, siccative oils also contain approximately 1% of other derivatives such as sterols, hydrocarbons, mineral salts, etc.

A siccative effect transforms an oily thin layer into an adherent thin solid film. This is the case for linseed oil, tung oil, etc. The unsaturated nature of the fatty acid is a determinant factor to induce the siccative phenomenon as well as the necessary presence of dioxygen. It has been shown [26] that a methylene group situated between two double bonds is much more effective than simple allylic sites. Following this observation, a schematic illustration of the chemical processes involved in the siccative phenomena has been proposed (Figure 8.34).

The formation of peroxydes and the phenomenon of autoxidation have been already studied by Moureu and Dufraisse around 1920 [30]. The reaction of dioxygen may be catalyzed by cobalt salts or related derivatives.

Alkyds

The 'alkyd' resins are formed from the combination of polyacids and polyols: such a resin is therefore a three-dimensional polyester. The starting materials are briefly listed below.

Acids or anhydrides
   Phthalic anhydride (m.p. 128 °C)

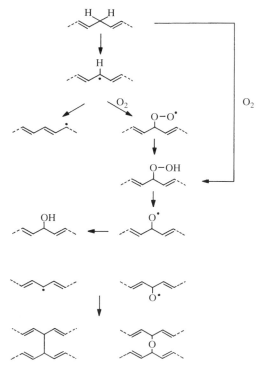

Figure 8.34 A few chemical reactions that can be involved in a siccative process

Maleic anhydride

$$\begin{array}{c} CH-CO \\ \parallel \diagdown O \\ CH-CO \diagup \end{array}$$

Acid		
Malonic	$HO_2C-CH_2-CO_2H$	135.6 °C
Succinic	$HO_2C-(CH_2)_2-CO_2H$	185 °C
Glutaric	$HO_2C-(CH_2)_3-CO_2H$	97.5 °C
Adipic	$HO_2C-(CH_2)_4-CO_2H$	151 °C
Pimelic	$HO_2C-(CH_2)_5-CO_2H$	105.5 °C
Suberic	$HO_2C-(CH_2)_6-CO_2H$	144 °C
Azelaic	$HO_2C-(CH_2)_7-CO_2H$	107 °C
Sebacic	$HO_2C-(CH_2)_8-CO_2H$	134 °C

Alcohols
 Glycerol

$$CH_2OH-CHOH-CH_2OH$$

INDUSTRIAL APPLICATIONS OF MOLECULAR MATERIALS

Pentaerythritol (m.p. 260.5 °C)

$$HOCH_2-\underset{\underset{CH_2OH}{|}}{\overset{\overset{CH_2OH}{|}}{C}}-CH_2OH$$

Sorbitol

$$HOCH_2-\underset{\underset{OH}{|}}{\overset{\overset{H}{|}}{C}}-\underset{\underset{OH}{|}}{\overset{\overset{H}{|}}{C}}-\underset{\underset{H}{|}}{\overset{\overset{OH}{|}}{C}}-\underset{\underset{OH}{|}}{\overset{\overset{H}{|}}{C}}-CH_2OH$$

Mannitol

$$HOCH_2-\underset{\underset{OH}{|}}{\overset{\overset{H}{|}}{C}}-\underset{\underset{OH}{|}}{\overset{\overset{H}{|}}{C}}-\underset{\underset{H}{|}}{\overset{\overset{OH}{|}}{C}}-\underset{\underset{H}{|}}{\overset{\overset{OH}{|}}{C}}-CH_2OH$$

Besides the bi- or polyfunctional acids and alcohols, fatty acids (usually as mixtures) are generally used to improve the technological qualities of the alkyd resin [31]. The physicochemical properties of the material depend on the molar mass of the polymer and on its degree of crosslinking.

The evolution of an alkyd resin depends on the altered or unaltered nature of its skeleton. Pure alkyds (with some saturated monoacids) form thin films by polycondensation reactions, generally in an oven. For modified alkyds, the film formation depends on the nature of the modification. An unsaturated fatty acid, such as linoleic acid, may be used as a functional group and then siccative properties may be found.

The dispersion of the pigment within the resin maintains the integrity of the primary particles but the aggregates break down the agglomerates. However, dispersed pigments may afterwards form loosely combined units known as *flocculates* (Figure 8.35). Flocculates are labile and can usually be broken down by stirring [15].

Latex

A latex is a colloidal dispersion of a polymeric material in a liquid, very often water. The polymer is synthesized in an emulsion. An amphiphilic molecule is used during the polymerization process in order to facilitate the solubilization of the constituents (monomer, initiator) and to stabilize the droplets of hydrophobic entities in water (Figure 8.36).

A typical latex recipe mentions [26]:

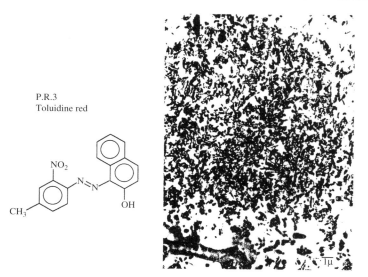

Figure 8.35 Pigment Red 3 (toluidine red) in an air-dried alkyd resin film. Electron micrograph of an ultramicrotome-cut thin layer of a pigment flocculate. (Reproduced by permission of Wiley-VCH from Ref. [15])

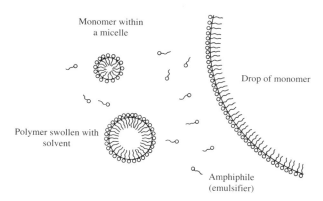

Figure 8.36 Schematic representation of an emulsion polymerization (Modified from Ref. [32]; cited in Ref. [26]) (typical latex size 0.1–0.5 μm)

A hydrophobic monomer	40–60% (v/v)
Water	40–60% (v/v)
Water soluble initiator	0.1–1% (w/w monomer)
Amphiphile (emulsifier)	1–5%

The monomers used depend on their applications (Table 8.3) [26].

The emulsifiers are also very diverse. Hoffmann (Bayer) at the beginning of the century used albumin or gelatine as protective colloids. Around 1920, oleates,

INDUSTRIAL APPLICATIONS OF MOLECULAR MATERIALS

Table 8.3 Some of the polymer latexes industrially used and their applications [26]

Polymer latex	Application
Styrene-butadiene, vinyl acetate, acrylic copolymers	Interior flat paints
Vinyl acetate, acrylic copolymers	Interior or exterior paints
Styrene-butadiene, acrylic copolymers	Industrial paints
Acrylic esters	Specially crosslinking paints

alkylarylsulfonates and abietates (see the next section on rosin) were used to form stable monomer emulsions that led to stable latexes. Hydrogen peroxide and other oxidizing agents (persulfates, perborates) were employed to increase the polymerization rate.

The latexes have recently gained more and more interests in industrial applications since they are water-based and do not require volatile organic solvents.

World Production and Market

The composition of a paint has been schematically described in the above section. A more detailed description is given in Figure 8.37.

The manufacture of paints is considered as 'mature' [27]. Paint companies become larger and larger and the total number of paint manufacturers decreases [27]. In 1992, the top ten companies produced about 30% of the world uses (Table 8.4).

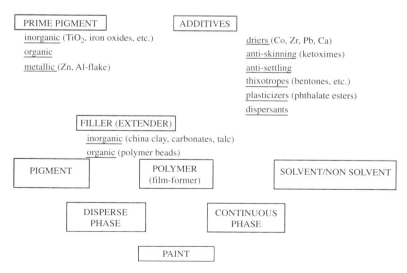

Figure 8.37 Components found in industrial paints. (After Ref. [27])

Table 8.4 The main world companies in 1991 classified according to their production in ML (millions of liters) [33]

Company	ML
1. ICI (UK)	805
2. Sherwin-Williams (USA)	533
3. PPG (USA)	515
4. BASF (Germany)	500
5. Akzo (Netherlands)	490
6. Nippon Paint (Japan)	350
7. Casco-Nobel (Sweden)	350
8. Courtalds (UK)	300
9. Kansai (Japan)	280
10. DuPont (USA)	265
11. Valspar (USA)	200
12. Sigma (Belgium)	180

8.3.4.2 Inks

An ink is a more or less viscous colored substance mainly used to write or print letters, drawings or images. Ink was fabricated in 2000 BC in China and Egypt from carbon black stabilized in water with gelatine, gum arabic, dextrin or other similar compounds. The invention of printing probably occurred in China [34]. This technique was introduced in Europe during the fifteenth century (J. Gutenberg 1440–1468). At that time, the ink was made of carbon black and mostly linseed oil. In 1818 P. Lorilleux created the first company fabricating inks in France.

Nowadays, the composition of inks is more diverse and depends on its final use. Either dyes (soluble) or pigments (insoluble) can be utilized depending on the application (Table 8.5).

A few examples of inks will be given in the following subsections.

Dyestuffs
The pigments used in inks are very often the same as for paints. However, in inks new soluble dyestuffs may also be employed. We will give below one example for each important class. Disperse dyes are used in both liquid and paste inks for heat transfer printing. Acid and basic dyes can be precipitated in the presence of inorganic substances or bases to yield pigments [36].

Table 8.5 The main components constituting an ink. (Modified from Ref. [35])

SOLIDS	Colorants	Pigments, dyestuffs (organic, inorganic)
	Support	Dispersants (colloids, gels)
		Crystalloids
	Additives	Emollients, tensioactive substitutes, fatty acids
LIQUIDS	Solvents	Hydrocarbons, others (alcohols)
	Binders	Oils (mineral, siccative) and resins
	Additives	Gums, resins, bitumen

INDUSTRIAL APPLICATIONS OF MOLECULAR MATERIALS

Acid dyes

The name 'acid dyes' originally applied to substances containing an acidic group (sulphonic, carboxylic, etc). However, such groups can also be present in dyes that do not belong to this class.

Eosine

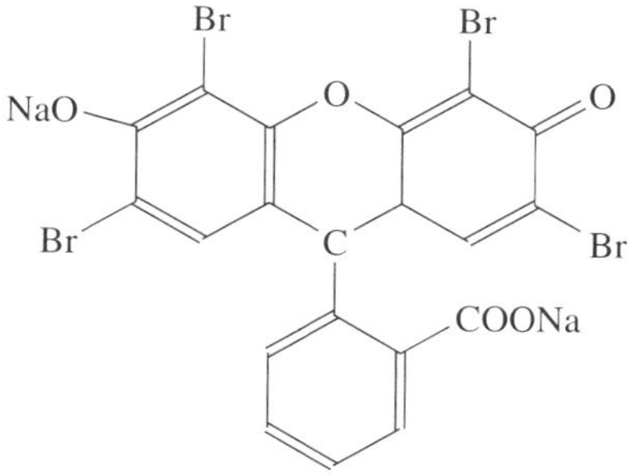

It gives a bright yellow-red hue and leads to a yellow-green fluorescence. It is soluble in water and related solvents. It is obtained by bromination of fluorescein and is used in water-based inks (invisible or fugitive inks).

Basic dyes

The first dyes produced from coal tar were basic dyestuffs. These cationic dyes have high tinctorial strength but their light fastness is poor. They are soluble in water and alcohols but they have a low solubility in most other organic solvents. These dyes are often used in conjunction with a mordant or laking agent like tannic acid. Tannic acid is produced from Turkish or Chinese nutgall. The chemistry of tannic acid (tannin, gallotannin, gallotannic acid) is complex [37]. Tannins may be divided into two groups: (a) derivatives of flavanols, (b) hydrolyzable tannins. These latter are the most important ones; they are esters of a sugar with one or more trihydroxybenzene carboxylic acid. One of these compounds is corilagin:

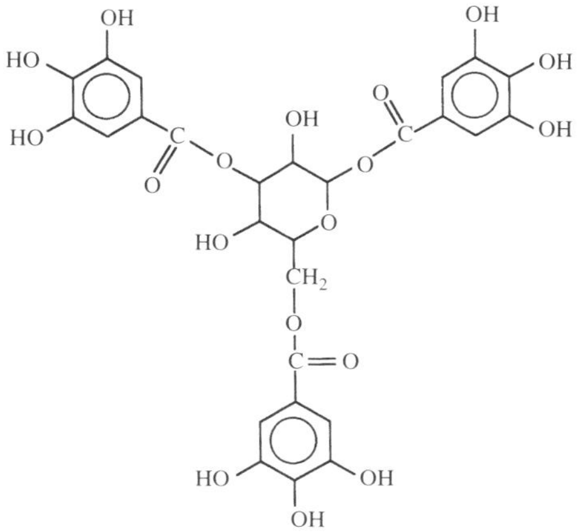

Victoria blue

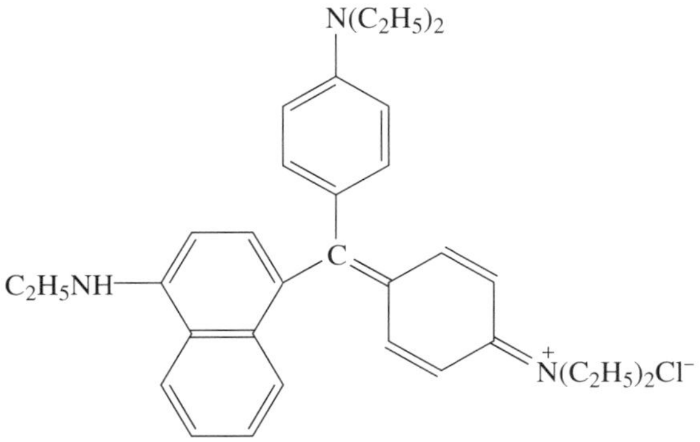

[Structure of Victoria blue: central C connected to a naphthalene bearing C$_2$H$_5$NH–, a phenyl ring bearing N(C$_2$H$_5$)$_2$, and a quinoid ring terminating in =$\overset{+}{N}$(C$_2$H$_5$)$_2$ Cl$^-$]

The basic dye 'Victoria blue' is bright blue. The compound is slightly soluble in cold water and has a poor light fastness. It is synthesized by condensing 4,4′-bis-(diethylamino)benzophenone with N-ethyl-1-naphthylamine. It is used as a dyestuff in flexographic inks. Different salts are used in all types of printing inks.

Solvent dyes
These dyes are soluble in an organic solvent. They include colorants described as fat dyes. The solubility is improved by many ways: base form of triarylmethane derivatives dissolved in fatty acids, acid dyes of the azochromium complexes, etc.

Induline (spirit soluble)

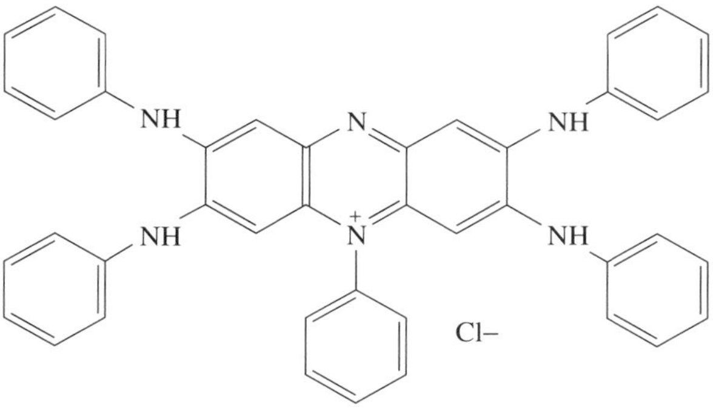

[Structure of Induline: a phenazinium core with four NH-phenyl substituents and an N-phenyl group on the central ring nitrogen, Cl$^-$ counterion]

This dye is reddish-blue and shows fair light fastness. It is stable up to 100 °C. It is insoluble in water but soluble in ethanol, wax, stearic acid and related compounds. It is used for stamping and gravure inks.

Disperse dyes
The name 'disperse dyes' was given because these colorants were sold as dispersions. Nearly all disperse dyes are amino derivatives of three main types: aminoazobenzene, aminoanthraquinone and nitrodiaryl amines. The dyes are

dispersed in a *vehicle* and are used for printing by various techniques. They are subsequently transferred under heat and pressure to the fabric in which they penetrate.

(C.I. disperse red 4)

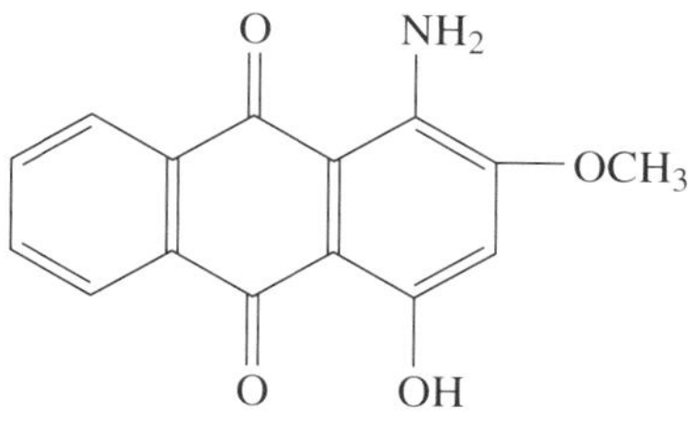

It is a bright scarlet dye with excellent light, washing, dry cleaning and perspiration fastness. It can be synthesized by methylating 4-aminoxanthopurpurin with dimethyl sulfate. It is used in inks for many conventional printing techniques, including the heat transfer process.

Rosin (Colophony)
Rosin is obtained from pine trees by distillation of the resin or solvent extraction of pine woods. The most important trading center of the oleoresin (rosin) was *Colophon* in present-day Turkey [28]. Two main products are isolated from pine tree resin [25]:

(a) wood turpentine obtained by distillation,
(b) rosin (not distillable residue).

90% of rosin is composed of five organic acids [38]. In the case of *Pinus Maritima*, the composition is as given in Figure 8.38.

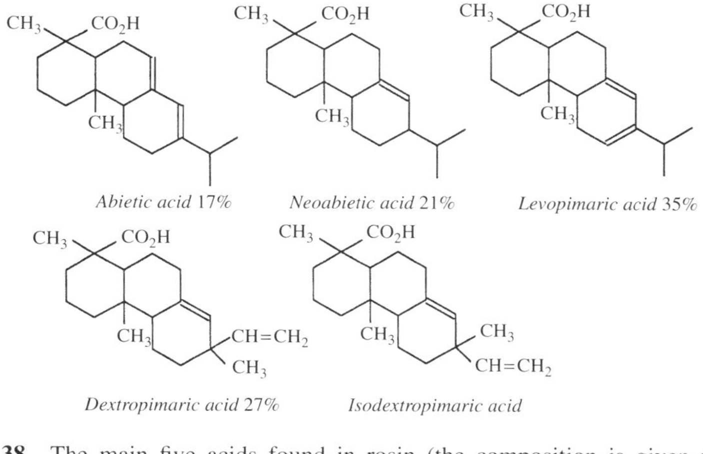

Figure 8.38 The main five acids found in rosin (the composition is given for *Pinus Maritima* as the natural source) [25]

Abietic acid (from Latin *abies*, meaning pine tree) can be crystallized from rosin solutions. It was isolated for the first time by Schultz in 1917 [38].

Different ester derivatives are formed which are used as varnishes, adhesives or paint systems [38]. Most of the abietic esters are highly viscous oils. Some of their physical properties are gathered in Table 8.6.

Rosin and related derivatives are susceptible to oxidation by air, leading to a darkening of the resin. To avoid this phenomenon, rosin is industrially hydrogenated [28] (Figure 8.39).

Table 8.6 Boiling points (b.p.) and melting points (m.p.) of various abietic esters. (After Ref. [38])

Esters of abietic acid	b.p. (pressure, mmHg)	m.p.
Me	210 °C (4 mm)	
Et	204–207 °C (4 mm)	
nBu	247–250 °C (3 mm)	
Ø	330–333 °C (4 mm)	
hexyl	299–302 °C (4 mm)	
cetyl (n-$C_{16}H_{33}$–)		42 °C
menthyl 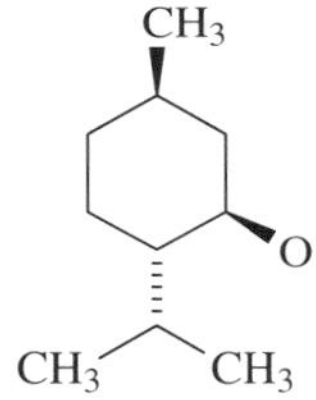		122–125 °C
bornyl 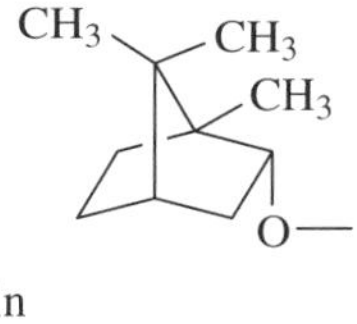		75–80 °C
glycerin (triabietic derivative)		150–162 °C

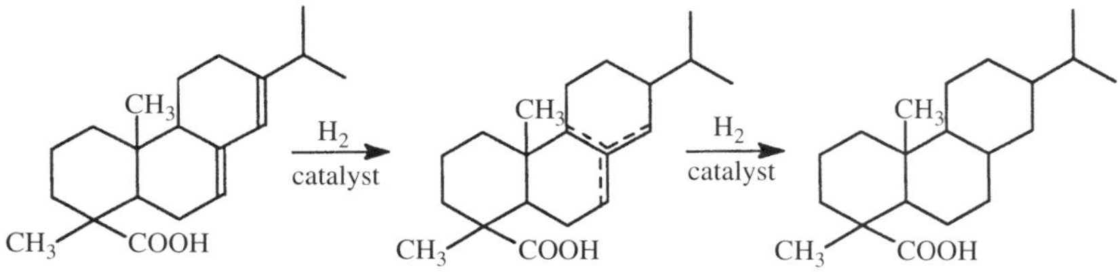

Figure 8.39 Hydrogenation of abietic acid. (After Ref. [28])

Abietic acid (2 double bonds) →H_2/catalyst→ Dihydroabietic acid (1 double bond) →H_2/catalyst→ Tetrahydroabietic acid (saturated)

INDUSTRIAL APPLICATIONS OF MOLECULAR MATERIALS

Increasing the degree of hydrogenation increases the oxidative stability but also changes the compatibility of the hydrogenated rosin in various polymers used in adhesives and chewing gum formulations. Adducts are obtained by reacting rosin with maleic anhydride. Levopimaric acid reacts at room temperature to give the adduct shown in Figure 8.40.

The other acids combine with maleic anhydride at higher temperatures (180 °C) [38]. The industrial abieto-maleic resins are obtained by esterification of the adduct with glycerin (ester gum). Various alcohols or even metallic oxides can also be used. Rosin can be condensed with many substances, but one of the most industrially important involves *resols*. These are obtained via condensation of phenol or substituted phenols with formol (Figure 8.41).

The resol is heated with rosin at temperatures around 300 °C to yield condensation products. These resins are widely used for printing inks (see, for example, Refs. [28] and [39]).

The oxidation of rosin, and in particular abietic acid, is fairly fast when the compound is dispersed. It then becomes insoluble in petroleum ether and chemical analysis shows the presence of peroxide groups [38] which further decompose into hydroxy groups. The oxidized rosin has a higher melting point than the genuine material. Rosin is also heat sensitive and often leads to a

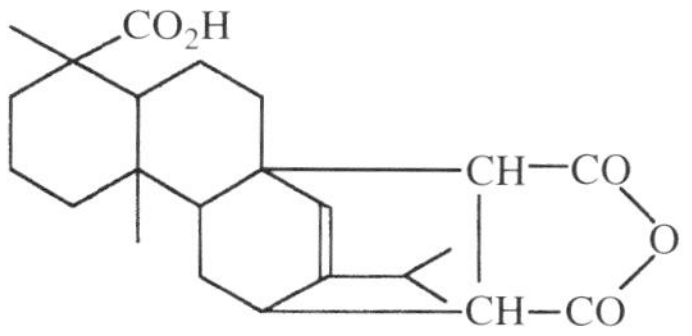

Figure 8.40 The adduct formed by the reaction of levopimaric acid and maleic anhydride

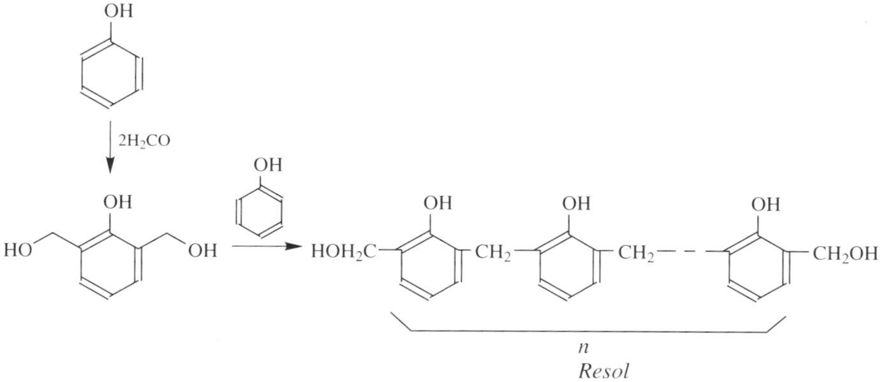

Figure 8.41 Production of resol from condensation of phenol with formol

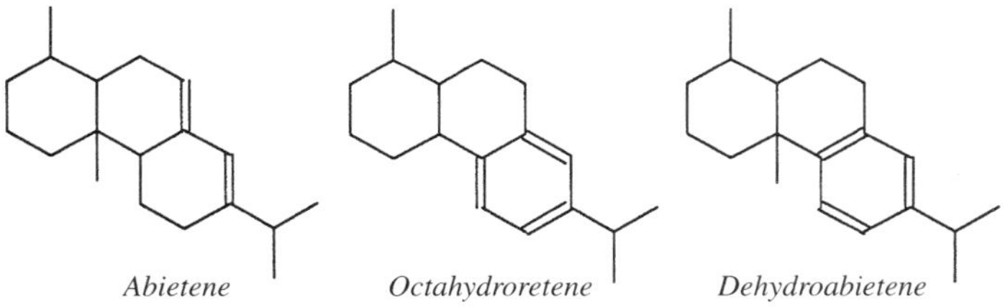

Abietene   *Octahydroretene*   *Dehydroabietene*

Figure 8.42 Three of the products obtained by pyrolysis of abietic acid. (After Ref. [38])

decarboxylation reaction. An oil is recovered by distillation (290–350 °C), which contains abietene, octahydroretene and dehydroabietene among other substances (Figure 8.42).

Rosin may also be transformed into the corresponding metal ion salts. The alkali derivatives of rosin are hydrophilic and soluble in water (and hydroxylic solvents) but insoluble in hydrocarbons. When rosin is treated with a water solution of NaOH (5%) at 100 °C, a limpid solution is obtained which forms a gel by cooling [38]. The soaps derived from rosin never become dry like conventional soaps based on fatty acids. Rosin soaps can be used to solubilize hydrocarbons and as emulsifiers. They can also be utilized to stabilize emulsions at low concentrations. They are incorporated in usual soaps (approximately 10%) to favor the formation of foams.

The salts deriving from ions other than alkali cations are, on the contrary, hydrophobic and soluble in organic solvents. In many cases, they are commercially available in solution in an organic solvent. The manganese and cobalt derivatives are employed as siccative agents (see also driers). In solution, they can be easily incorporated in paints and varnishes [38]. The calcium, zinc and aluminum derivatives are used in varnishes.

Nitrocellulose

More than 70% of all solvent-based liquid inks contain nitrocellulose. It is generally blended with other polymers in order to optimize many physicochemical properties such as adhesion, heat resistance, flexibility, etc. (Table 8.7) [40].

Nitrocellulose is obtained by treating cellulose with nitric acid under various conditions (Table 8.8):

Table 8.7 Typical ink formulation in % w/w [40]

Pigment	15%
Nitrocellulose	10%
Polyurethane	5%
Solvent	65%
Additives	5%

Table 8.8 Nitrocellulose: number of $-NO_2$ substituents depending on the conditions of preparation. (Modified from Ref. [41])

$R = -NO_2$	Number of substituents	Preparation
[structure, 500–2500]	1	HNO_3 (75%)
	2.4	HNO_3 (pure)
	~3	HNO_3–KNO_3
	0–2.9	HNO_3 (vapors)
	~3	HNO_3–AcOH

Other Components

Ink formulations are composed of many more products than those presently described. As an example, a typical Sheet-Fed paste ink formulation is shown in Table 8.9 [40].

Solvents are employed following the approximate empirical rule "like dissolves like" (see Chapter 5 concerning *'Le principe des affinités électives'*). These solvents can be petroleum distillates, aromatic hydrocarbons (toluene, xylene), alcohols (methanol, n-propanol, cyclohexanol, etc.), glycols, ketones (acetone, methylethylcetone), esters (ethyl acetate, n-butylacetate) [36].

Waxes are used to improve water repellency. They can be synthetic chain-branched polyethylene (to lower the melting point), polytetrafluoroethylene or fatty acid amides (stearamide). They can also be natural waxes consisting of mixtures of fatty acids, esters, alcohols and hydrocarbons [36].

The driers are used to promote the oxidation of drying oils: ink films dry hard in a few hours using these catalysts. Liquid driers are generally heavy metal ion derivatives of organic acids and soaps soluble in oils. Fatty acids, rosin and naphthenic derivatives are used. The colbalt ion is the most widely employed cation for making driers, followed by manganese and lead (which are less active) [36].

Table 8.9 Typical Sheet-Fed paste ink formulation [40]

	% w/w
Pigment	15
Rosin-modified phenolic	30
Alkyd	20
Modified hydrocarbon resin	~0
Wax	3
Driers	2
Solvents	29.3
Chelating agent	0.5
Antioxidant	0.2

Conventional chelating agents are ethylene diamine tetracetic acid (EDTA), nitrilotriacetic acid (NTA) and dimethylglyoxime:

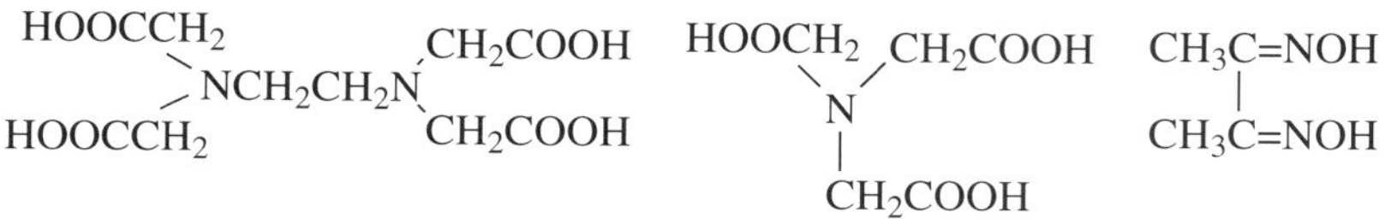

$$\begin{array}{cc} HOOCCH_2 & CH_2COOH \\ \diagdown NCH_2CH_2N \diagup & \\ HOOCCH_2 & CH_2COOH \end{array} \qquad \begin{array}{c} HOOCH_2 \diagdown \diagup CH_2COOH \\ N \\ | \\ CH_2COOH \end{array} \qquad \begin{array}{c} CH_3C=NOH \\ | \\ CH_3C=NOH \end{array}$$

EDTA   NTA   Dimethylglyoxine

Chelating agents are used to bind unwanted metal ions to form very stable (insoluble or soluble) chelates. They improve, for example, the wetting properties by sequestering iron, calcium or magnesium cations.

Antioxidants, like eugenol or ditert-butyl hydroxytoluene, react with the free radicals formed during the autoxidation process and therefore delay the initiation of the oxidative polymerization drying [36]:

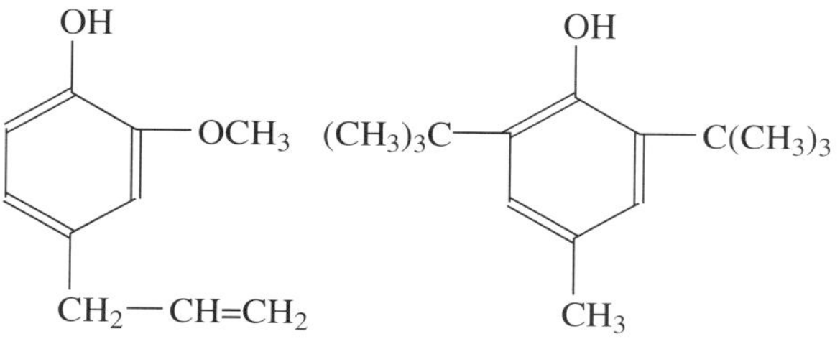

Eugenol   Di-tert-butylhydroxytoluene

Surfactants can adsorb on the surface of pigments by their polar head, the nonpolar part pointing outward into the vehicle. This can favour the dispersion of the pigment in the medium [36].

Plasticizers allow dried print to be more flexible and pliable. They can be considered as nonvolatile solvents. Di-n-butylphthalate, triethylcitrate and dicyclohexylphthalate are among the most popular plasticizers:

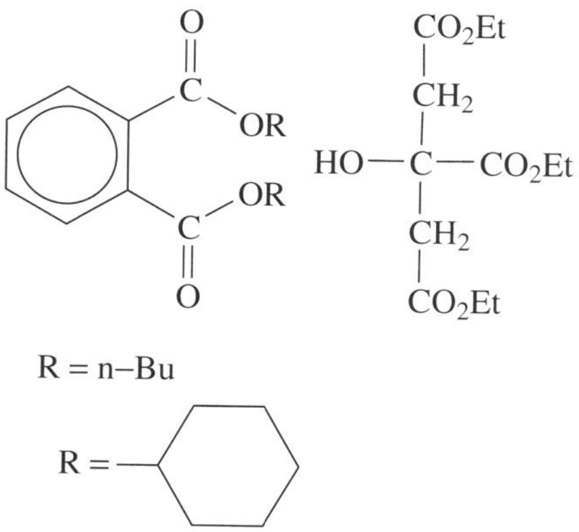

$R = n-Bu$

$R =$ —⬡

INDUSTRIAL APPLICATIONS OF MOLECULAR MATERIALS

Deodorants or reodorants, such as amyl and methyl salicylate, vanillin and natural essential oils, are also incorporated [36]:

Salicylates   *Vanillin*

R = Me —
R = C$_5$H$_{12}$ — (amyl)

World Market

The quantity of pigments used for making inks in the world was of the order of 0.6 Mtons in 1990 (the average pigment concentration in an ink is about 20%). The main manufacturers (about 50% of the world production) are Dai Nippon (Japan), BASF (Germany), Coates–Lorilleux (UK–France) and Toyo Ink (Japan) [42].

8.4 PHOTOCONDUCTORS AND PHOTOCOPYING MACHINES

8.4.1 Introduction

Photocopying machines are nowadays part of daily life. Xerography (from the Greek *xēros* meaning dry and *graphein* to write) was discovered by C. F. Carlson in 1938 and patented in 1942 [43] (Figure 8.43). A zinc plate was coated with a thin layer of sulfur. The sulfur surface was rubbed with a handkerchief in order to obtain an electrostatic charge. Then, it was recovered with a microscope slide

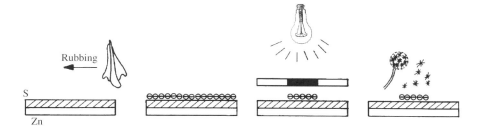

Figure 8.43 Schematic representation of the original Carlson discovery

on which was printed the notation "10–22–38 Astoria". The coated zinc plate and the glass slide were exposed to an incandescent lamp for a few seconds. The slide was removed and *lycopodium* powder was spread on the sulfur surface. By blowing on the surface, a duplicate of the notation appeared (for the original quotation, see Ref. [43]). The surface electrostatic charges seemed to have disappeared where the sulfur thin layer was illuminated.

In 1778, G. C. Lichtenbergh had made images by using surface electrostatic charges. He scattered dust on the surface of a resin previously exposed to an electric spark [43]. Many works related to 'electrophotography' were described in the 160 years separating Lichtenbergh from Carlson. However, the first commercial xerographic copier appeared in 1949 soon after Carlson's discovery (Haloid's Xerox Copier Model A) [43]. Before that time, copying machines were based on wet chemical processes.

In 1956, W. E. Bixby and O. A. Ullrich used selenium, which presents a much higher photoconductivity than sulfur (or anthracene) utilized in Carlson's original work [43].

The use of organic materials is related for the first time in 1957 by H. Hoegl, O. Süs and W. Neugebauer [43]. In 1969, R. C. Chittick, J. H. Alexander and H. F. Sterling (see Ref. [43]) investigated the photoconduction properties of amorphous silicon. However, nowadays molecular materials are the most widely used in industry.

8.4.2 Photoconduction in Molecular Materials

Numerous mechanisms for the photogeneration of carriers are possible. In all cases, it can be considered that a given molecular unit, A, is photochemically excited:

$$A, A, A \xrightarrow{h\nu} A, A^*, A$$

Energy transfer from molecular unit to molecular unit is then possible. At least two mechanisms are possible depending upon the magnitude of the interaction energy between the molecular units. At short distances, the orbital overlap is sufficient to induce the energy transfer (the Dexter mechanism). If the molecules are relatively far apart, the interaction between the transition moments dominates (the Förster mechanism). This last process is effective up to 50–100 Å whereas the Dexter mechanism relies on exchange interactions that require an orbital overlap operating at much shorter distances (5–10 Å) [44]:

$$A, A^*, A \longrightarrow A, A, A^*$$

In the absence of strong spin orbit coupling, the excitation of A to A^* greatly conserves the overall spin of the molecular unit. This precludes the migration of the triplet excited state $^3A^*$ by the Förster mechanism. However, the selection rule is relaxed in the case of the Dexter mechanism where orbital mixing is effective.

INDUSTRIAL APPLICATIONS OF MOLECULAR MATERIALS

At some stage, primary ionization occurs:

$$A^*, A \longrightarrow A^+, A^-$$

This can happen in the bulk material, on the surface of the material or in the vicinity of structural or chemical defects:

$$A^*, \text{defect} \longrightarrow A^+, (\text{defect})^-$$

Because of the low dielectric constant effective in molecular materials, the energy needed to separate the initial ion pair may be greater than the energy necessary to electronically excite one molecular unit:

$$A^+, A^- \longrightarrow A^+ \ldots A^- \quad \text{ion pair separation}$$

The Onsager approach is generally used to estimate the probability of the formation of free charge carriers as a function of the wavelength of the incident photon:

$$A \xrightarrow[h\nu]{} \underset{\substack{\text{not}\\\text{relaxed}}}{\text{`}A^*\text{'}} \longrightarrow A^+ + e^-$$

When the photon has excess energy compared to the one strictly necessary to excite one molecular unit, the energy difference can be used to break the ion pair interaction energy:

$$\Delta E = E_{\text{ip}} = -\frac{qq'}{\varepsilon r_0} \tag{8.12}$$

where

E_{ip} = ion pair interaction energy
q, q' = charges
ε = dielectric constant
r_0 = distance (Å)

r_0 is called the *thermalization length*. The condition for free carrier formation is that the coulombic attraction is less than the thermal energy kT. The thermalization length must therefore be such that

$$r_0 \geq -\frac{qq'}{\varepsilon kT} \tag{8.13}$$

Geminate recombination, the recombination of a charge carrier with its parent countercharge, governs the quantum yield of charge carrier formation in most molecular materials. The amount of geminate recombination is diminished if an external electrical field is applied to the material. Figure 8.44 illustrates such studies for thin films of amorphous selenium [45].

The thermalization length depends strongly on the excitation wavelength, from 70 Å at 400 nm to only 8.4 Å at 620 nm. Correspondingly, the number of carriers generated per photon absorbed (quantum efficiency) varies with the magnitude of the external electrical field applied, from about 0.5 at high fields (10^7 V/cm) to 10^{-4}–10^{-5} below 10^4 V/cm for an incident wavelength of 620 nm. These results found for amorphous selenium may be extrapolated to molecular materials.

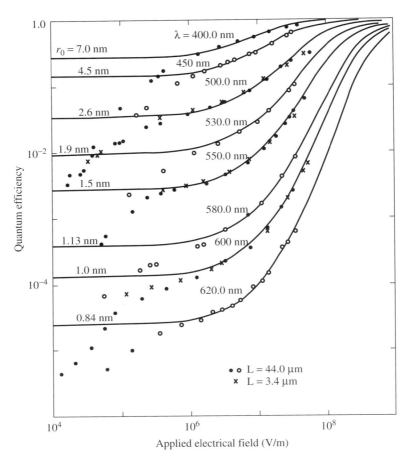

Figure 8.44 Quantum efficiency of photogeneration of charge carriers as a function of the wavelength of incident photons and of an external electrical field applied on thin films of amorphous selenium. Data concern films of two different thicknesses (3.4 and 44 µm). Solid lines follow Onsager's model for the initial separation r_0 indicated (Reproduced with permission from Ref. [45])

However, in some cases, the excess excitation energy can be dissipated by internal conversion only, the remainder being used to eject the charge carrier. In this case, the quantum efficiency for creation of charge carriers may be relatively insensitive to the incident radiation wavelength (see Ref. [44]).

8.4.3 Electrophotographic Processes [43, 46–48]

The electrophotographic process consists in transforming an image formed by a light source into a printed material. In this way, it can be compared with photography.

With standard light sources, approximately 10^{12} photons/cm^2 can be delivered. In order to obtain a dense black image, about 10^{17} light-absorbing molecules per cm^2 are necessary [48]. Photography therefore requires a 10^5-fold amplification. In silver halide based systems, the image is developed by reductive precipitation of silver in microcrystals that have received proper light exposure. The conversion of a 1 μm crystallite necessitates approximately a few dozen of photons. Since the whole crystallite is transformed, the amplification is given by the ratio of the metal Ag^0 formed over the minimum number of photons necessary. Each microcrystal is fully developed or not at all. Consequently, depending on grain size, 10^3–10^9-fold amplification may be obtained, the largest number corresponding to the poorest resolution [48].

In electrophotography, the primary quantum yield is 50–90% at high electrical fields — compared to about 1% for silver halides. With 10^{12} photons or electrons/cm^2, one must form a layer with absorbance corresponding to 10^{17} absorbing molecules per cm^2. The electrographic process must therefore have a gain of 10^5. This is achieved by the use of *toner* particles. A schematic scheme of the electrophotographic process is described in Figure 8.45.

In a first step (charge), a high voltage (5–15 kV) is applied between a wire and the substrate supporting the photoreceptor. Depending on the polarity, positive ions such as $(H_2O)_n H^+ (n=4-8)$ or negative ions, mainly CO_3^-, are deposited on the photoreceptor surface. In a second step (exposure), light is sent through the document; the black areas leave obscure the corresponding part of the photoreceptor. On the contrary, nonprinted parts yield to an irradiation of the photoconductor. Photogeneration of charge carriers occurs, the conductivity of the photoreceptor increases and neutralization of the positive and negative charges can take place. The surface dark conductivity of the photoreceptor being very low, redistribution of charges at the surface does not occur over the period of time considered. The surface charges that remain after exposure are therefore an image of the initial document. In the third step, development is realized. Light-absorbing entities are adsorbed on the charged areas via electrostatic interactions. However, a 1:1 ratio between adsorbed molecules and surface charges would not lead to sufficient coloration. The amplification is achieved by using toner particles. Dry toner is composed of a colorant (10%) within a resin binder. The diameter of the toner is about 5–20 μm. In two-component developers, the toner particles are themselves adsorbed on carrier (or bead) particles made of metal, glass or metal ferrite (Figure 8.46).

For the black color, carbon black is the most widely used material. The toner particles must be electrically charged in order to interact with the surface charges of the photoreceptor. One way to charge the toner particles is to pass them through high-voltage biased metallic nozzles to form an aerosol [48]. In the case where the toner particles are deposited on carriers the charging may be carried out triboelectrically. The toner powder is mixed with 50–100 times its weight of carrier particles possessing the desired triboelectric properties. Charges of the order of -50 to $+50$ μC/g can be obtained depending on the nature of the

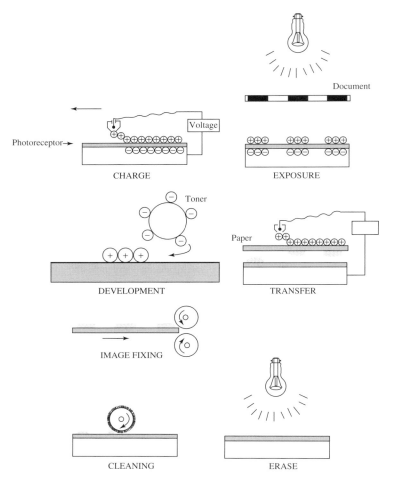

Figure 8.45 The main steps of the xerographic process. At the 'development' stage, part of the device (the photoreceptor) is magnified. (Modified from Ref. [43])

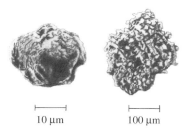

Figure 8.46 Schematic representation of a toner particle (10 μm scale) and of toner particles adsorbed on carrier (or bead) particles (100 μm scale). (From Ref. [43])

INDUSTRIAL APPLICATIONS OF MOLECULAR MATERIALS

pigment, the resin and the carrier [49]. It is also possible to employ in addition magnetized carrier beads [48]. In this last case, iron, cobalt or strontium ferrites are used [43].

The fourth step consists in the transfer of the toner to the final substrate, usually a sheet of paper. The transfer is electrostatically assisted by applying a corona discharge of proper sign on the back of the paper. In this way the transfer efficiency may be as high as 80–95% [43]. The fifth step is the 'image fixing'. The sheet of paper is passed between two heated rollers where fixing is achieved via pressure and heat. The choice of resin determines the fusing process. This is generally carried out in the range 120–130 °C. The photoreceptor must now be cleaned by removing all toner particles (sixth step) and all residual surface charges (seventh step) by overall illumination of the photoreceptor surface.

8.4.4 Xerographic Photoreceptors

As previously mentioned, in Carlson's original discovery, sulfur and anthracene were used as photoconductors followed rapidly by selenium. In 1957, the potential use of organic photoreceptors was reported. These can be either single-layer or dual-layer photoreceptors.

In the first case, the photoreceptor generally consists either of a pure molecular material (often vacuum deposited) or a colorant dispersed in a host polymer deposited by solvent coating. The photoconductive properties of mixed materials are given by the photochemical processes of the colorant particles, chemical reactions often arising at the interface between the particles and the host polymer (Figure 8.47).

The chemical and photochemical processes arising within colorant particles are of the same nature as those previously described for pure molecular materials. However, since the particles are of micrometric size, the influence of the surface must be higher than for large single crystals. Charge transport from particles to particles can occur whenever the distance between them is not too large. A simple model of charge transport was proposed by B. G. Bagley in 1970 [50] (Figure 8.48).

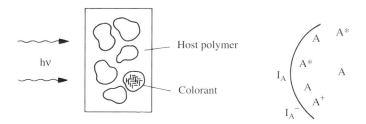

Figure 8.47 Illustration of a few photochemical processes arising in a colorant particle/host polymer system

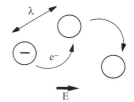

Figure 8.48 Schematic representation of a charge transfer between sites at a distance λ

Macroscopic charge transport involves multiple hoppings over barriers between sites at a distance λ. The energy barrier for hopping is ΔG. The hopping frequency, ν_1, is given by

$$\nu_1 = \nu_0 \exp\left(-\frac{\Delta G}{kT}\right) \qquad (8.14)$$

where

ν_0 = attempt frequency

When an electrical field is applied, the hopping probability is not the same *in* or *against* the direction of the field:

$$f = \nu_+ - \nu_- = \nu_1 \left[\exp\left(\frac{\lambda eE}{2kT}\right) - \exp\left(-\frac{\lambda eE}{2kT}\right)\right] \qquad (8.15)$$

where

E = electrical field

The term λeE represents the electrical energy difference before and after hopping and

$$f = 2\nu_1 \sinh\left(\frac{\lambda eE}{2kT}\right)$$

The drift mobility μ can be written as

$$\mu = \frac{\lambda f}{E}$$

$$\boxed{\mu = 2\frac{\lambda}{E}\nu_0 \exp\left(-\frac{\Delta G}{kT}\right) \sinh\left(\frac{\lambda eE}{2kT}\right)} \qquad (8.16)$$

At low electrical fields,

$$\sinh\left(\frac{\lambda eE}{2kT}\right) \to \frac{\lambda eE}{2kT}$$

$$\mu = \frac{\lambda^2 e \nu_0}{kT} \exp\left(-\frac{\Delta G}{kT}\right) \qquad (8.17)$$

INDUSTRIAL APPLICATIONS OF MOLECULAR MATERIALS

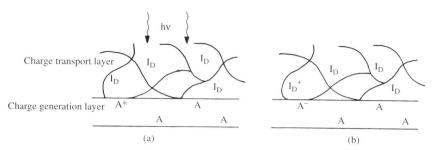

Figure 8.49 Dual-layer photoreceptor made of a charge generation layer (0.2–3 μm thick) and a charge transport layer (20–40 μm thick). (a) Excitation of the molecular unit of the charge generation layer. (b) Formation I_D, electron donor impurity; A, molecular unit

The mobility is in this case electrical field independent. At high electrical fields, the term sinh dominates.

More recent models have been proposed (see Chapter V of Ref. [43]) to take into account the different environments of the various hopping sites. It is considered that both site energies and intersite distances are nonuniform and subject to a distribution in energy.

In the mid 1980s [43], dual-layer photoreceptors were employed in industry. In this case, one layer (approximately 0.2–3 μm thick) is in charge of the absorption of the photons and the generation of charge carriers *(charge generation layer)*. The other layer — often consisting of an electron donor or an electron acceptor dispersed in a host polymer — transports one of the two generated charges *(charge transport layer)* (Figure 8.49).

8.4.5 Charge Transport Layer

Charge transport layers may be made of molecular units, aggregates or particles included in a polymer, pendant or main chain electroactive group polymers or organic glasses.

8.4.5.1 Polymer Binders

Some of the polymer binders that can be used are shown in Table 8.10.

8.4.5.2 Molecular Units Used in Charge Transport Layers

Electron donors derived from arylmethanes, arylamines, heterocycles, hydrazones or polysilylenes can be more or less easily oxidized and can therefore lead to hole transport. A few electron acceptors (trinitrofluorenone, 3,5'-dimethyl-3',5-di-*tert*-butyldiphenoquinone) can be alternatively employed (Table 8.11).

Table 8.10 Some of the polymer binders

Structure	Name
$(-CH_2-)_n$	Polyethylene (PE)
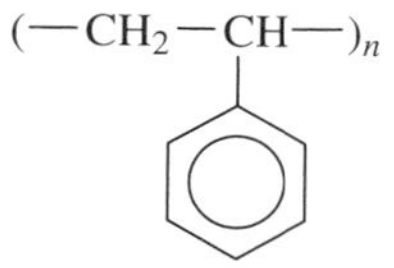 $(-CH_2-CH(C_6H_5)-)_n$	Polystyrene (PS)
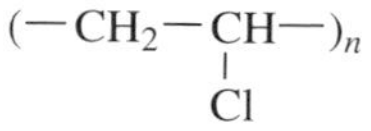 $(-CH_2-CHCl-)_n$	Poly(vinylchloride) (PVC)
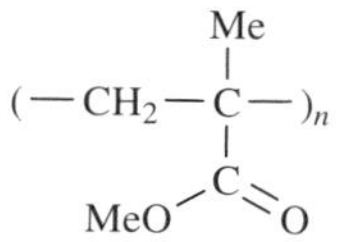 $(-CH_2-C(Me)(COOMe)-)_n$	Poly(methylmethacrylate) (PMMA)
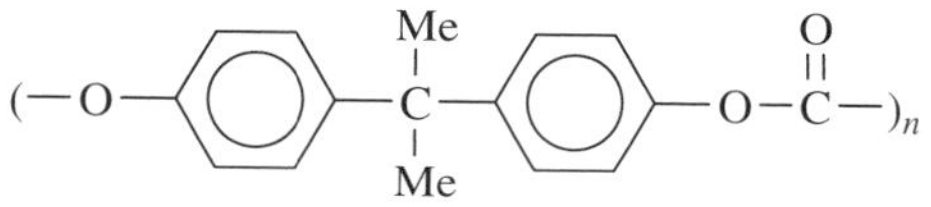 $(-O-C_6H_4-C(Me)_2-C_6H_4-O-C(=O)-)_n$	Bisphenol-A polycarbonate (PC)

8.4.5.3 Physicochemical Properties

In the Bässler formalism [43, 51], the low field mobility can be described as $\mu = \mu_0 \exp[-(T_0/T)^2]$, where T_0 is related to the width of the distribution of density of states and μ_0 is the mobility of a hypothetical disorder-free condensed phase extrapolated to a temperature $T \to \infty$.

Hole transport in MPMP/polystyrene and MPMP/polycarbonate couples have been studied (Figure 8.50) [52]. The concentration of MPMP is varied in the range 80–20%; correspondingly the intersite distance varies between 9 and 14 Å [43, 52]. In the same time μ_0 decreases by two orders of magnitude for both matrices (PC and PS), larger mobilities being obtained for the polycarbonate host polymer.

8.4.6 Charge Generation Layer

Many different compounds have been studied and commercially used for making charge generation layers:

Dyes: rhodamine and related molecules
Charge transfer complexes

INDUSTRIAL APPLICATIONS OF MOLECULAR MATERIALS

Table 8.11 Some of the molecular units used to form charge transport layers

Structure	Name
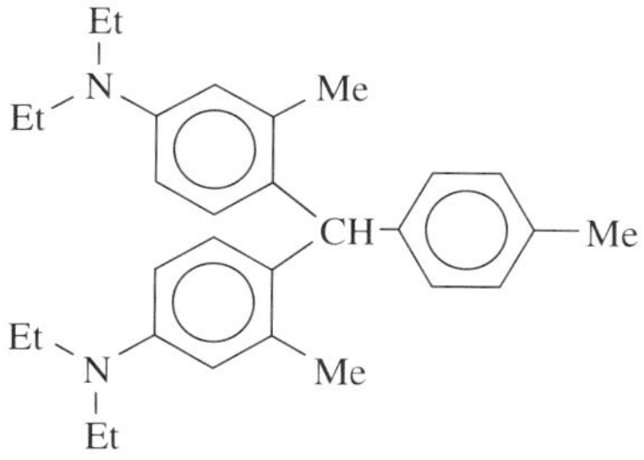 (structure shown)	bis-(4,N,N-Diethylamino-2-methylphenyl)-4-methylphenyl methane (MPMP)
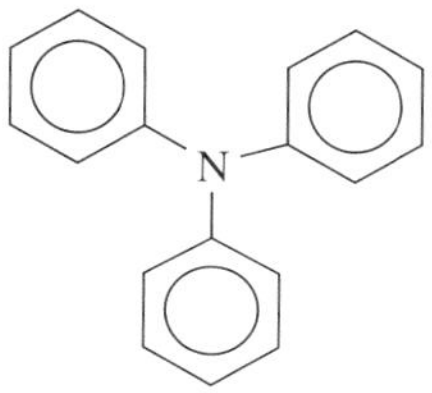 (structure shown)	Triphenylamine (TPA)
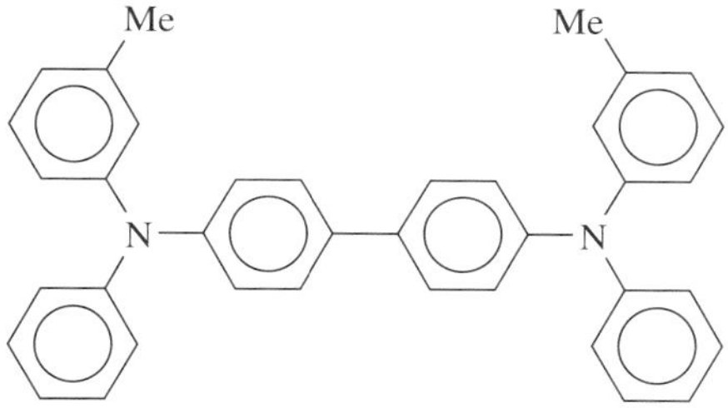 (structure shown)	N,N'-Diphenyl-N,N'-di(m-tolyl)-p-benzidine (TPD)
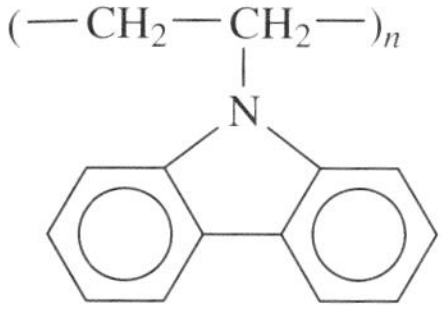 (structure shown)	Poly(N-vinylcarbazole) (PVK)
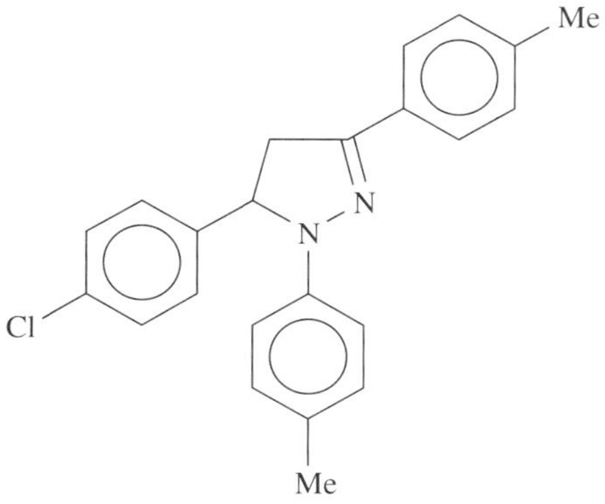 (structure shown)	1,3-di(4-Methylphenyl)-5-(4-chlorophenyl)-3-pyrazoline (TPPCl)

Table 8.11 (*continued*)

Structure	Name
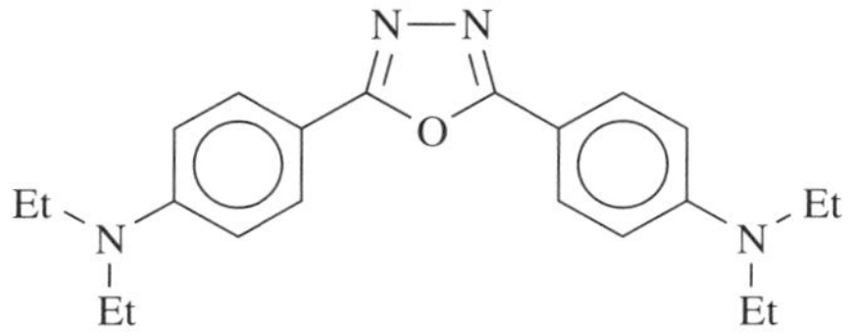 (2,5-bis(4-diethylaminophenyl)-1,3,4-oxadiazole structure)	2,5-bis(4-Diethylaminophenyl)-1,3,4-oxadiazole (OXD)
(p-diethylaminobenzaldehyde diphenylhydrazone structure)	*p*-Diethylaminobenzaldehyde diphenylhydrazone (DEH)
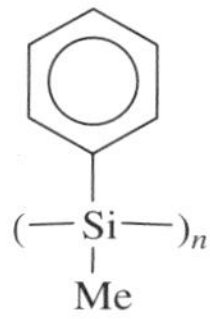 $(-Si-)_n$ with phenyl and Me substituents	Poly(methylphenylsilylene) (PMPS)

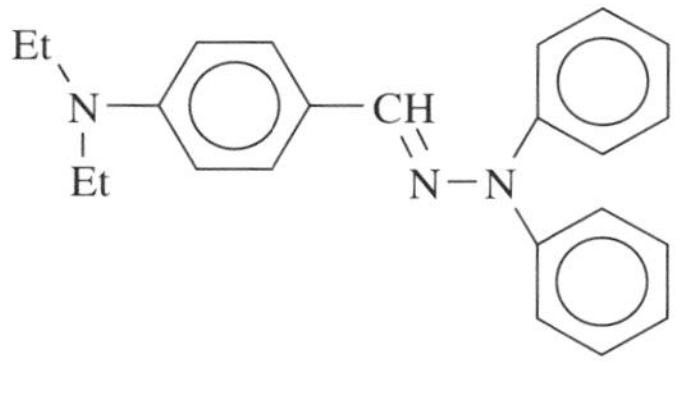

Pigments: azoderivatives, perylenes, phthalocyanines, squaraines, quinacridones, etc.

Polyacenes: anthracene, tetracene, pentacene, etc.

Polymers: poly(*N*-vinylcarbazole), poly(3-dodecylthiophene)

The first organic pigment used in a charge generation layer in a commercial photoreceptor was chlorodiane blue in 1975 [43] (Figure 8.51).

8.4.6.1 Azo Pigments

Bisazo derivatives are, among the azo pigments, the most widely studied as a generation layer [43]. Table 8.12 shows the various bisazo pigments used. The donor compound present in the transport layer has also been indicated.

8.4.6.2 Dye Polymer

Homogeneous films made of 4% 4-(4-dimethylaminophenyl)-2,6-diphenylthiapyrylium perchlorate in the polymer bisphenol-A polycarbonate have been prepared (cited in Ref. [43]) (Figure 8.52). The dye forms aggregates within the film in the presence of dichloromethane vapors. Electron microscopy shows the presence of fibers whose smallest diameter is about 70 Å. The aggregates occupy 10% of the overall volume [43]. Upon aggregation the optical absorption maximum wavelength is shifted from 580 to 690 nm. Such effects are also observed in the case of phthalocyanines.

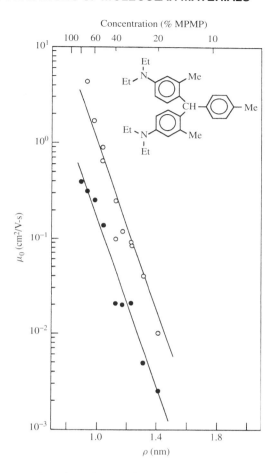

Figure 8.50 Concentration dependence of μ_0 for bis(4-N,N-diethylamino-2-methylphenyl)-4-methylphenylmethane (MPMP) included in polycarbonate (open circles) or poly(styrene) (solid circles): ρ = intersite distance. (after Refs. [43] and [52])

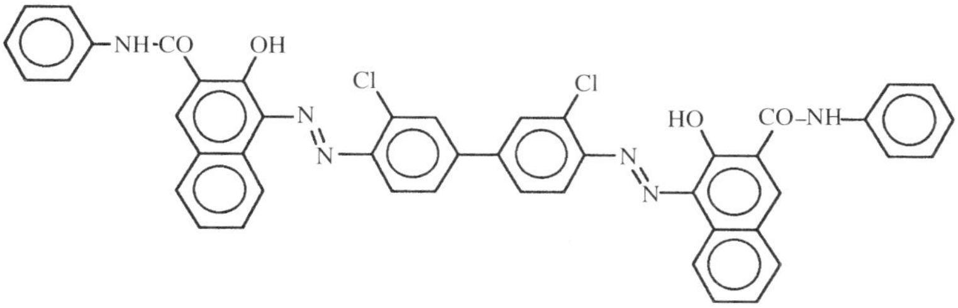

Figure 8.51 Molecular structure of chlorodiane blue

Table 8.12 A few of the bisazo pigments used in the generation layers. (After Ref. [43])

Pigment (generation layer)	Donor (transport layer)	Company
Chlorodiane blue 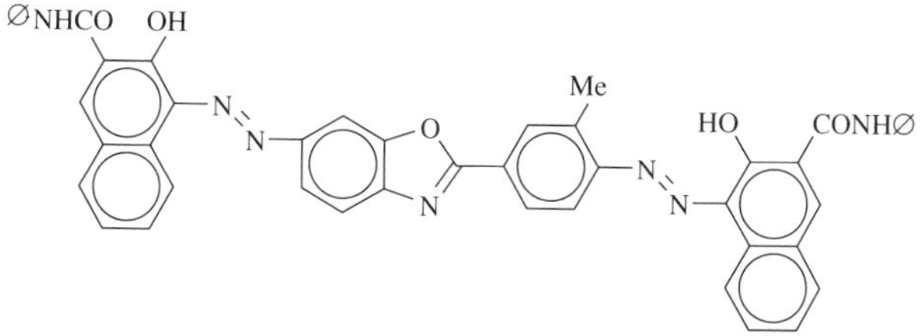	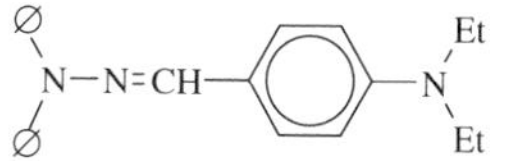	IBM
	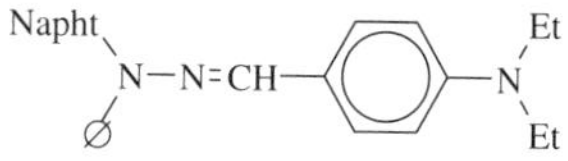	Canon
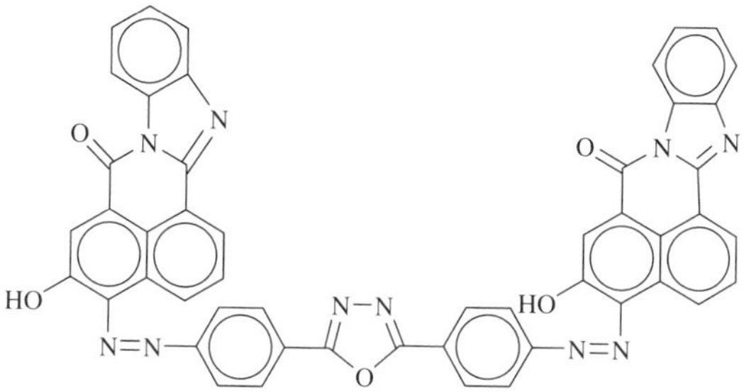	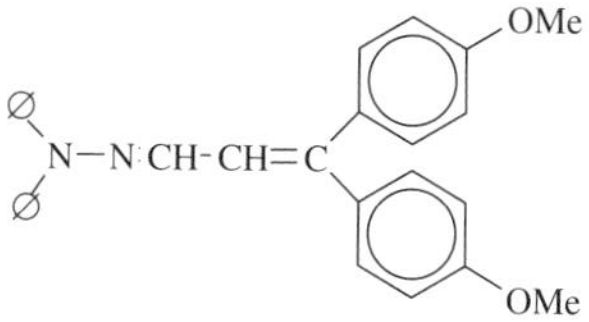	Mitsubishi
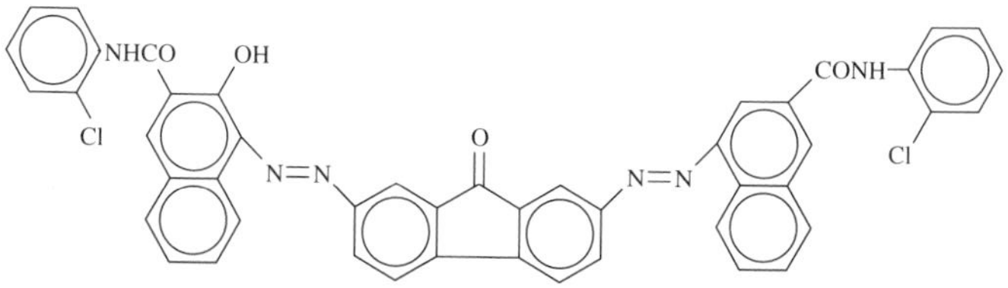	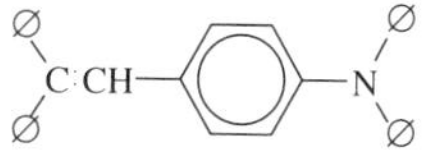	Ricoh

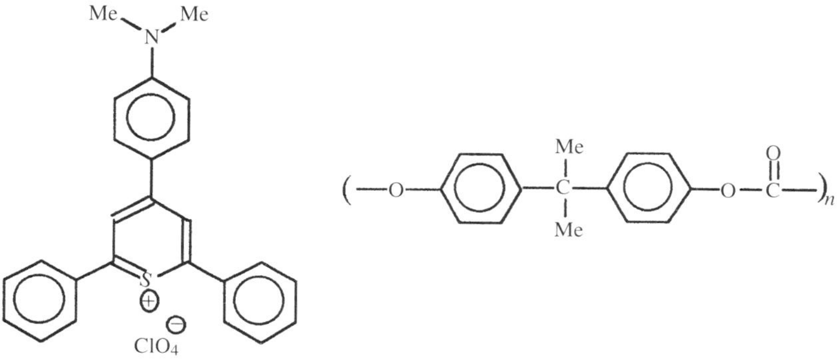

Figure 8.52 Chemical formula of 4-(4-dimethylaminophenyl)-2,6-diphenylthiapyrylium and bisphenol-A polycarbonate

8.4.6.3 Charge Transfer Complexes

H. Hoegl and coworkers in 1957 (cited in Ref. [43]) showed that the photoconductivity of poly(vinylcarbazole) (PVK) can be enhanced in the presence of Lewis acids (BF_3, $AlCl_3$, etc.) or molecular electron acceptors (nitroaromatics, cyanoaromatics, etc.). A commercial photoreceptor containing a mixture of PVK/2,4,7-trinitrofluorenone (TNF) charge transfer complex was described in 1969.

The association constant corresponding to the charge transfer:

$$A, I_A \rightleftharpoons A^+, I_A^-$$

$$K_{CT} = \frac{[A^+, I_A^-]}{[A][I_A]} \tag{8.18}$$

can be estimated from the redox properties of the constituents in solution (see Section 6.3) [53, 54]. If the absolute difference between the oxidation and reduction potentials is less than 0.25 V, partial charge transfer occurs ($K_{CT} \leq 10^4$).

In the case of PVK/TNF thin films the charge transfer complexes coexist with free TNF and PVK. The charge transport properties have been studied in detail (see Ref. [55]); for a review see Ref. [56]. The hole and electron drift mobilities were shown to be very sensitive to the TNF/PVK ratio (Figure 8.53) [55, 56].

The hole mobility decreases as the content of TNF increases; simultaneously, the electron mobility increases. The hole transport is reasonably associated with the carbazole subunit whereas the electron transport involves the TNF molecules (Figure 8.54).

In a PVK thin film containing increasing amounts of TNF, the intersite distance between carbazole moieties is approximately constant whereas the amount of carbazole cation increases. When the ratio (carbazole)$^+$/(carbazole) is too high, charge migration can no longer occur and the mobility decreases. With increasing TNF content, on the contrary, the mean distance between TNF (neutral or anionic)

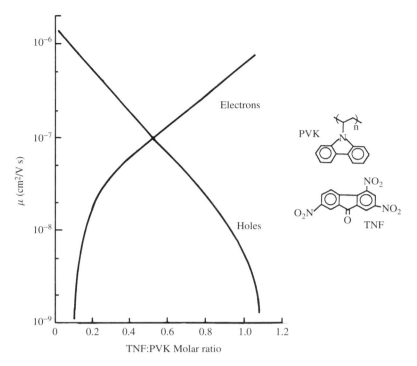

Figure 8.53 Hole and electron drift mobilities as a function of the TNF/PVK molar ratio. Temperature = 295 K; field = 5.0×10^5 V/cm. (After Ref. [55])

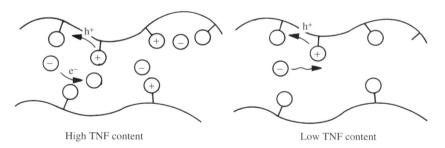

Figure 8.54 Illustration of the hole and electron migrations as a function of the TNF/PVK ratio (right: low TNF content; left: high TNF content)

decreases, facilitating electron migration. Electron mobility is exponentially dependent on the average separation distance of TNF dispersed either in PVK or in a polyester binder (cited in Ref. [56]).

The photogeneration of carriers in PVK/TNF has also been extensively studied (Figure 8.55) (Refs. [57] and [58]; cited in Ref. [56]).

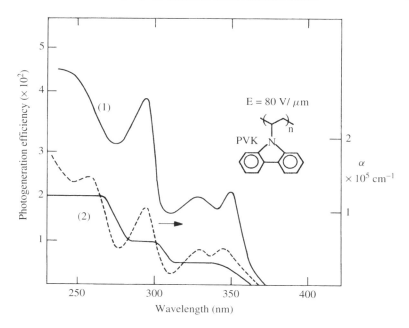

Figure 8.55 Photogeneration efficiency at a field of 8×10^5 V/cm for commercial (1) or purified (2) poly(vinylcarbazole). The corresponding optical absorption spectrum of PVK is also shown (------). (After Refs. [56] to [58])

Commercial PVK leads to higher photogeneration efficiencies than purified PVK, demonstrating the role of impurities or defects in the photochemical generation of carriers. In this case, the absorption spectrum and the photogeneration curve behave similarly as a function of the wavelength. This is not true for purified PVK in which, in the same absorption band, the photogeneration efficiency is roughly independent of the wavelength.

In the late 1970s [43], due to possible carcinogenicity of TNF, this type of molecular material was abandoned.

8.4.6.4 Perylene Pigments

Perylene diimides are synthesized by reacting perylene tetracarboxylic acid or anhydride with a monofunctional or bifunctional primary amine (Figure 8.56). The methyl perylene derivative was investigated in the late 1970s [43]. The thin film was prepared by vapor deposition.

8.4.6.5 Squaraine Pigments

This class of molecules can be considered as a zwitterion (Figure 8.57). A large number of related derivatives can be synthesized on a large scale [43].

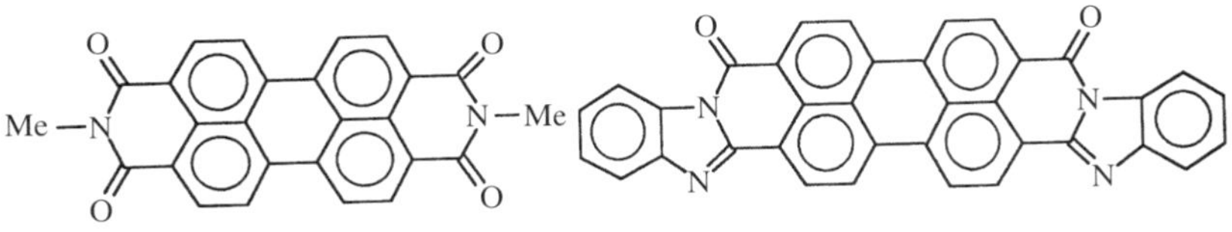

Figure 8.56 Two examples of perylene diimides

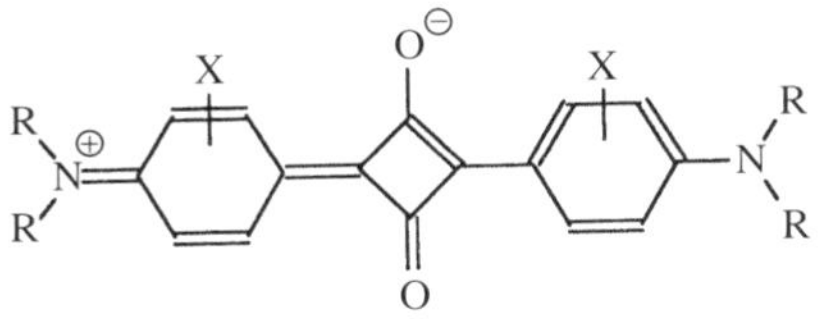

Figure 8.57 Molecular structure of the squaraine molecule

8.4.6.6 Phthalocyanine Pigments

Various phthalocyanines differing by the nature of the central metallic ion and the substitution of the macrocyclic ring have been used as photoreceptors (Figure 8.58).

The optical absorption spectra of molecular compounds highly depend on the crystalline form considered. In the case of metal-free phthalocyanine (PcH_2) at least four different forms (α, β, τ, x) have been characterized. They differ significantly in their optical spectra (Figure 8.59).

In most cases, photogeneration studies are performed on vapor-deposited films which are made of a mixture of α, β and amorphous PcH_2. The x-form of PcH_2 is used as dispersions for making both single- and dual-layer photoreceptors. In this last case, the transport layer is PVK [43]. PcCu and PcMg dispersed in various media have also been studied as materials for making photoreceptors.

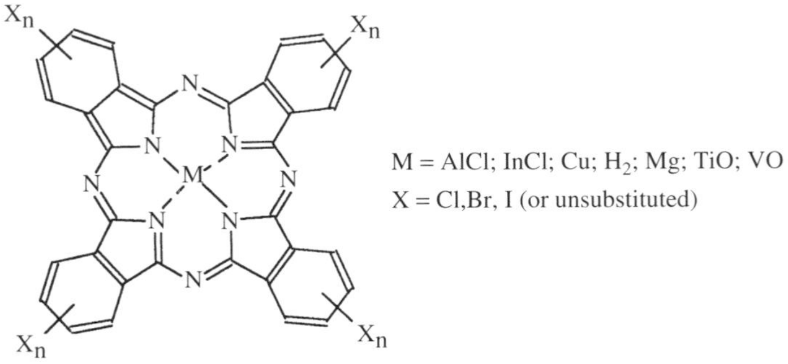

M = AlCl; InCl; Cu; H_2; Mg; TiO; VO
X = Cl, Br, I (or unsubstituted)

Figure 8.58 Examples of phthalocyanine derivatives used as photoreceptors

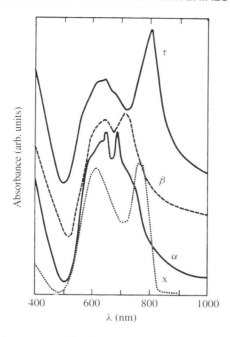

Figure 8.59 Absorption spectra of various crystalline forms of PcH$_2$. (After Refs. [59] and [60]; cited in Ref. [43], courtesy of Marcel Dekker, Inc., NY, 1993)

PcMX (M = Al, In) and PcMO (M = Ti, V) derivatives have been used either vapor deposited or solvent casted dispersed in a polymer. They have, in general, a good photoresponse in the range 650–850 nm [43].

In determining the photoconductive properties of PcM, one must pay attention to the effect of ambience (see Ref. [44] and references therein). For example, the α-form of PcH$_2$ has been found to bind O$_2$ more easily than the β-form [61], yielding a higher photoconductivity. This can also be demonstrated in the case of PcZn. The photoresponse of PcZn in the visible is diminished 40-fold after 24 h at 10^{-5} torr in order to remove O$_2$ [62].

8.4.7 Comparative Xerographic Performances

The various steps of the xerographic process may be followed by plotting the potential arising on the photoconductor as a function of time [58] (Figure 8.60).

In the first step, corona charging allows a potential difference V_0 to be obtained on the photoreceptor. Spontaneous discharge occurs due to a nonnegligible dark conductivity of the photoreceptor and/or to charge injection. Photodischarge occurs under irradiation.

The light exposure energy ($\xi_{1/2}$) required to generate a discharge from V_i to $V_i/2$ (in erg/cm^2) is one of the key parameters of the device together with the

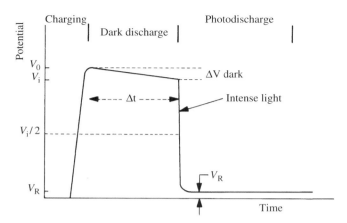

Figure 8.60 Potential effective on the photoconductor as a function of the various xerographic steps (photodischarge curve) V_0 = initial potential after corona charging, ΔV_{dark} = dark decay, V_R = residual potential due to carriers in deep traps). (Redrawn from Ref. [63])

Table 8.13 Dark discharge (ΔV_{dark}) light exposure to discharge ($\xi_{1/2}$) and percentage of discharge under irradiation at 10 erg/cm^2 for PcM dispersed in a polymer or PcM vapor-deposited photogeneration layers. (After Refs. [64] and [65]; cited in Ref. [43])

Charge generation layer	Transport layer	Dark discharge ΔV_{dark} (V/s)	Exposure to discharge $\xi_{1/2}^a$ (erg/cm^2)	Discharge at 10 erg/cm^2 (%)
PcM pigment dispersed in a polyester				
PcVO	TPD in PC[b]	30	4.1	71
τ-PcH$_2$	—	20	5.0	70
PcTiO	—	30	5.6	69
PcGe(OH)$_2$	—	60	7.0	65
PcInCl	—	60	8.0	60
PcAlCl	—	35	10.0	55
(Cl)PcAlCl	—	25	11.0	50
PcMg	—	80	30	24
PcM vapor deposited				
PcInCl	TPD in PC	45	2.6	88
PcAlCl	—	40	6.0	72
PcPb	—	130	10	49
PcVO	—	30	16	36
PcGeCl$_2$	—	100	>100	10
PcGaF	—	26	>100	4
PcSnCl$_2$	—	60	>100	2
PcSn	—	50	>100	1

[a] Exposure to discharge at 830 nm down to $V_i/2$.
[b] TPD: N,N'-diphenyl-N,N'-di(m-tolyl)-p-benzidine (see Table 8.11); PC: polycarbonate.

percentage of discharge under a given light exposure energy. The third important parameter is the amount of dark discharge (Figure 8.60). Comparative performances of a few devices are given in Table 8.13 for phthalocyanine derivatives.

It is important to note that vapor-deposited films are approximately one-half the thickness of the PcM dispersed in polymer coatings. By comparing the ΔV_{dark} values, it can thus be concluded that the dark decays are in the same range for both types of materials. Vapor-deposited PcInCl shows remarkable values for both $\xi_{1/2}$ and the percentage of discharge upon irradiation at $10\,erg/cm^2$. It is worth stressing that the fatigue observed in photoreceptors is due partly to the formation of ozone during the irradiation together with the effects of humidity [43].

8.5 LIQUID CRYSTAL DISPLAYS

8.5.1 Introduction

The history of liquid crystals is exemplary to illustrate the time and the effort necessary to transform a discovery into a widespread industrial application. The discovery of liquid crystals is attributed to the Austrian botanist F. Reinitzer (1888) [66] who observed an intermediate phase when he studied the melting of cholesteryl benzoate. Two years later, a physicist, O. Lehman [67], found that the intermediate phase studied by F. Reinitzer possessed some of the properties of crystals while remaining liquid. In particular, he found evidence of birefringence properties; the material was named, for this reason, a *liquid crystal*. G. Friedel around 1920 [68] interpreted the textures of the liquid crystalline phases (the material being observed by optical microscopy with crossed polarizers).

However it was only in 1963 that R. Williams [69] published an article describing an electro-optical device by showing that light scattering could be induced by an electrical field applied on a nematic-type material. In 1968, G. H. Heilmeier and L. A. Zanoni discovered the first usable display effect using dynamic scattering [70]. In 1964, J. L. Fergason proposed the use of cholesteric liquid crystals in which the pitch of the helix changes with temperature; this effect was later used to detect breast cancer [71]. In 1968, G. H. Heilmeier, L. A. Zanoni and L. A. Barton [72] described the use of liquid crystals containing dichroic or pleochroic dyes.

In 1971, M. Schadt and W. Helfrich [73] discovered the twisted nematic effect which led to the most important type of device for liquid crystal displays. In 1974, D. L. White and G. N. Taylor proposed adding a dichroic dye within the twisted nematic liquid crystal [74]. In 1973, F. J. Kahn proposed a display based on smectic liquid crystals instead of nematic ones. In this case the two states (OFF/ON) are provided by an aligned single domain of the liquid crystal (fully transparent) or a polydomain with milky appearance due to light scattering [75]. In 1975, a physicist (R. B. Meyer) and a group of chemists (L. Liébert, L. Strzelecki and P. Keller) predicted from symmetry arguments the possibility of obtaining a polar smectic-C phase using optically active mesogens. They then synthesized the appropriate molecules and characterized a polar mesophase (SmC*) [76, 77]. In 1980, N. A. Clark and S. T. Lagerwall [78] demonstrated the possibility of making devices using polar S_C^*.

Approximately one century has separated the discovery of liquid crystals to their widespread use in industry. Nowadays the display market is expanding rapidly in size from tens of millions of dollars in 1985 to nearly ten billion dollars around 1995 [79].

8.5.2 Dynamic Scattering Displays

This technique was used for the first generation of liquid crystal displays [80]. A nematic liquid crystalline phase having a negative dielectric anisotropy, $\Delta\varepsilon$, is used:

$$\Delta\varepsilon = \varepsilon_\| - \varepsilon_\perp \tag{8.19}$$

where

$\varepsilon_\|, \varepsilon_\perp$ = dielectric constants parallel and perpendicular to the molecular axis, respectively

It is usually doped with an ionic solute to increase the ionic conductivity of the material. A cell is constituted in which the mesophase is oriented with the molecular axis parallel to the substrate (planar arrangement). The two sides of the cell are covered with semitransparent conducting materials (ITO). In the absence of voltage between the two electrodes (OFF state), the display is transparent. When a sufficient voltage is applied, typically around 10 V, scattering takes place, giving a milky-white appearance. This phenomenon is due to charge injection from the electrodes to (or from) the liquid crystal molecule or dopant:

Injection: A = molecular unit

electrode (−) / A A A / (+)
              A A
              A A

Ionic migration:

(−) / A⊖ → A A / (+)
      A A

Discharge:

(−) / A A⊖ / (+)
      A A
      A

Hole injection could also be considered. The ionic solute (dopant) can also play a role in the charge injection processes. Finally, in the ON state, an electrodynamic, vortex-like, flow takes place which is responsible for the scattering (Figure 8.61).

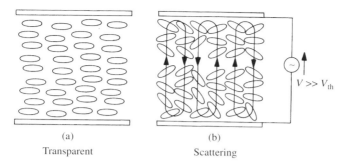

Figure 8.61 Dynamic scattering cell: (a) no applied voltage (OFF state), the liquid crystal has a planar orientation; (b) $V > V_{th}$, where V_{th} = threshold voltage. The applied voltage generates flow loops in the liquid crystal layer. (Redrawn from Ref. [80])

This device does not need the use of polarizers. However, the power consumption per unit of active area is high (100–1000 μW/cm^2) due to the high operating voltage (10–60 V) and the nonnegligible current through the cell. The turn on time depends on the voltage but is of the order of 10 ms; the turn-off time is about 100–200 ms. The viewing angle is poor. The lifetime of the device is of the order of 10 000 h, significantly less than for the other devices (see subsequent sections). These numbers are given for reflective displays [80].

8.5.3 Twisted Nematic Liquid Crystal Displays

A nematic liquid crystalline material is sandwiched between two semitransparent conductive electrodes (ITO). The electrodes are covered by a thin film of polyimide which is rubbed in one direction. The nematic molecules align themselves parallel to the buffing direction and are slightly inclined (surface tilt angle) relative to the glass plate (1–3°). The rubbing directions of the upper and lower glass plates are perpendicular to each other and, consequently, the material is twisted by 90° over the thickness of the mesophase layer [81] (Figure 8.62).

The twisting is reinforced by adding a very small amount of a cholesteric dopant. The product of the birefringence $\Delta n = n_e - n_o$ (n_e = extraordinary refractive index, n_o = ordinary refractive index) with the cell spacing d (optical path difference) is large compared with half of the wavelength of the incident light. The cell spacing is usually of the order of 5–10 μm. When an electrical field is applied (ON state), the molecules tend to align parallel to the field (positive dielectric anisotropy). In the ON state, the polarization of the incident light remains unchanged within the liquid crystalline material and the light is consequently absorbed by passing through the analyzer (crossed polarization) (Figure 8.63).

For intermediate voltages between V_{90} and V_{10} (V_{90} = 90% transmission, V_{10} = 10% transmission), the molecules within the cell are oriented at a significant angle to the normal of the electrodes. This generates a principal

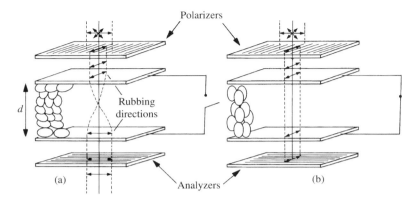

Figure 8.62 Twisted nematic liquid crystal display: (a) no applied voltage — transparent state; (b) a voltage is applied inducing a homeotropic orientation d = cell spacing. (Redrawn from Ref. [82])

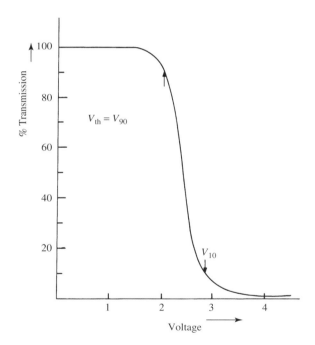

Figure 8.63 Percentage of transmission as a function of the voltage for a twisted nematic liquid crystal display (crossed polarizers). (Modified from Ref. [83])

viewing cone. The threshold voltage ($V_{90} = V_{th}$) varies with viewing angle and with temperature. Typical operating characteristics (reflective displays) are V_{th} = 1–2 V, operating voltage = 3–20 V, power consumption per unit of active area = 1–10 μW/cm² (10–1000 times smaller than for dynamic scattering displays), turn-on time = 0.2–100 ms (voltage dependent), turn-off time = 30–100 ms, viewing angle = good($\pm 40°$), lifetime >50 000 hours [78].

The cell can be directly 'addressed': each element has its own electrical circuit in order to apply uncorrelated voltages on each cell. This is not a practical way to make displays when the number of elementary cells becomes large. The solution is to change to a so-called 'multiplex addressing' (Figure 8.64). The ITO electrode on one glass is divided into M columns and, on the other plate, to N rows, giving M times N picture elements (pixels).

By multiplex addressing, the information is applied parallel to the columns and time sequentially to the rows. However, in this system, although the maximum voltage is applied on the cell corresponding to the intersection of the row and the column switched ON, the other cells also undergo part of the previous applied voltage. It is therefore necessary to use a transistor associated with each pixel in order to induce a nonlinearity (see for example, Ref. [84]) (Figure 8.65).

When a voltage is applied on a row, all the transistors of this line are switched ON and the voltage of the columns are transferred through the transistor to the electrode of the cell. When the transistor is OFF, the state of polarization of the cell is conserved (at least until leak electrical currents ruin the voltage difference between the electrodes). The problem was to realize an electronic circuit possessing as many transistors as image dots (of the order of 10^6 transistors without defects for a conventional TV screen).

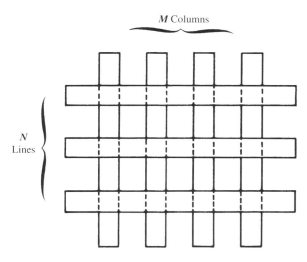

Figure 8.64 Schematic representation of a multiplex addressing constisting of M columns on the back electrode and N rows on the front electrode

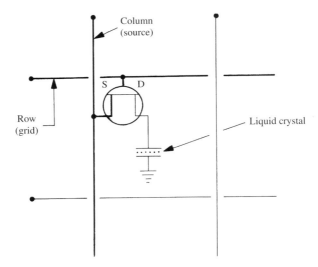

Figure 8.65 Equivalent electrical circuit associated with each picture element (pixel). The source of the transistor is connected to a column, the grid to a row. Depending on the grid voltage the transistor is in an OFF or ON state

In practical terms, one applies a voltage, alternatively positive and negative, to avoid ionic currents, in approximately 20 ms. A short voltage signal allows all the transistors of a given line to switch from OFF to ON. In this case two voltages are used for the lines: $V_{G_{ON}}$ and $V_{G_{OFF}}$. The screen is addressed line by line; if there are N lines, the time available to address one line is of the order of $t = 20\,\text{ms}/N$. This time varies in the range $t = 15\text{--}30\,\mu\text{s}$ [84]. During this period of time the capacitance associated with the liquid crystal, C_{LC}, must be charged through the resistance of the transistor, R^{trans}. Therefore,

$$R_{ON}^{\text{trans}} \times C_{LC} \ll t$$

As an example, if a pixel (200 μm × 200 μm) is addressed in 64 μs, $R_{ON}^{\text{trans}} < 10^7\,\Omega$. The time the information will remain on the transistor is related to R_{OFF}^{trans}, which reflects the leak currents. It should be noted that $R_{OFF}^{\text{trans}} > 10^{12}\,\Omega$ [84] if the charge is maintained 20 ms.

A more detailed description of an element of a liquid crystal display based on a twisted nematic is shown in Figure 8.66. The electrical characteristics of the transistor determine the qualities of the display. Hydrogenated amorphous silicon (hydrogen content ~10%) can be deposited at temperatures lower than 300 °C by various methods: radiofrequency glow, reactive sputtering, photochemical-vapor deposition. Under these conditions, the optical and electrical parameters that can be obtained are listed in Table 8.14 [85].

The dark conductivity of undoped a-SiH made from the decomposition of SiH_4 varies with the substrate temperature on which the thin film is deposited (T_{subs}): for $T_{\text{subs}} = 100\,°\text{C}$, $\sigma \leq 10^{-11}\,\Omega^{-1}\,\text{cm}^{-1}$; for $T_{\text{subs}} = 550\,°\text{C}$, $\sigma \leq 10^{-5}\,\Omega^{-1}\,\text{cm}^{-1}$

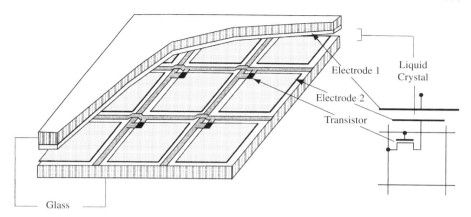

Figure 8.66 Schematic structure of a picture element in a twisted nematic liquid crystal display. (Modified from Ref. [84])

Table 8.14 Electrical characteristics of a thin film of hydrogenated amorphous silicon (a-SiH). (After Ref. [85])

Optical bandgap	1.70–1.72 eV
Conductivity	10^{-11}–$10^{-12}\,\Omega^{-1}\,cm^{-1}$
Activation energy	~0.8 eV
H content	~10 at%
Drift mobility	$\mu_e \sim 1\,cm^2/V\,s$ ($\mu_h \sim 0.01$)
Dangling bond density	10^{15}–$10^{16}\,cm^{-3}$

[86]. The intensity of the drain-to-source current as a function of the gate voltage is also a determining parameter (see Figure 6.61). The steepness of the electrooptical characteristics of the liquid crystal cell (see Figure 8.63) depends on the elastic properties of the mesophase. The basic elastic deformations (splay, twist, bend) that can be defined for a nematic liquid crystal are shown in Figure 8.67. The ratio K_3/K_1 determines the steepness of the electro-optical characteristics of the liquid crystal cell (see Ref. [87] for a review).

The liquid crystalline phase used for fabricating a twisted nematic liquid crystal display must have a reasonable range of thermal stability around room

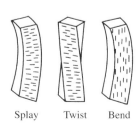

Figure 8.67 Representation of the three elastic deformations: splay (K_1), twist (K_2), bend (K_3). (Modified from Ref. [81])

temperature. The molecular unit must also be chemically and photochemically stable. Generally mixtures of mesogens are used. In many cases biphenyl or terphenyl subunits are employed as rigid cores (Figure 8.68) [88].

The breakthrough in the mesogen design and synthesis was made by G. W. Gray and coworkers in the early 1970s. They used highly stable biphenyl and terphenyl subunits as rigid cores for their mesogens. A mixture of approximately 65% 5CB and 35% 8OCB yields a nematic phase stable in the range 5–50 °C [88] (Figure 8.69).

The mixtures used in liquid crystal displays are eutectic mixtures consisting, very often, of four to ten components. Phenylcyclohexane and biphenylcyclohexane subunits are employed to extend further the domain of stability of the mesophase. A mixture, whose code name is NP1694, demonstrates a nematic range from −20 to 86 °C.

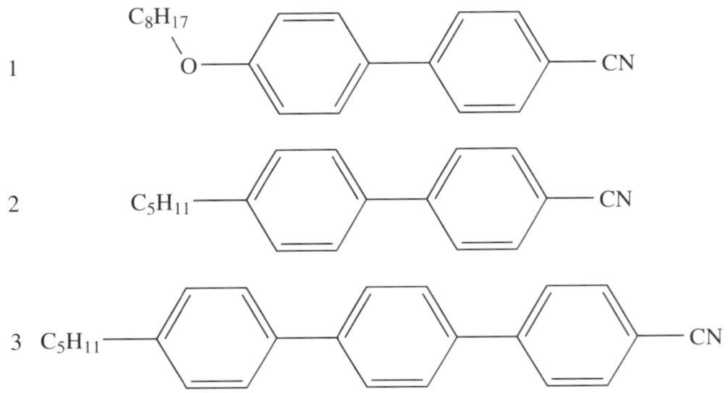

Figure 8.68 Three of the mesogens widely used in liquid crystal displays [88]: 1, *p*-n-octyloxy-*p'*-cyanobiphenyl (8OCB); 2, *p*-n-pentyl-*p'*-cyanobiphenyl (5CB); 3, *p*-n-pentyl-*p'*-cyano terphenyl

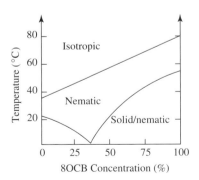

Figure 8.69 Transition temperatures for mixtures of pentyl-cyanobiphenyl (5CB) and octyloxy-cyanobiphenyl (8OCB) [88]

INDUSTRIAL APPLICATIONS OF MOLECULAR MATERIALS

In France, huge efforts have been made since the early discoveries of liquid crystal displays. The first appearance of pocket-sized liquid crystal TV sets (3inch) occurred in 1985. Around this period of time a flat screen was fabricated with an area of $10 \times 13\,\text{cm}^2$ consisting of 250 lines $\times$ 320 columns [89].

8.5.4 Dichroic Liquid Crystal Displays

Dichroic liquid crystal displays were discovered by G. H. Heilmeier earlier than twisted nematic displays. However, their industrial fate was more laborious [90].

A small amount of dye, usually with an elongated structure, is dissolved in a nematic liquid crystal. The dye molecules consequently align within the liquid crystalline material and will follow the orientation of this material when an electrical field is applied (guest–host interaction). The dyes, because of their shape and electronic structure absorb light depending on the relative orientation of the molecular long axis and the electrical field associated with the photon. This phenomenon is called *dichroism* when two colors are involved and *pleochroism* for several (Figure 8.70).

Surface treatment allows the planar orientation of a nematic phase, which includes the dye. In this state, the light is strongly absorbed. The application of an electrical field reorients the guest–host matrix, resulting in a less absorbing state (Figure 8.71).

The liquid crystals used are based on the cyanobiphenyl derivatives first synthesized by G. W. Gray and coworkers [92, 93] and launched on the market by BDH in England. The dichroic dyes must present very important characteristics:

(a) solubility of the dye in the liquid crystal,
(b) chemical and photochemical stabilities,
(c) good dichroic ratio,
(d) appropriate order parameter of the dye within the liquid crystal.

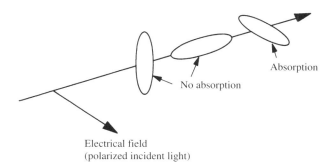

Figure 8.70 Optical properties of a dichroic dye molecule. The transition moment is considered to be aligned with the long molecular axis. (Modified from Ref. [80])

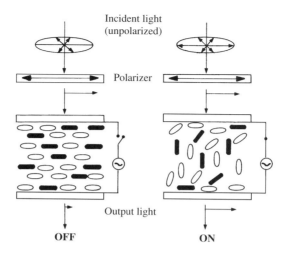

Figure 8.71 Principle of a dichroic display (Heilmeier display): OFF, no electrical field, planar orientation of the nematic liquid crystal; ON, with electrical field. (Modified from Ref. [91]

Figure 8.72 The orientation of the various individual molecular units relative to the director

The preferred direction of a liquid crystal is called the *director*. At some time, each molecule is oriented at some angle θ relative to the director (Figure 8.72). In the case where there is a perfect orientational order the average angle is zero. On the contrary, with no orientational order, the angle is between 0 and 90°. In three dimensions there are innumerable orientations of molecules making an angle of 90° to the director but only one orientation affords an angle of 0° (Figure 8.73). If one takes the average when there is no orientational order [88], the result will be $\theta = 57°$. Smaller angles indicate some orientation of the mesogens.

Instead of averaging θ, one can average the function $\mathcal{P} = (3\cos^2\theta - 1)/2$. When $\theta = 0$, $\cos\theta = 1$, the value of the *order parameter* $\mathcal{P} = 1$ corresponds to a perfect orientational order. With no orientational order, the average function is zero, with an average angle such that $\langle\cos^2\theta\rangle = 1/3$.

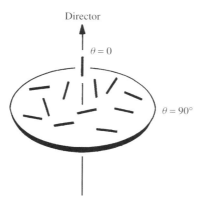

Figure 8.73 Molecules having a relative angle to the director of $\theta = 0°$ or $\theta = 90°$

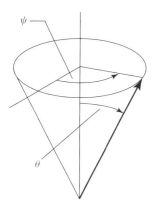

Figure 8.74 The two components influencing the average angle of the long molecular axes direction relative to the director

Two components are present in the order parameter: the fluctuation (Ψ) concerning the precession around the director direction and the fluctuations of the angle θ while maintaining Ψ approximately constant (Figure 8.74).

The order parameter $\mathcal{P}$ depends on the temperature and on the nature of the liquid crystal. Close to the clearing point, the order parameter drops to values around 0.4 to reach zero in the isotropic phase. For mesogens used in displays, $\mathcal{P}$ is of the order of 0.6–0.7. The order parameters of the guest–host liquid crystals as well as the absorption wavelength of the dye are among the most important parameters for a display. The main dyes employed are listed in Table 8.15.

8.5.5 Ferroelectric Liquid Crystal Displays

8.5.5.1 Orientation of S_C^* Mesophases by Surfaces

The generalities have been given in Section 7.2.4 [94]. N. A. Clark and S. T. Lagerwall, in 1978 [95, 96], started studies to unwind the helix of the S_C^*

Table 8.15 Some of the dichroic dyes used in displays. The order parameter has been determined at the maximum of absorption of the dye. (Modified from Ref. [90])

Class	Structure	λ_{max}	order parameter
Diazo	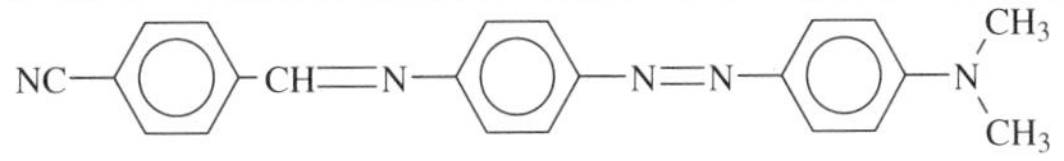	455	0.67
		391	791
	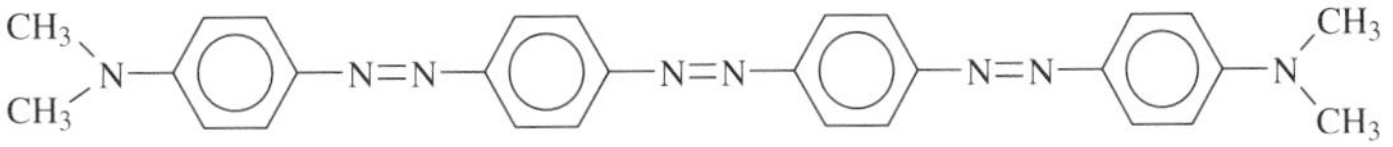	523	790
	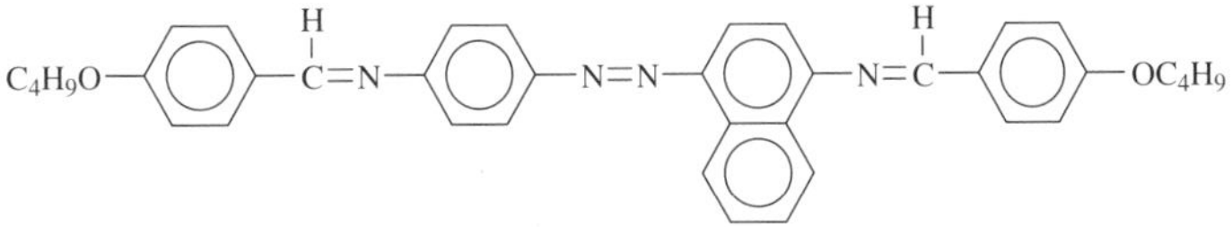	455	787
	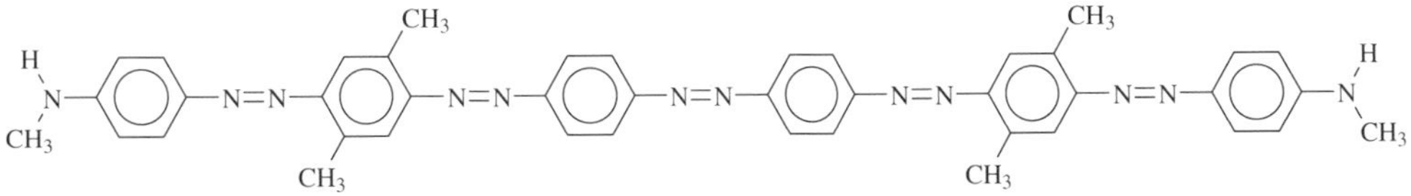	595	0.80

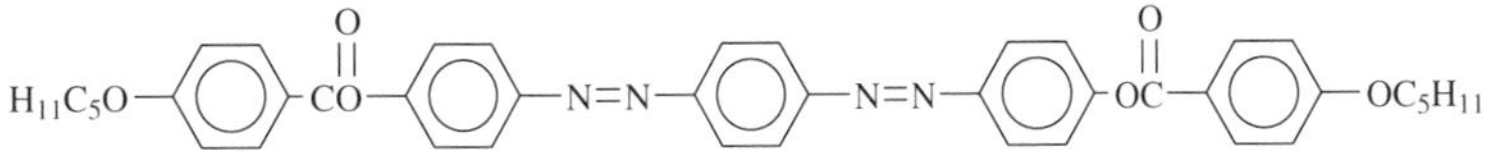

Anthraquinone

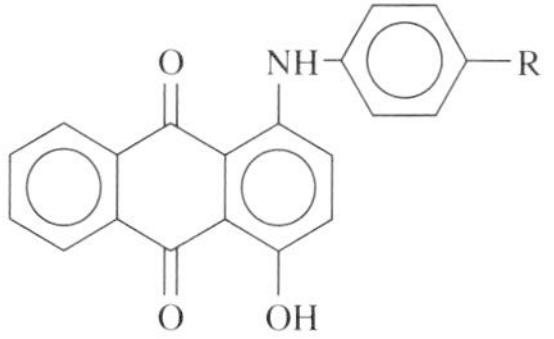

Structure		
R=OC$_9$H$_{19}$	596	0.65
R=N(CH$_3$)$_2$	612	0.63
(bis-Schiff base with C$_8$H$_{17}$)		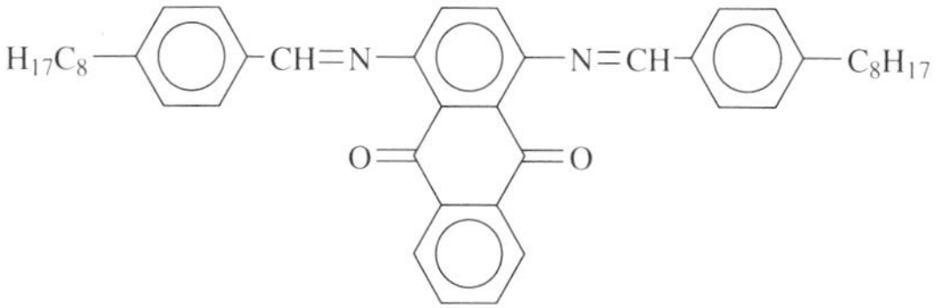 0.91
R = C$_2$H$_5$	554	0.67
R = OC$_5$H$_{11}$	557	0.68
R = N(CH$_3$)$_2$	546	0.65
R = —N=N—C$_6$H$_5$	524	0.71
R = C$_6$H$_4$C$_6$H$_5$		0.90

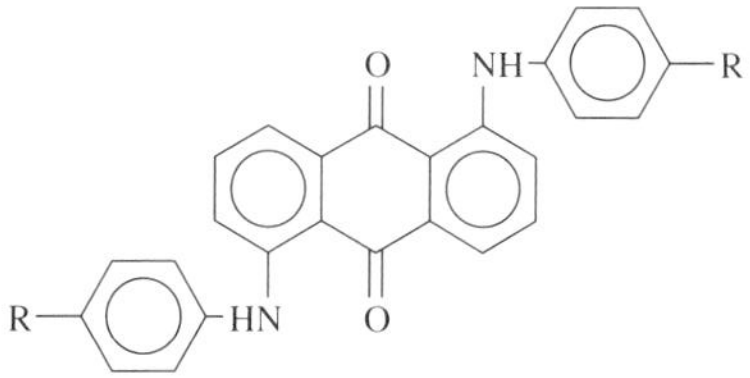

(continued overleaf)

Table 8.15 (continued)

Class	Structure	λ_{max}	order parameter
	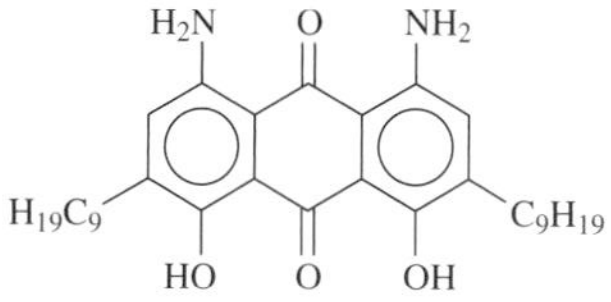		0.7
	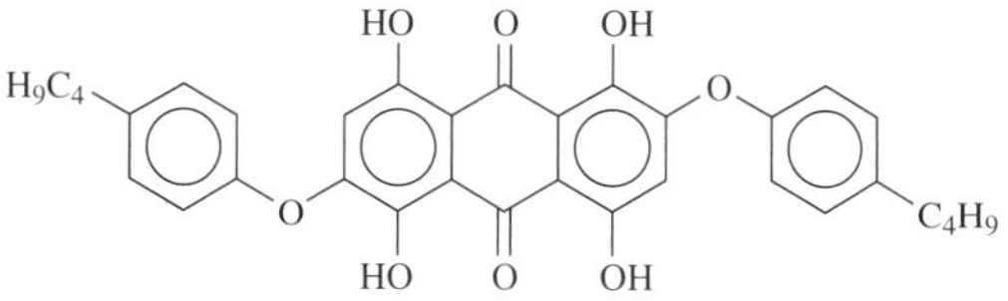	589	0.76
	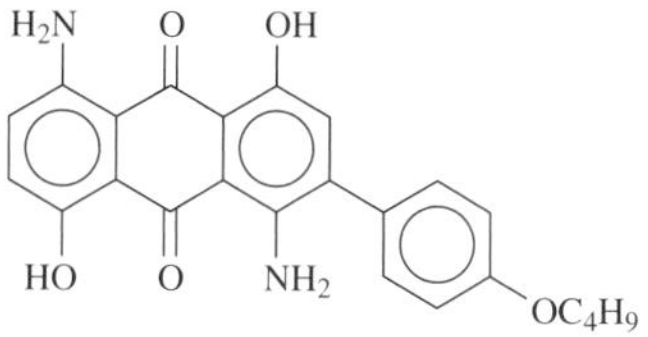		~0.7
	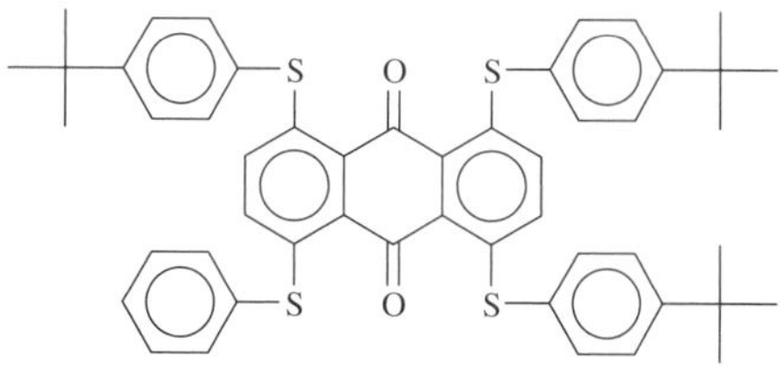	550	0.78

Class	Structure	λ (nm)	
Perylene	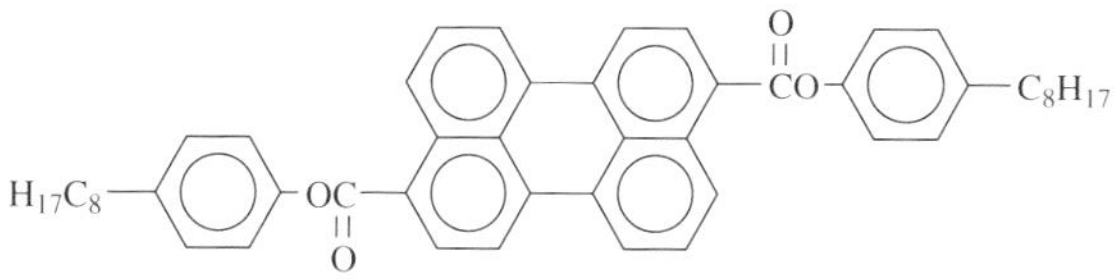	475	0.72
Tetrazine	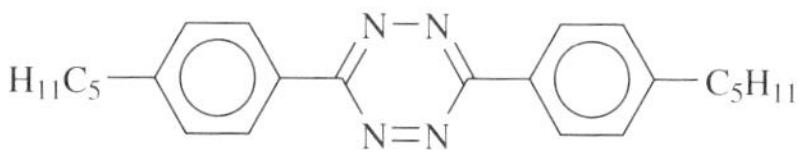	555	−0.39
Imidazole	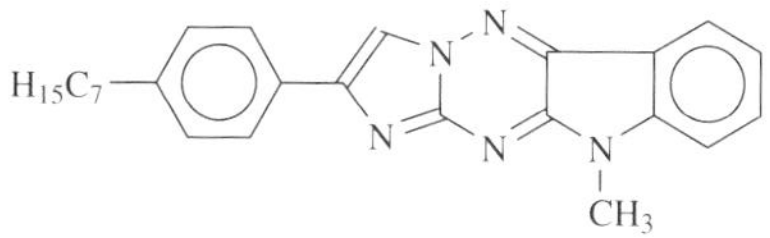	400	0.66
	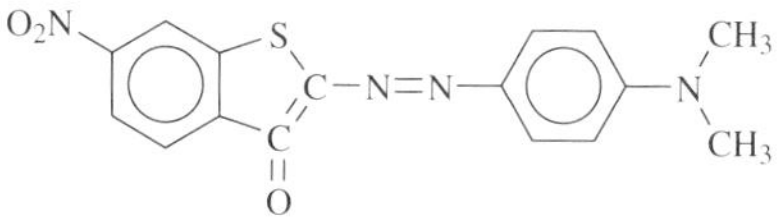	594	0.60
Methine	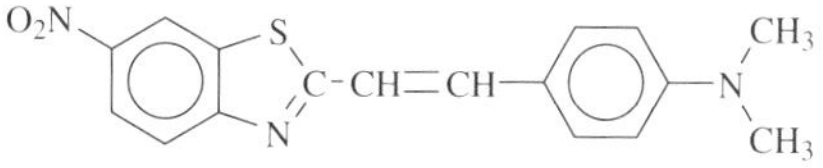	457	0.70
Organometallic	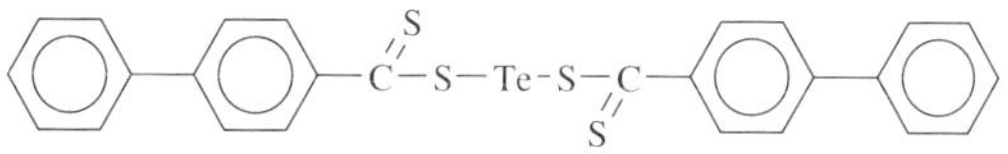	455	

phase by using electrical fields and surface electrode treatments. The main point was to obtain thin samples; results were, however, finally obtained without any spacer. Although the initial cells were always shorted, further works permitted correctly aligned layers to be obtained, with a fairly uniform thickness in the range 0.5–2 μm (Figure 8.75).

In many cases, a polymer (polyethylene, poly(vinylalcohol), poly (hexamethylene-adipamide), etc.) is deposited on the electrodes and rubbed in one direction [97]. Partial polymeric chain orientations or formation of microgrooves occur during buffing [97]. The long axis of the molecular units will tend to align with the buffing direction (Figure 8.76).

Smectic C* mesophases, even on a polymer coated and rubbed surface, can often exhibit random alignment. In the presence of an applied electrical field, large areas of homogeneously aligned samples can be obtained. A schematic representation of the domains that can arise in S_C^* mesophases together with the corresponding polarity has been given by N. A. Clark and S. T. Lagerwall [96] (Figure 8.77).

The switching from one state (polarity up) to the other one (polarity down) can be followed by optical microscopy. The domain wall motion can then be observed (Figure 8.78) [96].

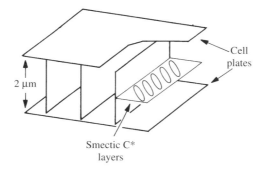

Figure 8.75 Orientation of smectic C* thin films sandwiched between two electrodes

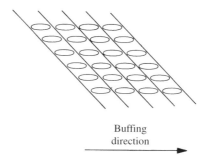

Figure 8.76 Orientation of the S_C^* layers relative to the buffing direction

INDUSTRIAL APPLICATIONS OF MOLECULAR MATERIALS

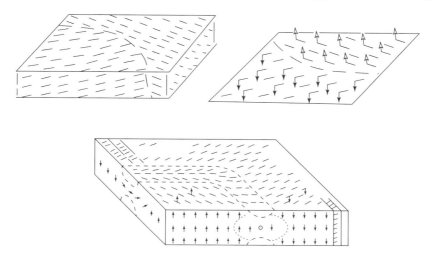

Figure 8.77 Domains in surface stabilized ferroelectric liquid crystal thin films: bistable case when the helical structure is quenched. (Reproduced by permission of Gordon and Breach Publishers from Ref. [96])

The optical microscopy photograph with crossed polarizers is obtained by slowly applying increasing voltages from 0 to 1.5 V. Observations have been carried out in the case of the smectic F* mesophase. A S_F^* shows short-range hexagonal order, long-range orientational order and the columns tilted to the edge of the hexagonal array (Figure 8.79) [97].

When the voltage is slowly increased, the size of one of the domains increases. The transition from one bistable state to the other is ensured by moving domain walls. These can encounter *pinning* sites, which can be overcome when sufficiently high energy is given to the system.

In a conventional S_C^* liquid crystal derived from an epoxide containing mesogen, a polarity of approximately 45 nC/cm^2 has been measured, which corresponds to a contribution of 0.2 debye per molecule of mesogen. (Figure 8.80). This last value may be compared with the epoxide molecular dipole moment of 2 debye. The difference is due to orientational movements of the mesogen around its long molecular axis in the liquid crystalline phase [96].

The magnitude of the spontaneous polarization has been tentatively related to the structure of the molecular unit. It has been shown that the polarization is primarily controlled by the chiral moiety, a stronger molecular dipole leading to a greater spontaneous polarization (Figure 8.81).

8.5.5.2 Chronology of the Ferroelectric Liquid Crystal Display Technological Advances

Most of the following indications have been taken from Ref. [98].

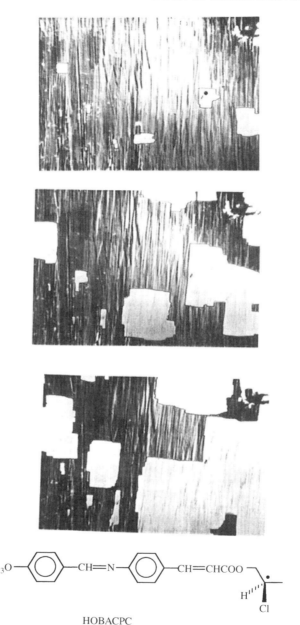

HOBACPC

Figure 8.78 Thin film (1.5 μm) of a smectic F* mesophase of HOBACPC showing the size increase of one domain. (Reproduced by permission of Gordon and Breach Publishers from Ref. [96])

INDUSTRIAL APPLICATIONS OF MOLECULAR MATERIALS

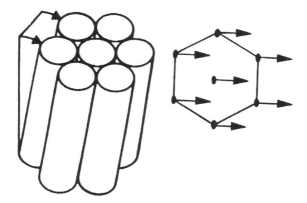

Figure 8.79 Structure of the S_F^* showing the tilt towards the edge of the hexagonal array. (Modified from Ref. [97])

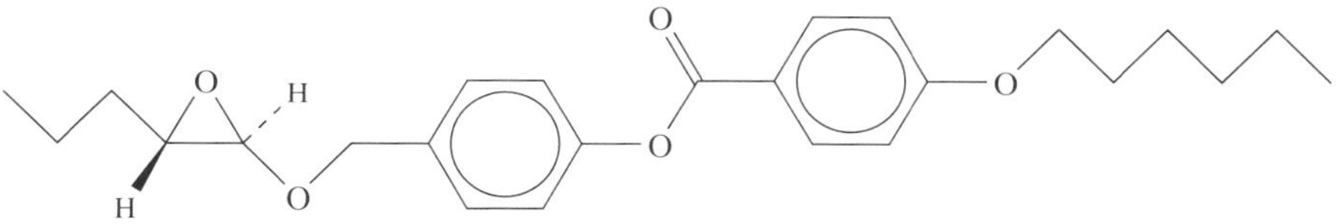

Figure 8.80 One of the epoxy-based mesogens

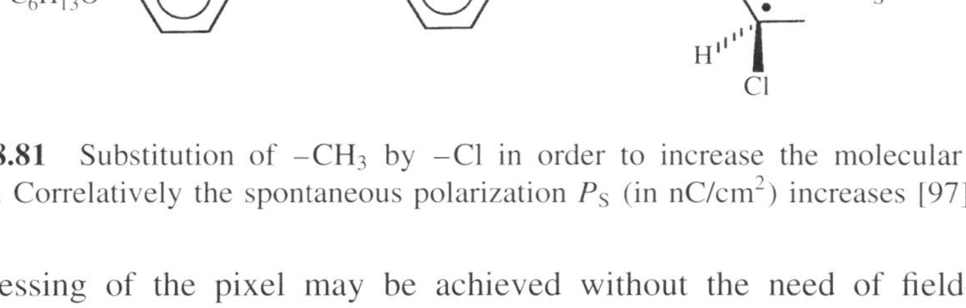

Figure 8.81 Substitution of $-CH_3$ by $-Cl$ in order to increase the molecular dipole moment. Correlatively the spontaneous polarization P_S (in nC/cm^2) increases [97]

Addressing of the pixel may be achieved without the need of field-effect transistors as long as the number of lines is not too large. In this case, the corresponding electrical circuit is as shown in Figure 8.82.

In Figure 8.82, the selected row is at voltage zero whereas unselected rows are at $2V/3$ (arbitrary units). The columns are at V or $V/3$ when selected or unselected, respectively. The addressed capacitor undergoes of $-V$ voltage. Addressing of columns and rows is carried out using pulses. There is a threshold

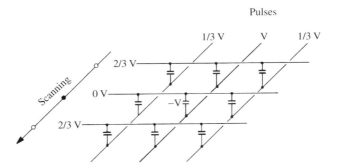

Figure 8.82 Electrical interdependence of matrix elements in a matrix (capacitance couplings). (Modified from Ref. [98])

required to switch an element from one state to another which depends both on time (τ) and voltage (V). For high values of field (i.e. 5×10^4 V/cm), there is a critical value of the product $(\tau V)_{th}$ to obtain switching [98]. In Figure 8.82, the addressed capacitor is at voltage $-V$ while the other ones are at $\pm v/3$. The selection is 3:1 and is optimized.

AT & T Bell Laboratories in 1985 fabricated a display with 60×60 pixels. The first company that attempted the development of ferroelectric liquid crystal (FLC) screens was Seiko Instruments and Electronics in May 1985 (640×400 elements). This company demonstrated that it was possible to obtain uniform alignment and uniform spacing over a fairly large area (12-inch panel; pixel of 0.4×0.4 mm^2).

In 1985, ferroelectric liquid crystal (FLC) chemicals were available. Seiko Instruments and Electronics and Teikoku Inc. developed FLC mixtures usable in the temperature range $+3-50\,°C$. By the end of 1986, other Japanese companies presented prototypes. Efforts started in England in 1986 (JOERS/Alvey Project). By the end of 1986, a 64×64 panel was fabricated which developed afterwards (1987) into a 720×400 lines screen (12 inch). Philips (1988) fabricated a 108×92 pixel screen driven by amorphous silicon thin field-effect transistors. LETI (1987) proposed a way to obtain grey levels. The LETI screen comprised 320×256 pixels (size, $310\,\mu m^2$) and showed long-term memory.

The greatest research and development efforts in ferroelectric liquid crystals has been made by Canon Inc. in Japan. This company launched an ambitious program in 1985 and in 1988 was the first to develop a commercial product.

Figure 8.83 (Plate 8) shows a display commercially available in 1998.

8.6 REFERENCES

1. *Le Grand Larousse Encyclopédique*, Savon, p. 9375, Librairie Larousse, Paris (1985).
2. J. Rouanet and C. Trezain, in *l'Actualité Chimique*, **15**, March–May (1996).

3. W. K. Purves, G. H. Orians and H. C. Heller, *The Science of Biology*, 4th life edn, Sinauer and W. H. Freeman, New York (1995).
4. BASF-France, same as Ref. [2], pp. 58–62.
5. B. Brancq, same as Ref. [2], pp. 63–65.
6. F. Garcia and C. Parlant, same as Ref. [2], pp. 66–71.
7. R. D. Vold and G. S. Hattiangdi, *Ind. Eng. Chem.*, **41**, 2311 (1949).
8. A. Skoulios, *Ann. Phys.*, **3**, 421 (1978).
9. J. Simon, J.-J. André and A. Skoulios, *Nouv. J. de Chimie*, **10**, 295 (1986).
10. P. J. Flory, *Statistical Mechanics of Chain Molecules*, Interscience and John Wiley & Sons, Chichester (1969).
11. P. A. Winsor, *Chem. Rev.*, **68**, 1 (1968).
12. B. Bahadur, in Vol. II, p. 28 of Ref. [90].
13. H. Zollinger, *Color Chemistry*, VCH, Weinheim (1991).
14. B. and G. Delluc, *Connaître Lascaux*, Photo R. Delvert, Sud-Ouest, Luçon (1989).
15. W. Herbst and K. Hunger, *Industrial Organic Pigments*, VCH, Weinheim (1993).
16. P. F. Gordon and P. Gregory, *Organic Chemistry in Colour*, Springer-Verlag, Berlin (1983).
17. P. Gramet, *Découvrir les animaux du littoral (Aquarelles: A. Jourcin)*, Ouest France, Rennes (1978).
18. W. Mueller-Kaul (Hoechst AG), C. Ratcliff (Hoechst Celanese), *American Ink Maker*, p. 39 (November 1992).
19. M. Pope and C. E. Swenberg, *Electronic Processes in Organic Crystals*, Clarendon Press, Oxford (1982).
20. M. Kasha, *Spectroscopy of the Excited State* (ed. B. Di Bartole), Plenum Press, New York (1976).
21. A. T. Davidson, *J. Chem. Phys.*, **77**, 168 (1982).
22. H. J. Hesse, W. Fuhs, G. Weiser and L. von Szentpaly, *Phys. State Solid*, **B76**, 817 (1976).
23. G. Geissler, *The Origins of Colors*, Hoechst, Frankfurt.
24. A. Doroszkowski, in *Technological Applications of Dispersions*, Vol. 52 (ed. R. B. McKay), Marcel Dekker, New York (1994).
25. G. Champetier and H. Rabaté, *Chimie des peintures, vernis et pigments*, Vol. I and II, Dunod, Paris (1956).
26. A. Marrion, *The Chemistry and Physics of Coatings*, Royal Society of Chemistry, Cambridge (1994).
27. R. B. McKay, in *Technological Applications of Dispersions*, Marcel Dekker, New York (1994).
28. Documentation, *'Hercules' Rosin Resins Selection Guide*, Rijswijk, The Netherlands.
29. J. Petit, Vol. I, Chapter II of Ref. [25].
30. Vol. I, p. 32 of Ref. [25].
31. J. Petit, Vol. I, Chapter IX of Ref. [25].
32. J. W. Vanderhoft *et al.*, *Adv. Chem.*, **34**, 32 (1962); cited in Ref. [26].
33. H. Jotischky, *The Paint Industry: The Economic Framework*, Paint Research Association (1991); cited in Ref. [27].
34. *Encyclopaedia Universalis, Thesaurus Index (Encres)*, Paris (1990).
35. M. Fontaine, Vol. II, Chapter XXV of Ref. [25].
36. D. E. Bisset, C. Goodacre, H. A. Idle, R. H. Leach and C. H. Williams (eds.), *The Printing Ink Manuel*, 3rd edn, Van Nostrand Reinhold, Wokingham (1979).
37. M. Windholz (ed.), *The Merck Index*, 10th edn, Merck, Rahway (1983).
38. R. Lombard, Vol. I, Chapter XI of Ref. [25].
39. Documentation, *Raw Materials for the Printed Image*, (Lawter International Inc.,) Illinois.
40. G. Battersby, Chapter 5 of Ref. [24].

41. P. Clément, Vol. I Chapter III of Ref. [25].
42. C. Delmas, *Europhtal*, documentation (1995).
43. P. M. Borsenberger and D. S. Weiss, *Organic Photoreceptors for Imaging Systems*, Marcel Dekker, New York (1993).
44. J. Simon and J.-J. André, *Molecular Semiconductors*, Springer-Verlag, Berlin (1985).
45. D. M. Pai and R. C. Enck, *Phys. Rev.*, **B11**, 5163 (1975).
46. J. Mort and D. M. Pai, *Photoconductivity and Related Phenomena*, Elsevier Scientific, Amsterdam (1976).
47. H. Meier, *Organic Semiconductors Monographs in Modern Chemistry*, Vol. 2, Verlag Chemie, Weinheim (1974).
48. J. W. Weigl, *Angew. Chem. Int. Ed. Engl.*, **16**, 374 (1977).
49. See, for example, H.-T. Macholdt and A. Sieber, *J. Imaging Technol.*, **14**, 89 (1988).
50. B. G. Bagley, *Solid State Commun.*, **8**, 345 (1970).
51. P. M. Borsenberger, L. T. Pautmeier and H. Bässler, *Phys. Rev.*, **B46**, 12 145 (1992) and references therein.
52. P. M. Borsenberger, *Phys. Status Solidi*, **B173**, 671 (1992).
53. E. Silinsh, M. Bouvet and J. Simon, *Molec. Mater.*, **5**, 1 and 255 (1995).
54. P. Turek, P. Petit, J. Simon, R. Even, B. Boudjema, G. Guillaud and M. Maitrot, *J. Am. Chem. Soc.*, **109**, 5119 (1987).
55. W. D. Gill, *J. Appl. Phys.*, **43**, 5033 (1972).
56. J. M. Pearson, *Pure Appl. Chem.*, **49**, 463 (1977).
57. D. M. Pai, *J. Chem. Phys.*, **50** 2285 (1970).
58. P. J. Regensburger, *Photochem. Photobiol.*, **8**, 429 (1968).
59. A. Kakuta, Y. Mori, S. Takano, M. Sawada and I. Shibuya, *J. Imaging Technol.*, **11**, 7 (1987).
60. R. O. Loutfy, *Can. J. Chem.*, **59**, 549 (1981).
61. C. Y. Liang and E. G. Scalco, *J. Electrochem. Soc.*, **110**, 779 (1963).
62. P. Day and R. J. P. Williams, *J. Chem. Phys.*, **37**, 567 (1962).
63. M. K. Engel, personal communication (1995).
64. Advances Printing of Paper Summaries: Fourth International Congress on *Advances in Non-Impact Printing Technology*, 20–25 March 1988, p. 52, Society of Imaging Science and Technology (1988).
65. R. O. Loutfy, A.-M. Hor, C. -K. Hsiao, G. Baranyi and P. Kazmaier, *Pure Appl. Chem.*, **60**, 1047 (1988).
66. F. Reinitzer, *Monatsh. Chem.*, **9**, 421 (1888).
67. O. Lehmann, *Z. Kristallogr. Mineral*, **18**, 464 (1890).
68. G. Friedel, *Ann. Physique*, **18**, 273 (1922).
69. R. Williams, *Nature Lond.*, **199**, 273 (1963).
70. G. H. Heilmeier and L. A. Zanoni, *Appl. Phys. Lett.*, **13**, 91 (1968).
71. I. W. Davison, K. L. Ewing, J. Fergason, M. Chapman, A. Can and C. C. Voorhis, *Cancer*, **29**, 1123 (1972).
72. G. H. Heilmeier, L. A. Zanoni and L. A. Barton, *Proc. Inst. Elec. Electron Engrs*, **56**, 1162 (1968).
73. M. Schadt and W. Helfrich, *Appl. Phys. Lett.*, **18**, 127 (1971).
74. D. L. White and G. N. Taylor, *J. Appl. Phys.*, **45**, 4718 (1974).
75. F. J. Kahn, *Appl. Phys. Lett.*, **22**, 111 (1973).
76. R. B. Meyer, L. Liébert, L. Strzelecki and P. Keller, *Journal de Physique (Lettres)*, **36**, L-69 (1975).
77. P. Keller, L. Liébert and L. Strzelecki, *J. de Physique Colloques C3*, **37**, 3 (1976).
78. N. A. Clark and S. T. Lagerwall, *Appl. Phys. Lett.*, **36**, 899 (1980).
79. J. S. Im and A. Chiang, *MRS Bull.*, p. 27 (March 1996).
80. I. A. Shanks, *Contemp. Phys.*, **23**, 65 (1982).
81. B. S. Scheuble, *Kontakte (Darmstadt)*, 34 (1989).

82. R. Zimmermann, Thèse de doctorat d'état, Université L. Pasteur, Strasbourg (1984).
83. B. Bahadur, *Molec. Cryst. Liquid Cryst.*, **109**, 3 (1984).
84. M. Le Contellec (CNET-Lannion B), *Images de la Physique*, p. 121, CNRS (1992).
85. Jun-Ichi Hanna and I. Shimizu, *MRS Bull.*, **35** (March 1996).
86. D. E. Carlson and C. R. Wronski, in *Amorphous Semiconductors*, (ed. M. H. Brodsky), Topics in Applied Physics **36**, Springer-Verlag, Heidelberg (1979).
87. M. J. Thompson, *J. Vac. Sci. Technol.*, **B2**, 827 (1984).
88. P. J. Collings, *Liquid Crystals: Nature's Delicate Phase of Matter*, Adam Hilger, Bristol (1990).
89. CNET-Lannion B booklet printed in 1987.
90. B. Bahadur, *Liquid Crystals*, Vols. I–III, World Scientific, Singapore (1992).
91. B. Bahadur, Vol. III, p. 69 of Ref. [90].
92. G. W. Gray, K. J. Harrison and J. A. Nash, *Electron. Lett.*, **9**, 130 (1974).
93. G. W. Gray, *Advances in Liquid Crystal Materials for Applications*, BDH Publication (1978).
94. C. Escher, *Kontakte (Darmstadt) (2)*, **3** (1986).
95. J. W. Goodby *et al.*, *Ferroelectric Liquid Crystals: Ferroelectricity and Related Phenomena*, Vol. 7, Gordon & Breach., Philadelphia (1991).
96. N. A. Clark and S. T. Lagerwall, in p. 1 of Ref. [95].
97. J. W. Goodby in p. 99 of Ref. [95].
98. N. A. Clark and S. T. Lagerwall, in p. 409 of Ref. [95].
99. See also Canon, Display Technology, pp. 12–13 (1998).
100. V. Luzzati, A. Tardieu and T. Gulik-Krzywicki, *Nature (Lond.)*, **217**, 1028 (1968).
101. V. Luzzati and P. Spegt, *Nature (Lond.)*, **215**, 701 (1967).

Appendixes

1. Main Symmetry Point Groups: Notation and Symbols — 429
2. Tables of Characters of the Main Point Symmetry Groups — 433
3. Group–Subgroup Relationships for the Crystallographic and Infinite Groups — 443
4. Two-Dimensional (Monocolor) Space Groups — 444
5. Isohedral Tilings — 450
6. Isohedral Tilings Derived from the Topological Class [3^6] — 458
7. Isohedral Tilings Corresponding to a Given Site Symmetry — 466
8. Piezoelectrical and Nonlinear Optical Tensor Coefficients — 468
9. Irreducible Representations of the Group K_h — 471
10. The Main Dyes and Pigments — 475

Hinoki ni naro
Y. Inoue (1907–1992)
Asunaro

Appendix 1
Main Symmetry Point Groups: Notation and Symbols

Reference

T. Hahn (ed.), *International Tables for Crystallography*, The International Union of Crystallography, Kluwer Academic, Dordrecht (1989).

Definitions

Lattice: periodic array of points, each of them having an identical surrounding relative to all other points.

Unit cell: parallelepiped formed by three noncoplanar lattice vectors.

Primitive unit cell: unit cell with only one lattice point at each corner and therefore one point per unit cell.

Space group: set of symmetry operations (including translations) transforming a periodic crystal into itself.

Point group of the crystal (crystalline class): symmetry group in which there is an invariant point determined by the various symmetry operations (excluding translations). Glide plane and screw axes are then considered as plain planes or rotation axes (symmetry of orientation).

Site symmetry: it is the point symmetry group formed by the symmetry operations that are common to both the molecule (or more generally the site) and to the crystal space group. The elements of symmetry which include translations (glide plane, screw axes) are not taken into account.

Asymmetric unit: set of atoms that cannot be transformed into each other by using a symmetry operation of the space group.

Main Symbols

Table A1.1 Symmetry planes normal to the plane of projection (three dimensions) and symmetry lines in the plane of the figure (two dimensions)

Symmetry plane or symmetry line	Graphical symbol	Glide vector in units of lattice translation vectors parallel and normal to the projection plane	Printed symbol
Reflection plane, mirror plane	───────	None	m
Reflection line, mirror line (two dimensions)	───────	None	m
'Axial' glide plane	-----------	$\frac{1}{2}$ along line parallel to projection plane	a, b or c
Glide line (two dimensions)	-----------	$\frac{1}{2}$ along line in plane	g
'Axial' glide plane	············	$\frac{1}{2}$ normal to projection plane	a, b, or c
'Diagonal' glide plane	—·—·—·—	$\frac{1}{2}$ along line parallel to projection plane, combined with $\frac{1}{2}$ normal to projection plane	n

Table A1.2 Symmetry planes parallel to the plane of projection

Symmetry plane	Graphical symbol	Glide vector in units of lattice translation vectors parallel to the projection plane	Printed symbol
Reflection plane, mirror plane	⌐ ⌐	None	m
'Axial' glide plane	↓⌐ ⌐→	$\frac{1}{2}$ in the direction of the arrow	a, b or c
'Axial' glide plane	⌐→↓	$\frac{1}{2}$ in either of the directions of the two arrows	a, b or c
'Diagonal' glide plane	⌐↘	$\frac{1}{2}$ in the direction of the arrow	n

APPENDIXES

Table A1.3 Symmetry axes normal to the plane of projection (three dimensions) and symmetry points in the plane of the figure (two dimensions)

Symmetry axis or symmetry point	Graphical symbol	Screw vector of a right-handed screw rotation in units of the shortest lattice translation vector parallel to the axis	Printed symbol
Identity	None	None	1
Twofold rotation axis (two dimensions)	●	None	2
Twofold screw axis: '2 sub 1'	⟋	$\frac{1}{2}$	2_1
Threefold rotation axis (two dimensions)	▲	None	3
Threefold screw axis: '3 sub 1'	▲	$\frac{1}{3}$	3_1
Fourfold rotation axis (two dimensions)	◆	None	4
Fourfold screw axis: '4 sub 1'	◆	$\frac{1}{4}$	4_1
Sixfold rotation axis (two dimensions)	⬢	None	6
Sixfold screw axis: '6 sub 1'	⬢	$\frac{1}{6}$	6_1
Centre of symmetry, inversion centre: '1 bar'	○	None	$\bar{1}$
Inversion axis: '3 bar'	△	None	$\bar{3}$
Inversion axis: '4 bar'	⬦	None	$\bar{4}$
Inversion axis: '6 bar'	⬣	None	$\bar{6}$
Twofold rotation axis with centre of symmetry	●	None	$2/m$
Twofold screw axis with centre of symmetry	⟋	$\frac{1}{2}$	$2_1/m$
Fourfold rotation axis with centre of symmetry	◆	None	$4/m$

Table A1.4 Symmetry axes parallel to the plane of projection

Symmetry axis	Graphical symbol	Screw vector of a right-handed screw rotation in units of the shortest lattice translation vector parallel to the axis	Printed symbol
Twofold rotation axis	← →	None	2
Twofold screw axis: '2 sub 1'	⟵ ⟶	$\frac{1}{2}$	2_1
Fourfold rotation axis	⊢ ⊣	None	4
Fourfold screw axis: '4 sub 1'	⊢ ⊣	$\frac{1}{4}$	4_1

Main Point Symmetry Groups (^3D): the 32 Crystal Classes

Shoenflies	Shubnikov	International Tables, short symbol
C_1	1	1
C_i	$\tilde{2}$	$\bar{1}$
C_2	2	2
C_s	m	m
C_{2h}	2:m	2/m
D_2	2:2	222
C_{2v}	2·m	mm2
D_{2h}	m·2:m	mmm
C_4	4	4
S_4	$\tilde{4}$	$\bar{4}$
C_{4h}	4:m	4/m
D_4	4:2	422
C_{4v}	4·m	4mm
D_{2d}	$\tilde{4}$:2	$\bar{4}$2m
D_{4h}	m·4:m	4/mmm
C_3	3	3
S_6	$\tilde{6}$	$\bar{3}$
D_3	3:2	32
C_{3v}	3·m	3m
D_{3d}	$\tilde{6}$·m	$\bar{3}$m
C_6	6	6
C_{3h}	3:m	$\bar{6}$
C_{6h}	6:m	6/m
D_6	6:2	622
C_{6v}	6·m	6mm
D_{3h}	m·3:m	$\bar{6}$m2
D_{6h}	m·6:m	6/mmm
T	3/2	23
T_h	$\tilde{6}$/2	m$\bar{3}$
O	3/4	432
T_d	3/$\tilde{4}$	$\bar{4}$3m
O_h	$\tilde{6}$/4	m$\bar{3}$m

Appendix 2
Tables of Characters of the Main Point Symmetry Groups

Reference

O. Kahn and M. F. Koenig, *Données fondamentales pour la Chimie*, Hermann, Paris (1972).

The nomenclature proposed by R. S. Mulliken has been used:

(a) One-dimensional representations symmetric with respect to rotation $2\pi/n$ around the main axis (C_n) are designated A; those antisymmetric B.
(b) Subscripts 1 and 2 attached to A or B designate representations symmetric or antisymmetric with respect to a C_2 perpendicular to the principal axis or, if missing, a vertical plane of symmetry.
(c) Primes and double primes indicate a symmetry or antisymmetry relative to σ_h.
(d) g (gerade: even) or u (ungerade: odd) is used to indicate the presence or not of an inversion centre.
(e) E and T indicate two-dimensional and three-dimensional representations.

Nonaxial Groups

C_1	E
A	1

C_s	E	σ_h		
A'	1	1	x, y, R_z	x^2, y^2, z^2, xy
A''	1	-1	z, R_x, R_y	yz, xz

C_i	E	i		
A_g	1	1	R_x, R_y, R_z	$x^2, y^2, z^2, xy, xz, yz$
A_u	1	-1	x, y, z	

C_n Groups

C_2	E	C_2		
A	1	1	z, R_z	x^2, y^2, z^2, xy
B	1	-1	x, y, R_x, R_y	yz, xz

C_3	E	C_3	C_3^2		$\varepsilon = \exp(2\pi i/3)$
A	1	1	1	z, R_z	$x^2 + y^2, z^2$
E	$\begin{cases} 1 \\ 1 \end{cases}$	$\begin{matrix} \varepsilon \\ \varepsilon^* \end{matrix}$	$\begin{matrix} \varepsilon^* \\ \varepsilon \end{matrix}$	$(x, y), (R_x, R_y)$	$(x^2 - y^2, xy)(yz, xz)$

C_4	E	C_4	C_2	C_4^3		
A	1	1	1	1	z, R_z	$x^2 + y^2, z^2$
B	1	-1	1	-1		$x^2 - y^2, xy$
E	$\begin{cases} 1 \\ 1 \end{cases}$	$\begin{matrix} i \\ -i \end{matrix}$	$\begin{matrix} -1 \\ -1 \end{matrix}$	$\begin{matrix} -i \\ i \end{matrix}$	$(x, y), (R_x, R_y)$	(xz, yz)

C_5	E	C_5	C_5^2	C_5^3	C_5^4		$\varepsilon = \exp(2\pi i/5)$
A	1	1	1	1	1	z, R_z	$x^2 + y^2, z^2$
E_1	$\begin{cases} 1 \\ 1 \end{cases}$	$\begin{matrix} \varepsilon \\ \varepsilon^* \end{matrix}$	$\begin{matrix} \varepsilon^2 \\ \varepsilon^{2*} \end{matrix}$	$\begin{matrix} \varepsilon^{2*} \\ \varepsilon^2 \end{matrix}$	$\begin{matrix} \varepsilon^* \\ \varepsilon \end{matrix}$	$(x, y), (R_x, R_y)$	(xz, yz)
E_2	$\begin{cases} 1 \\ 1 \end{cases}$	$\begin{matrix} \varepsilon^2 \\ \varepsilon^{2*} \end{matrix}$	$\begin{matrix} \varepsilon^* \\ \varepsilon \end{matrix}$	$\begin{matrix} \varepsilon \\ \varepsilon^* \end{matrix}$	$\begin{matrix} \varepsilon^{2*} \\ \varepsilon^2 \end{matrix}$		$(x^2 - y^2, xy)$

C_6	E	C_6	C_3	C_2	C_3^2	C_6^5		$\varepsilon = \exp(2\pi i/6)$
A	1	1	1	1	1	1	z, R_z	$x^2 + y^2, z^2$
B	1	−1	1	−1	1	−1		
E_1	1	ε	$-\varepsilon^*$	−1	$-\varepsilon$	ε^*	$(x, y), (R_x, R_y)$,	(xz, yz)
	1	ε^*	$-\varepsilon$	−1	$-\varepsilon^*$	ε		
E_2	1	$-\varepsilon^*$	$-\varepsilon$	1	$-\varepsilon^*$	$-\varepsilon$		$(x^2 - y^2, xy)$
	1	$-\varepsilon$	$-\varepsilon^*$	1	$-\varepsilon$	$-\varepsilon^*$		

D_n Groups

D_2	E	$C_2(z)$	$C_2(y)$	$C_2(x)$		
A	1	1	1	1		x^2, y^2, z^2
B_1	1	1	−1	−1	z, R_z	xy
B_2	1	−1	1	−1	y, R_y	xz
B_3	1	−1	−1	1	x, R_x	yz

D_3	E	$2C_3$	$3C_2$		
A_1	1	1	1		$x^2 + y^2, z^2$
A_2	1	1	−1	z, R_z	
E	2	−1	0	$(x, y), (R_x, R_y)$	$(x^2 - y^2, xy) (xz, yz)$

D_4	E	$2C_4$	$C_2(=C_4^2)$	$2C_2'$	$2C_2''$		
A_1	1	1	1	1	1		$x^2 + y^2, z^2$
A_2	1	1	1	−1	−1	z, R_z	
B_1	1	−1	1	1	−1		$x^2 - y^2$
B_2	1	−1	1	−1	1		xy
E	2	0	−2	0	0	$(x, y), (R_x, R_y)$	(xz, yz)

D_5	E	$2C_5$	$2C_5^2$	$5C_2$		
A_1	1	1	1	1		$x^2 + y^2, z^2$
A_2	1	1	1	−1	z, R_z	
E_1	2	$-\tfrac{1}{2}(1-\sqrt{5})$	$-\tfrac{1}{2}(1+\sqrt{5})$	0	$(x, y) (R_x, R_y)$	(xz, yz)
E_2	2	$-\tfrac{1}{2}(1+\sqrt{5})$	$-\tfrac{1}{2}(1-\sqrt{5})$	0		$(x^2 - y^2, xy)$

D_6	E	$2C_6$	$2C_3$	C_2	$3C_2'$	$3C_2''$		
A_1	1	1	1	1	1	1		$x^2 + y^2, z^2$
A_2	1	1	1	1	−1	−1	z, R_z	
B_1	1	−1	1	−1	1	−1		
B_2	1	−1	1	−1	−1	1		
E_1	2	1	−1	−2	0	0	$(x, y), (R_x, R_y)$	(xz, yz)
E_2	2	−1	−1	2	0	0		$(x^2 - y^2, xy)$

C_{nv} Groups

C_{2v}	E	C_2	$\sigma_v(xz)$	$\sigma'_v(yz)$		
A_1	1	1	1	1	z	x^2, y^2, z^2
A_2	1	1	−1	−1	R_z	xy
B_1	1	−1	1	−1	x, R_y	xz
B_2	1	−1	−1	1	y, R_x	yz

C_{3v}	E	$2C_3$	$3\sigma_v$		
A_1	1	1	1	z	x^2+y^2, z^2
A_2	1	1	−1	R_z	
E	2	−1	0	$(x, y), (R_x, R_y)$	$(x^2-y^2, xy)(xz, yz)$

C_{4v}	E	$2C_4$	C_2	$2\sigma_v$	$2\sigma_d$		
A_1	1	1	1	1	1	z	x^2+y^2, z^2
A_2	1	1	1	−1	−1	R_z	
B_1	1	−1	1	1	−1		x^2-y^2
B_2	1	−1	1	−1	1		xy
E	2	0	−2	0	0	$(x, y), (R_x, R_y)$	(xz, yz)

C_{5v}	E	$2C_5$	$2C_5^2$	$5\sigma_v$		
A_1	1	1	1	1	z	x^2+y^2, z^2
A_2	1	1	1	−1	R_z	
E_1	2	$-\frac{1}{2}(1-\sqrt{5})$	$-\frac{1}{2}(1+\sqrt{5})$	0	$(x, y), (R_x, R_y)$	(xz, yz)
E_2	2	$-\frac{1}{2}(1+\sqrt{5})$	$-\frac{1}{2}(1-\sqrt{5})$	0		(x^2-y^2, xy)

C_{6v}	E	$2C_6$	$2C_3$	C_2	$3\sigma_v$	$3\sigma_d$		
A_1	1	1	1	1	1	1	z	x^2+y^2, z^2
A_2	1	1	1	1	−1	−1	R_z	
B_1	1	−1	1	−1	1	−1		
B_2	1	−1	1	−1	−1	1		
E_1	2	1	−1	−2	0	0	$(x, y), (R_x, R_y)$	(xz, yz)
E_2	2	−1	−1	2	0	0		(x^2-y^2, xy)

C_{nh} Groups

C_{2h}	E	C_2	i	σ_h		
A_g	1	1	1	1	R_z	x^2, y^2, z^2, xy
B_g	1	−1	1	−1	R_x, R_y	xz, yz
A_u	1	1	−1	−1	z	
B_u	1	−1	−1	1	x, y	

C_{3h}	E	C_3	C_3^2	σ_h	S_3	S_3^5		$\varepsilon = \exp(2\pi i/3)$
A'	1	1	1	1	1	1	R_z	x^2+y^2, z^2
E'	$\begin{cases} 1 \\ 1 \end{cases}$	$\begin{matrix}\varepsilon \\ \varepsilon^*\end{matrix}$	$\begin{matrix}\varepsilon^* \\ \varepsilon\end{matrix}$	$\begin{matrix}1 \\ 1\end{matrix}$	$\begin{matrix}\varepsilon \\ \varepsilon^*\end{matrix}$	$\begin{matrix}\varepsilon^* \\ \varepsilon\end{matrix}$	(x,y)	(x^2-y^2, xy)
A''	1	1	1	-1	-1	-1	z	
E''	$\begin{cases} 1 \\ 1 \end{cases}$	$\begin{matrix}\varepsilon \\ \varepsilon^*\end{matrix}$	$\begin{matrix}\varepsilon^* \\ \varepsilon\end{matrix}$	$\begin{matrix}-1 \\ -1\end{matrix}$	$\begin{matrix}-\varepsilon \\ -\varepsilon^*\end{matrix}$	$\begin{matrix}-\varepsilon^* \\ -\varepsilon\end{matrix}$	(R_x, R_y)	(xz, yz)

C_{4h}	E	C_4	C_2	C_4^3	i	S_4^3	σ_h	S_4		
A_g	1	1	1	1	1	1	1	1	R_z	x^2+y^2, z^2
B_g	1	-1	1	-1	1	-1	1	-1		x^2-y^2, xy
E_g	$\begin{cases}1\\1\end{cases}$	$\begin{matrix}i\\-i\end{matrix}$	$\begin{matrix}-1\\-1\end{matrix}$	$\begin{matrix}-i\\i\end{matrix}$	$\begin{matrix}1\\1\end{matrix}$	$\begin{matrix}i\\-i\end{matrix}$	$\begin{matrix}-1\\-1\end{matrix}$	$\begin{matrix}-i\\i\end{matrix}$	(R_x, R_y)	(xz, yz)
A_u	1	1	1	1	-1	-1	-1	-1	z	
B_u	1	-1	1	-1	-1	1	-1	1		
E_u	$\begin{cases}1\\1\end{cases}$	$\begin{matrix}i\\-i\end{matrix}$	$\begin{matrix}-1\\-1\end{matrix}$	$\begin{matrix}-i\\i\end{matrix}$	$\begin{matrix}-1\\-1\end{matrix}$	$\begin{matrix}-i\\i\end{matrix}$	$\begin{matrix}1\\1\end{matrix}$	$\begin{matrix}i\\-i\end{matrix}$	(x, y)	

D_{nh} Groups

D_{2h}	E	$C_2(z)$	$C_2(y)$	$C_2(x)$	i	$\sigma(xy)$	$\sigma(xz)$	$\sigma(yz)$		
A_g	1	1	1	1	1	1	1	1		x^2, y^2, z^2
B_{1g}	1	1	-1	-1	1	1	-1	-1	R_z	xy
B_{2g}	1	-1	1	-1	1	-1	1	-1	R_y	xz
B_{3g}	1	-1	-1	1	1	-1	-1	1	R_x	yz
A_u	1	1	1	1	-1	-1	-1	-1		
B_{1u}	1	1	-1	-1	-1	-1	1	1	z	
B_{2u}	1	-1	1	-1	-1	1	-1	1	y	
B_{3u}	1	-1	-1	1	-1	1	1	-1	x	

D_{3h}	E	$2C_3$	$3C_2$	σ_h	$2S_3$	$3\sigma_v$		
A_1'	1	1	1	1	1	1		x^2+y^2, z^2
A_2'	1	1	-1	1	1	-1	R_z	
E'	2	-1	0	2	-1	0	(x, y)	(x^2-y^2, xy)
A_1''	1	1	1	-1	-1	-1		
A_2''	1	1	-1	-1	-1	1	z	
E''	2	-1	0	-2	1	0	(R_x, R_y)	(xz, yz)

D_{4h}	E	$2C_4$	C_2	$2C_2'$	$2C_2''$	i	$2S_4$	σ_h	$2\sigma_v$	$2\sigma_d$		
A_{1g}	1	1	1	1	1	1	1	1	1	1		x^2+y^2, z^2
A_{2g}	1	1	1	−1	−1	1	1	1	−1	−1	R_z	
B_{1g}	1	−1	1	1	−1	1	−1	1	1	−1		x^2-y^2
B_{2g}	1	−1	1	−1	1	1	−1	1	−1	1		xy
E_g	2	0	−2	0	0	2	0	−2	0	0	(R_x, R_y)	(xz, yz)
A_{1u}	1	1	1	1	1	−1	−1	−1	−1	−1		
A_{2u}	1	1	1	−1	−1	−1	−1	−1	1	1	z	
B_{1u}	1	−1	1	1	−1	−1	1	−1	−1	1		
B_{2u}	1	−1	1	−1	1	−1	1	−1	1	−1		
E_u	2	0	−2	0	0	−2	0	2	0	0	(x, y)	

D_{5h}	E	$2C_5$	$2C_5^2$	$5C_2$	σ_h	$2S_5$	$2S_5^3$	$5\sigma_v$		
A_1'	1	1	1	1	1	1	1	1		x^2+y^2, z^2
A_2'	1	1	1	−1	1	1	1	−1	R_z	
E_1'	2	$-\frac{1}{2}(1-\sqrt{5})$	$-\frac{1}{2}(1+\sqrt{5})$	0	2	$-\frac{1}{2}(1-\sqrt{5})$	$-\frac{1}{2}(1+\sqrt{5})$	0	(x, y)	
E_2'	2	$-\frac{1}{2}(1+\sqrt{5})$	$-\frac{1}{2}(1-\sqrt{5})$	0	2	$-\frac{1}{2}(1+\sqrt{5})$	$-\frac{1}{2}(1-\sqrt{5})$	0		(x^2-y^2, xy)
A_1''	1	1	1	1	−1	−1	−1	−1		
A_2''	1	1	1	−1	−1	−1	−1	1	z	
E_1''	2	$-\frac{1}{2}(1-\sqrt{5})$	$-\frac{1}{2}(1+\sqrt{5})$	0	−2	$-\frac{1}{2}(1-\sqrt{5})$	$\frac{1}{2}(1+\sqrt{5})$	0	(R_x, R_y)	(xz, yz)
E_2''	2	$-\frac{1}{2}(1+\sqrt{5})$	$-\frac{1}{2}(1-\sqrt{5})$	0	−2	$-\frac{1}{2}(1+\sqrt{5})$	$\frac{1}{2}(1-\sqrt{5})$	0		

D_{nd} Groups

D_{2d}	E	$2S_4$	C_2	$2C_2'$	$2\sigma_d$			
A_1	1	1	1	1	1			x^2+y^2, z^2
A_2	1	1	1	−1	−1	R_z		
B_1	1	−1	1	1	−1			x^2-y^2
B_2	1	−1	1	−1	1	z		xy
E	2	0	−2	0	0	$(x, y), (R_x, R_y)$		(xz, yz)

D_{3d}	E	$2C_3$	$3C_2$	i	$2S_6$	$3\sigma_d$		
A_{1g}	1	1	1	1	1	1		x^2+y^2, z^2
A_{2g}	1	1	−1	1	1	−1	R_z	
E_g	2	−1	0	2	−1	0	(R_x, R_y)	$(x^2-y^2, xy)(xz, yz)$
A_{1u}	1	1	1	−1	−1	−1		
A_{2u}	1	1	−1	−1	−1	1	z	
E_u	2	−1	0	−2	1	0	(x, y)	

D_{4d}	E	$2S_8$	$2C_4$	$2S_8^3$	C_2	$4C_2'$	$4\sigma_d$		
A_1	1	1	1	1	1	1	1		x^2+y^2, z^2
A_2	1	1	1	1	1	-1	-1	R_z	
B_1	1	-1	1	-1	1	1	-1		
B_2	1	-1	1	-1	1	-1	1	z	
E_1	2	$\sqrt{2}$	0	$-\sqrt{2}$	-2	0	0	(x, y)	
E_2	2	0	-2	0	2	0	0		(x^2-y^2, xy)
E_3	2	$-\sqrt{2}$	0	$\sqrt{2}$	-2	0	0	(R_x, R_y)	(xz, yz)

D_{5d}	E	$2C_5$	$2C_5^2$	$5C_2$	i	$2S_{10}^3$	$2S_{10}$	$5\sigma_d$		
A_{1g}	1	1	1	1	1	1	1	1		x^2+y^2, z^2
A_{2g}	1	1	1	-1	1	1	1	-1	R_z	
E_{1g}	2	$-\tfrac{1}{2}(1-\sqrt{5})$	$-\tfrac{1}{2}(1+\sqrt{5})$	0	2	$-\tfrac{1}{2}(1-\sqrt{5})$	$-\tfrac{1}{2}(1+\sqrt{5})$	0	(R_x, R_y)	(xz, yz)
E_{2g}	2	$-\tfrac{1}{2}(1+\sqrt{5})$	$-\tfrac{1}{2}(1-\sqrt{5})$	0	2	$-\tfrac{1}{2}(1+\sqrt{5})$	$-\tfrac{1}{2}(1-\sqrt{5})$	0		(x^2-y^2, xy)
A_{1u}	1	1	1	1	-1	-1	-1	-1		
A_{2u}	1	1	1	-1	-1	-1	-1	1	z	
E_{1u}	2	$-\tfrac{1}{2}(1-\sqrt{5})$	$-\tfrac{1}{2}(1+\sqrt{5})$	0	-2	$\tfrac{1}{2}(1-\sqrt{5})$	$\tfrac{1}{2}(1+\sqrt{5})$	0	(x, y)	
E_{2u}	2	$-\tfrac{1}{2}(1+\sqrt{5})$	$-\tfrac{1}{2}(1-\sqrt{5})$	0	-2	$\tfrac{1}{2}(1+\sqrt{5})$	$\tfrac{1}{2}(1-\sqrt{5})$	0		

S_n Groups

S_4	E	S_4	C_2	S_4^3			
A	1	1	1	1	R_z		x^2+y^2, z^2
B	1	-1	1	-1	z		(x^2-y^2, xy)
E	$\begin{Bmatrix}1\\1\end{Bmatrix}$	$\begin{matrix}i\\-i\end{matrix}$	$\begin{matrix}-1\\-1\end{matrix}$	$\begin{matrix}-i\\i\end{matrix}$	$(x, y), (R_x, R_y)$		(xz, yz)

S_6	E	C_3	C_3^2	i	S_6^5	S_6		$\varepsilon = \exp(2\pi i/3)$
A_g	1	1	1	1	1	1	R_z	x^2+y^2, z^2
E_g	$\begin{Bmatrix}1\\1\end{Bmatrix}$	$\begin{matrix}\varepsilon\\\varepsilon^*\end{matrix}$	$\begin{matrix}\varepsilon^*\\\varepsilon\end{matrix}$	$\begin{matrix}1\\1\end{matrix}$	$\begin{matrix}\varepsilon\\\varepsilon^*\end{matrix}$	$\begin{matrix}\varepsilon^*\\\varepsilon\end{matrix}$	(R_x, R_y)	$(x^2-y^2, xy), (xz, yz)$
A_u	1	1	1	-1	-1	-1	z	
E_u	$\begin{Bmatrix}1\\1\end{Bmatrix}$	$\begin{matrix}\varepsilon\\\varepsilon^*\end{matrix}$	$\begin{matrix}\varepsilon^*\\\varepsilon\end{matrix}$	$\begin{matrix}-1\\-1\end{matrix}$	$\begin{matrix}-\varepsilon\\-\varepsilon^*\end{matrix}$	$\begin{matrix}-\varepsilon^*\\-\varepsilon\end{matrix}$	(x, y)	

S_8	E	S_8	C_4	S_8^3	C_2	S_8^5	C_4^3	S_8^7		$\varepsilon = \exp(2\pi i/8)$
A	1	1	1	1	1	1	1	1	R_z	x^2+y^2, z^2
B	1	-1	1	-1	1	-1	1	-1	z	
E_1	$\begin{Bmatrix}1\\1\end{Bmatrix}$	$\begin{matrix}\varepsilon\\\varepsilon^*\end{matrix}$	$\begin{matrix}i\\-i\end{matrix}$	$\begin{matrix}-\varepsilon^*\\-\varepsilon\end{matrix}$	$\begin{matrix}-1\\-1\end{matrix}$	$\begin{matrix}-\varepsilon\\-\varepsilon^*\end{matrix}$	$\begin{matrix}-i\\i\end{matrix}$	$\begin{matrix}\varepsilon^*\\\varepsilon\end{matrix}$	$(x, y), (R_x, R_y)$	
E_2	$\begin{Bmatrix}1\\1\end{Bmatrix}$	$\begin{matrix}i\\-i\end{matrix}$	$\begin{matrix}-1\\-1\end{matrix}$	$\begin{matrix}-i\\i\end{matrix}$	$\begin{matrix}1\\1\end{matrix}$	$\begin{matrix}i\\-i\end{matrix}$	$\begin{matrix}-1\\-1\end{matrix}$	$\begin{matrix}-i\\i\end{matrix}$		(x^2-y^2, xy)
E_3	$\begin{Bmatrix}1\\1\end{Bmatrix}$	$\begin{matrix}-\varepsilon^*\\-\varepsilon\end{matrix}$	$\begin{matrix}-i\\i\end{matrix}$	$\begin{matrix}\varepsilon\\\varepsilon^*\end{matrix}$	$\begin{matrix}-1\\-1\end{matrix}$	$\begin{matrix}\varepsilon^*\\\varepsilon\end{matrix}$	$\begin{matrix}i\\-i\end{matrix}$	$\begin{matrix}-\varepsilon\\-\varepsilon^*\end{matrix}$		(xz, yz)

Cubic Groups

T	E	$4C_3$	$4C_3^2$	$3C_2$			$\varepsilon = \exp(2\pi i/3)$
A	1	1	1	1			$x^2 + y^2 + z^2$
E	$\left\{\begin{array}{l}1\\1\end{array}\right.$	$\begin{array}{c}\varepsilon\\\varepsilon^*\end{array}$	$\begin{array}{c}\varepsilon^*\\\varepsilon\end{array}$	$\left.\begin{array}{l}1\\1\end{array}\right\}$			$(2z^2 - x^2 - y^2, x^2 - y^2)$
T	3	0	0	-1	$(x, y, z), (R_x, R_y, R_z)$		(xy, yz, xz)

T_h	E	$4C_3$	$4C_3^2$	$3C_2$	i	$4S_6$	$4S_6^5$	$3\sigma_h$			$\varepsilon = \exp(2\pi i/3)$
A_g	1	1	1	1	1	1	1	1			$x^2 + y^2, z^2$
E_g	$\left\{\begin{array}{l}1\\1\end{array}\right.$	$\begin{array}{c}\varepsilon\\\varepsilon^*\end{array}$	$\begin{array}{c}\varepsilon^*\\\varepsilon\end{array}$	$\begin{array}{c}1\\1\end{array}$	$\begin{array}{c}1\\1\end{array}$	$\begin{array}{c}\varepsilon\\\varepsilon^*\end{array}$	$\begin{array}{c}\varepsilon^*\\\varepsilon\end{array}$	$\left.\begin{array}{l}1\\1\end{array}\right\}$			$(2z^2 - x^2 - y^2, x^2 - y^2)$
T_g	3	0	0	-1	3	0	0	-1	(R_x, R_y, R_z)		(xy, yz, xz)
A_u	1	1	1	1	-1	-1	-1	-1			
E_u	$\left\{\begin{array}{l}1\\1\end{array}\right.$	$\begin{array}{c}\varepsilon\\\varepsilon^*\end{array}$	$\begin{array}{c}\varepsilon^*\\\varepsilon\end{array}$	$\begin{array}{c}1\\1\end{array}$	$\begin{array}{c}-1\\-1\end{array}$	$\begin{array}{c}-\varepsilon\\-\varepsilon^*\end{array}$	$\begin{array}{c}-\varepsilon^*\\-\varepsilon\end{array}$	$\left.\begin{array}{l}-1\\-1\end{array}\right\}$			
T_u	3	0	0	-1	-3	0	0	1	(x, y, z)		

O	E	$8C_3$	$6C_2'$	$6C_4$	$3C_2$		
A_1	1	1	1	1	1		$x^2 + y^2 + z^2$
A_2	1	1	-1	-1	1		
E	2	-1	0	0	2		$(2z^2 - x^2 - y^2, x^2 - y^2)$
T_1	3	0	-1	1	-1	$(x, y, z), (R_x, R_y, R_z)$	
T_2	3	0	1	-1	-1		(xy, yz, xz)

T_d	E	$8C_3$	$3C_2$	$6S_4$	$6\sigma_d$		
A_1	1	1	1	1	1		$x^2 + y^2 + z^2$
A_2	1	1	1	-1	-1		
E	2	-1	2	0	0		$(2z^2 - x^2 - y^2, x^2 - y^2)$
T_1	3	0	-1	1	-1	(R_x, R_y, R_z)	
T_2	3	0	-1	-1	1	(x, y, z)	(xy, yz, xz)

O_h	E	$8C_3$	$6C_2$	$6C_4$	$3C_2$ $(=C_4^2)$	i	$6S_4$	$8S_6$	$3\sigma_h$	$6\sigma_d$		
A_{1g}	1	1	1	1	1	1	1	1	1	1		$x^2 + y^2 + z^2$
A_{2g}	1	1	-1	-1	1	1	-1	1	1	-1		
E_g	2	-1	0	0	2	2	0	-1	2	0		$(2z^2 - x^2 - y^2, x^2 - y^2)$
T_{1g}	3	0	-1	1	-1	3	1	0	-1	-1	(R_x, R_y, R_z)	
T_{2g}	3	0	1	-1	-1	3	-1	0	-1	1		(xy, yz, xz)
A_{1u}	1	1	1	1	1	-1	-1	-1	-1	-1		
A_{2u}	1	1	-1	-1	1	-1	1	-1	-1	1		
E_u	2	-1	0	0	2	-2	0	1	-2	0		
T_{1u}	3	0	-1	1	-1	-3	-1	0	1	1	(x, y, z)	
T_{2u}	3	0	1	-1	-1	-3	1	0	1	-1		

Icosahedral Groups

I	E	$12C_5$	$12C_5^2$	$20C_3$	$15C_2$		
A	1	1	1	1	1		$x^2+y^2+z^2$
T_1	3	$\frac{1}{2}(1+\sqrt{5})$	$\frac{1}{2}(1-\sqrt{5})$	0	-1	(x,y,z), (R_x, R_y, R_z)	
T_2	3	$\frac{1}{2}(1-\sqrt{5})$	$\frac{1}{2}(1+\sqrt{5})$	0	-1		
G	4	-1	-1	1	0		
H	5	0	0	-1	1		$(2z^2-x^2-y^2,$ $x^2-y^2,$ $xy, yz, xz)$

I_h	E	$12C_5$	$12C_5^2$	$20C_3$	$15C_2$	i	$12S_{10}$	$12S_{10}^3$	$20S_6$	15σ		
A_g	1	1	1	1	1	1	1	1	1	1		$x^2+y^2+z^2$
T_{1g}	3	$\frac{1}{2}(1+\sqrt{5})$	$\frac{1}{2}(1-\sqrt{5})$	0	-1	3	$\frac{1}{2}(1-\sqrt{5})$	$\frac{1}{2}(1+\sqrt{5})$	0	-1	(R_x, R_y, R_z)	
T_{2g}	3	$\frac{1}{2}(1-\sqrt{5})$	$\frac{1}{2}(1+\sqrt{5})$	0	-1	3	$\frac{1}{2}(1+\sqrt{5})$	$\frac{1}{2}(1-\sqrt{5})$	0	-1		
G_g	4	-1	-1	1	0	4	-1	-1	1	0		
H_g	5	0	0	-1	1	5	0	0	-1	1		$(2z^2-x^2-y^2,$ $x^2-y^2,$ $xy, yz, xz)$
A_u	1	1	1	1	1	-1	-1	-1	-1	-1		
T_{1u}	3	$\frac{1}{2}(1+\sqrt{5})$	$\frac{1}{2}(1-\sqrt{5})$	0	-1	-3	$-\frac{1}{2}(1-\sqrt{5})$	$-\frac{1}{2}(1+\sqrt{5})$	0	1	(x,y,z)	
T_{2u}	3	$\frac{1}{2}(1-\sqrt{5})$	$\frac{1}{2}(1+\sqrt{5})$	0	-1	-3	$-\frac{1}{2}(1+\sqrt{5})$	$-\frac{1}{2}(1-\sqrt{5})$	0	1		
G_u	4	-1	-1	1	0	-4	1	1	-1	0		
H_u	5	0	0	-1	1	-5	0	0	1	-1		

Linear Groups

$C_{\infty v}$	E	$2C_\Phi$	$\cdots$	$\infty\sigma_v$		
Σ^+	1	1	$\cdots$	1	z	x^2+y^2, z^2
Σ^-	1	1	$\cdots$	-1	R_z	
Π	2	$2\cos\Phi$	$\cdots$	0	$(x,y), (R_x, R_y)$	(xz, yz)
Δ	2	$2\cos 2\Phi$	$\cdots$	0		(x^2-y^2, xy)
Φ	2	$2\cos 3\Phi$	$\cdots$	0		
$\cdots$	$\cdots$	$\cdots$	$\cdots$	$\cdots$		

D_∞	E	$2C_\Phi$	$\cdots$	∞C_2		
Σ^+	1	1	$\cdots$	1		x^2+y^2, z^2
Σ^-	1	1	$\cdots$	-1	R_z, z	
Π	2	$2\cos\Phi$	$\cdots$	0	$(x,y), (R_x, R_y)$	(xz, yz)
Δ	2	$2\cos 2\Phi$	$\cdots$	0		(x^2-y^2, xy)
Φ	2	$2\cos 3\Phi$	$\cdots$	0		
.	.	.	$\cdots$	.		
.	.	.	$\cdots$	.		

$D_{\infty h}$	E	$2C_\Phi$	$\cdots$	$\infty\sigma_v$	i	$2S_\Phi$	$\cdots$	∞C_2		
Σ_g^+	1	1	$\cdots$	1	1	1	$\cdots$	1		$x^2+y^2+z^2$
Σ_g^-	1	1	$\cdots$	-1	1	1	$\cdots$	-1	(R_z)	
Π_g	2	$2\cos\Phi$	$\cdots$	0	2	$2\cos\Phi$	$\cdots$	0	(R_x, R_y)	(xz, yz)
Δ_g	2	$2\cos 2\Phi$	$\cdots$	0	2	$2\cos 2\Phi$	$\cdots$	0		x^2-y^2, xy
$\cdots$	$\cdots$	$\cdots$	$\cdots$	$\cdots$	$\cdots$	$\cdots$	$\cdots$	$\cdots$		
Σ_u^+	1	1	$\cdots$	1	-1	-1	$\cdots$	-1	z	
Σ_u^-	1	1	$\cdots$	-1	-1	-1	$\cdots$	1		
Π_u	2	$2\cos\Phi$	$\cdots$	0	-2	$-2\cos\Phi$	$\cdots$	0	(x, y)	
Δ_u	2	$2\cos 2\Phi$	$\cdots$	0	-2	$-2\cos 2\Phi$	$\cdots$	0		
$\cdots$	$\cdots$	$\cdots$	$\cdots$	$\cdots$	$\cdots$	$\cdots$	$\cdots$	$\cdots$		

Group of Rotations

K	E	∞C_Φ	$\cdots$		
S	1	1	$\cdots$		$x^2+y^2+z^2$
P	3	$1+2\cos\Phi$	$\cdots$	$(x, y, z), (R_x, R_y, R_z)$	
D	5	$1+2\cos\Phi+2\cos 2\Phi$	$\cdots$		$(2z^2-x^2-y^2,$ $x^2-y^2,$ $xy, yz, xz)$
F	7	$1+2\cos\Phi+2\cos 2\Phi+2\cos 3\Phi$	$\cdots$		
$\cdots$	$\cdots$	$\cdots$			

Appendix 3
Group–Subgroup Relationships for the Crystallographic and Infinite Groups

Reference

Y. Shirotine and M. Chaskolskaïa, *Fondements de la Physique des Cristaux*, Ed. Mir, Moscow (1984).

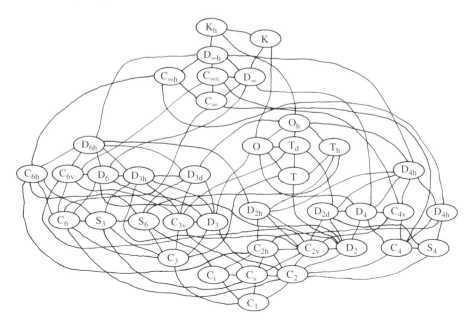

Appendix 4
Two-Dimensional (Monocolor) Space Groups

Description of the 2D space groups

Oblique System

symbol: p1   **crystal class: 1**

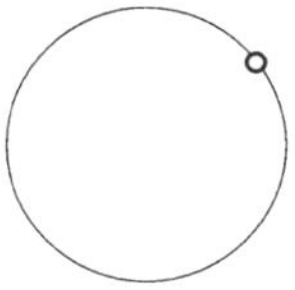

	number of molecules per unit cell:	molecular site symmetry:
a	(a) 1	(a) 1
b	(b) 2 (different)	(b) 1

APPENDIXES

symbol: p2

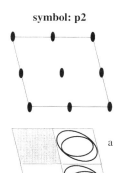

crystal class: 2

number of molecules per unit cell:
(a) 1
(b) 2 (identical)

molecular site symmetry:
(a) 2
(b) 1

Rectangular System

symbol: pm

crystal class: m

number of molecules per unit cell:
(a) 1
(b) 2

molecular site symmetry:
(a) m
(b) 1

symbol: pg

number of molecules per unit cell:
(a), (b), (c) : 2

molecular site symmetry:
(a), (b), (c) : 1

symbol: cm

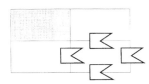

number of molecules per unit cell:
2

molecular site symmetry:
m

crystal class: 2mm

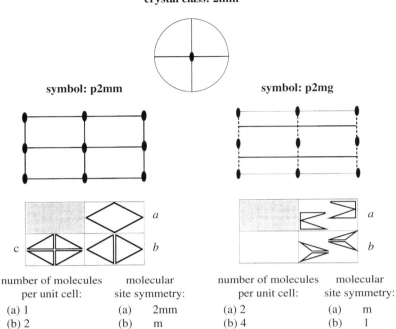

symbol: p2mm

number of molecules per unit cell:
(a) 1
(b) 2
(c) 4

molecular site symmetry:
(a) 2mm
(b) m
(c) 1

symbol: p2mg

number of molecules per unit cell:
(a) 2
(b) 4

molecular site symmetry:
(a) m
(b) 1

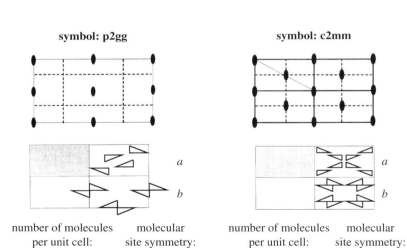

symbol: p2gg

number of molecules per unit cell:
(a) 4
(b) 2

molecular site symmetry:
(a) 1
(b) 2

symbol: c2mm

number of molecules per unit cell:
(a) 8
(b) 4

molecular site symmetry:
(a) 1
(b) 2

APPENDIXES

Square System

symbol: p4

crystal class: 4

number of molecules per unit cell:
(a) 4
(b) 2
(c) 1

molecular site symmetry:
(a) 1
(b) 2
(c) 4

symbol: p4mm

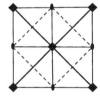 

crystal class: 4mm

number of molecules per unit cell:
(a) 8
(b) 4
(c) 1

molecular site symmetry:
(a) 1
(b) m
(c) 4

symbol: p4gm

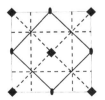

number of molecules per unit cell:
2

molecular site symmetry:
4

Hexagonal System

symbol: p3 **crystal class: 3**

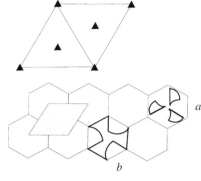

a number of molecules   molecular
   per unit cell: site symmetry:
   (a) 3   (a)   1
   (b) 1   (b)   3

b

symbol: p3m1 **crystal class: 3m**

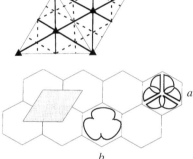

a number of molecules   molecular
   per unit cell: site symmetry:
   (a) 6   (a)   1
   (b) 1   (b)   3m

b

symbol: p31m

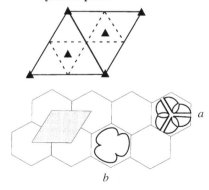

a number of molecules   molecular
   per unit cell: site symmetry:
   (a) 6   (a)   1
   (b) 1   (b)   3m

b

APPENDIXES

symbol: p6

crystal class: 6

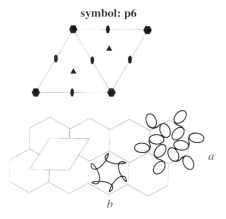

number of molecules per unit cell:
(a) 3
(b) 1

molecular site symmetry:
(a) 2
(b) 6

symbol: p6mm

crystal class: 6mm

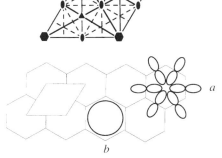

number of molecules per unit cell:
(a) 3
(b) 1

molecular site symmetry:
(a) 2mm
(b) 6mm

Appendix 5
Isohedral Tilings

Reference

B. Grünbaum, G. C. Shephard, *Tilings and Patterns*, W. H. Freeman, New York (1987).

Below are indicated the 81 unmarked isohedral tilings. For each tiling an arrow ↑ has been used to indicate the relative orientations of the tiles. The total of 93 isohedral tilings is reached with the 12 types of 'marked' tiles, which cannot be represented only by the contour of the tile and the singular points (for more details see the reference cited).

APPENDIXES 451

[3^6]

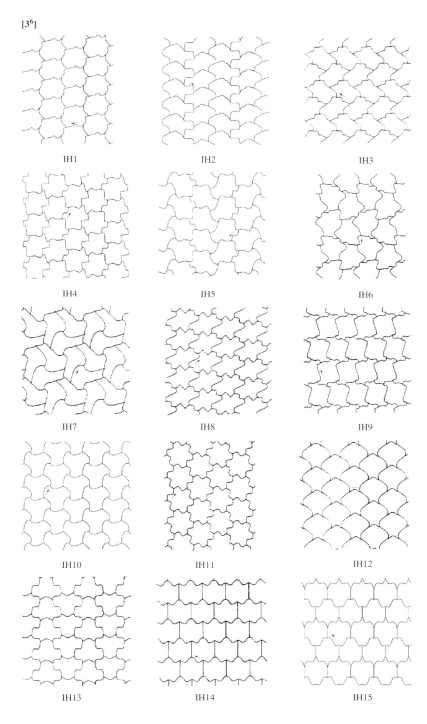

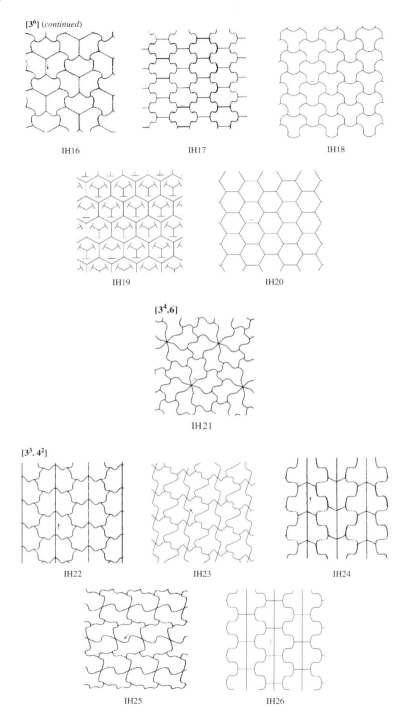

APPENDIXES

[$3^2.4.3.4$]

[3.4.6.4]

[3.6.3.6]

[3.12^2]

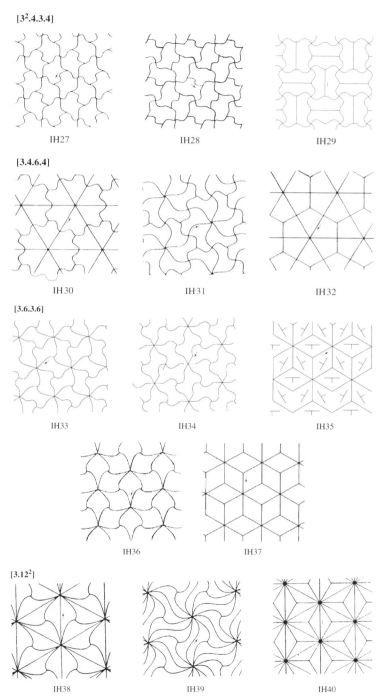

IH27 IH28 IH29
IH30 IH31 IH32
IH33 IH34 IH35
IH36 IH37
IH38 IH39 IH40

454   *DESIGN OF MOLECULAR MATERIALS*

$[4^4]$

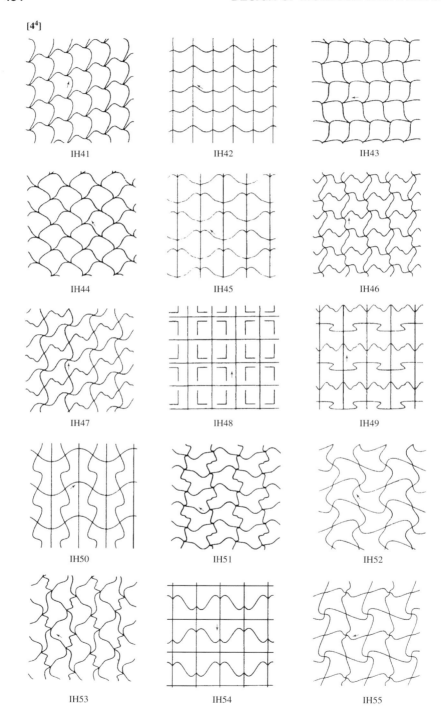

IH41   IH42   IH43

IH44   IH45   IH46

IH47   IH48   IH49

IH50   IH51   IH52

IH53   IH54   IH55

APPENDIXES

[4⁴] (*continued*)

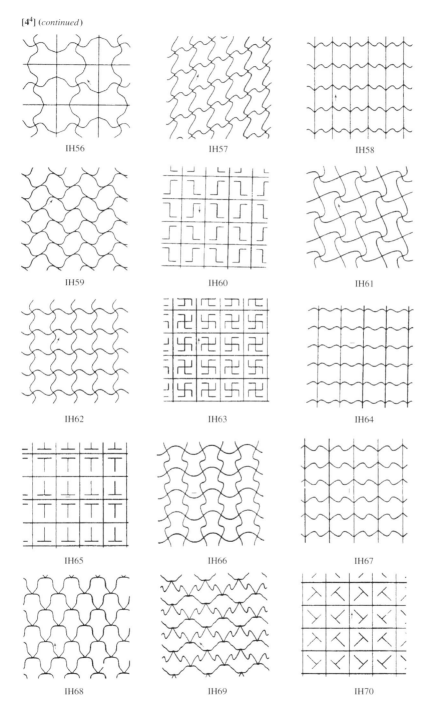

456   *DESIGN OF MOLECULAR MATERIALS*

[4^4] (*continued*)

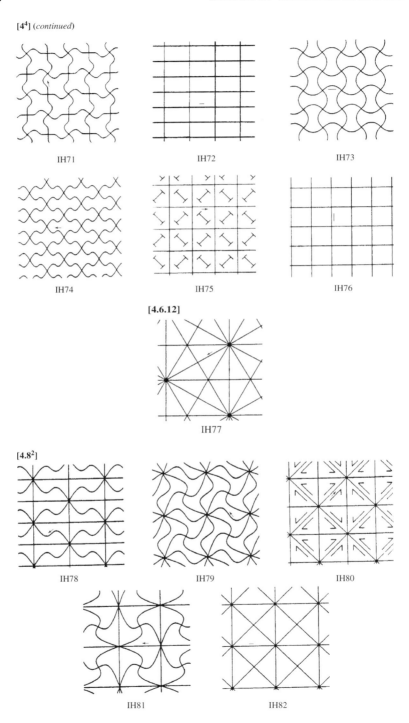

APPENDIXES

[6³]

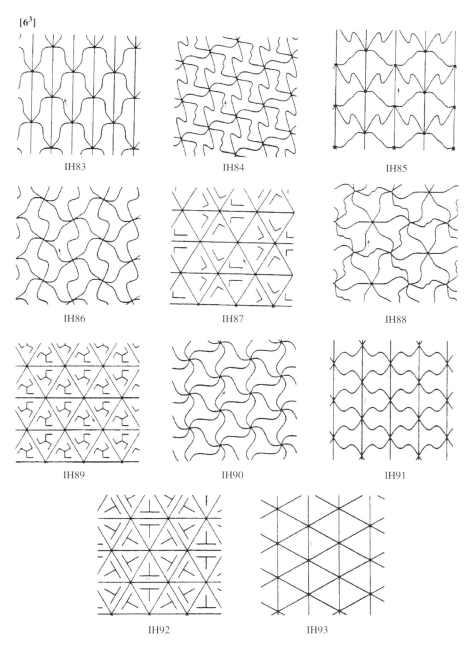

Appendix 6
Isohedral Tilings Derived from the Topological Class [3^6]

References

The following appendix has been adapted from B. Grünbaum and G. C. Shepard, *Tilings and Patterns*, W. H. Freeman, New York (1987). The notation is, however, different.

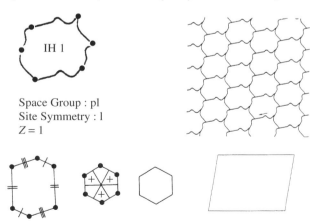

IH 1

Space Group : p1
Site Symmetry : 1
Z = 1

APPENDIXES

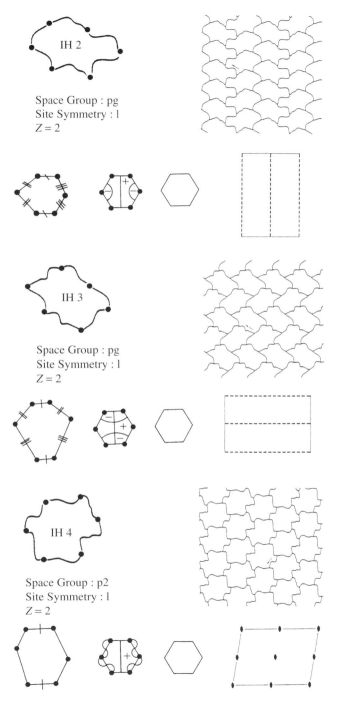

IH 2

Space Group : pg
Site Symmetry : 1
Z = 2

IH 3

Space Group : pg
Site Symmetry : 1
Z = 2

IH 4

Space Group : p2
Site Symmetry : 1
Z = 2

DESIGN OF MOLECULAR MATERIALS

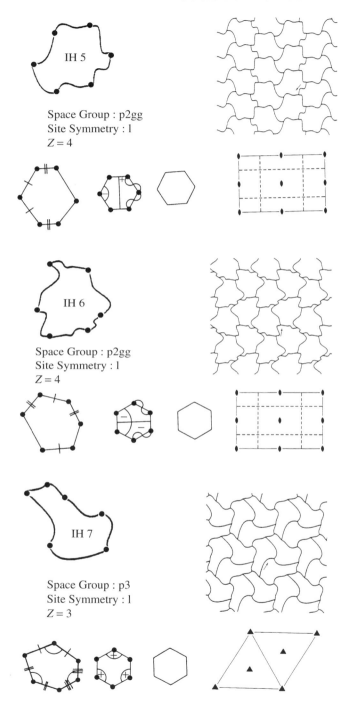

IH 5

Space Group : p2gg
Site Symmetry : 1
$Z = 4$

IH 6

Space Group : p2gg
Site Symmetry : 1
$Z = 4$

IH 7

Space Group : p3
Site Symmetry : 1
$Z = 3$

APPENDIXES

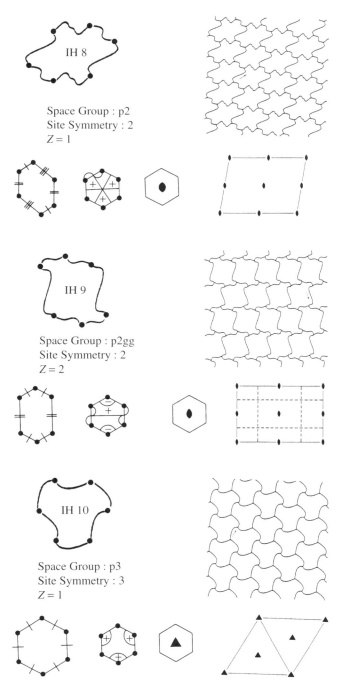

IH 8

Space Group : p2
Site Symmetry : 2
$Z = 1$

IH 9

Space Group : p2gg
Site Symmetry : 2
$Z = 2$

IH 10

Space Group : p3
Site Symmetry : 3
$Z = 1$

462   **DESIGN OF MOLECULAR MATERIALS**

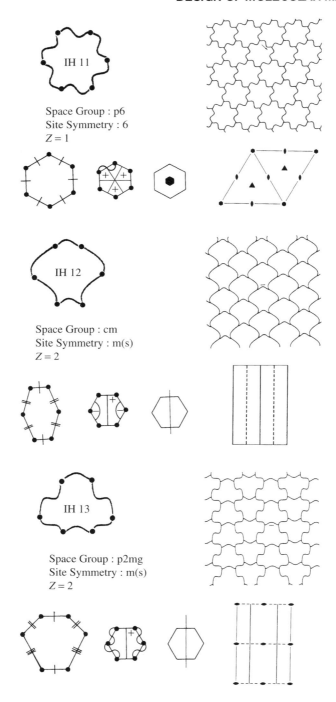

IH 11

Space Group : p6
Site Symmetry : 6
$Z = 1$

IH 12

Space Group : cm
Site Symmetry : m(s)
$Z = 2$

IH 13

Space Group : p2mg
Site Symmetry : m(s)
$Z = 2$

APPENDIXES

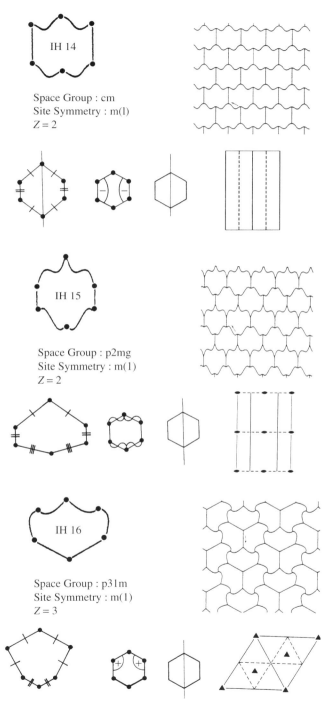

IH 14

Space Group : cm
Site Symmetry : m(1)
Z = 2

IH 15

Space Group : p2mg
Site Symmetry : m(1)
Z = 2

IH 16

Space Group : p31m
Site Symmetry : m(1)
Z = 3

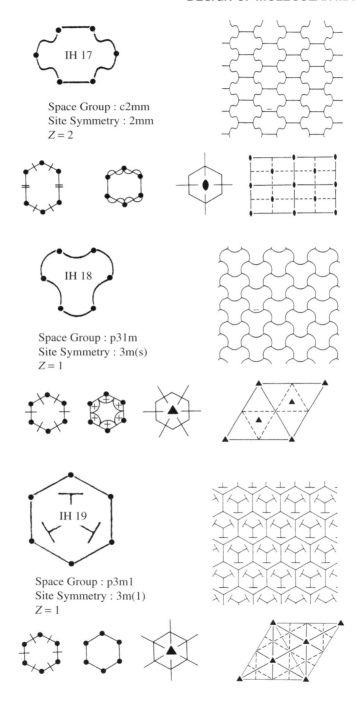

IH 17
Space Group : c2mm
Site Symmetry : 2mm
Z = 2

IH 18
Space Group : p31m
Site Symmetry : 3m(s)
Z = 1

IH 19
Space Group : p3m1
Site Symmetry : 3m(1)
Z = 1

APPENDIXES

Space Group : p6mm
Site Symmetry : 6mm
$Z = 1$

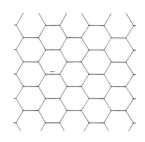

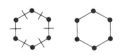

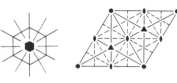

Appendix 7
Isohedral Tilings Corresponding to a Given Site Symmetry (other than 1 or C_1; unmarked tiles)

	[3⁶]	[3³.4²]	[3².4.3.4]	[3.4.6.4]	[3.6.3.6]	[3.12²]	[4⁴]	[4.8²]	[6³]
6mm	**20**								
6	*11*								
4mm							**76**		
4							*62*		
3m	18(s)								**93**
3	*10*								*90*
2mm(s)	**17**				**37**		**72**, **73**		
2mm(l)	*17*				**37**		*74*		
2	8, 9				*34*		57, 58, 59, 61		
m(s)	12, 13	**26**	**29**	**32**		**40**	64, 66, 67	**82**	**91**
m(l)	14, 15, 16	*26*	*29*	*32*	*36*	*40*	68, 69, 71	**82**	*91*

[3⁶4] and [4.6.12]: no tiling with site symmetry other than 1.
Bold face: convex.
Italic: nonconvex.
Upright: convex or nonconvex.
s and l indicate two different series of mirror plane.

Appendix 8
Piezoelectrical and Nonlinear Optical Tensor Coefficients

$$\begin{pmatrix} P_x \\ P_y \\ P_z \end{pmatrix} = \begin{pmatrix} d_{11} & d_{12} & d_{13} & d_{14} & d_{15} & d_{16} \\ d_{21} & d_{22} & d_{23} & d_{24} & d_{25} & d_{26} \\ d_{31} & d_{32} & d_{33} & d_{34} & d_{35} & d_{36} \end{pmatrix} \begin{pmatrix} E_x E_x \\ E_y E_y \\ E_z E_z \\ 2E_y E_z \\ 2E_x E_z \\ 2E_x E_y \end{pmatrix}$$

In the case of piezoelectricity, $E_i E_j$ must be replaced by the stress component σ_{ij}.

Key to notation:
- ·   Zero modulus
- •   Nonzero modulus
- •—•   Equal moduli
- •—○   Moduli numerically equal, but opposite in sign

Centrosymmetrical Classes:
 All moduli vanish.

Triclinic

Class : 1, C₁

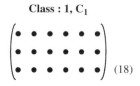

(18)

APPENDIXES

Monoclinic

Class : 2, C_2 — $C_2 \| z$ (8)

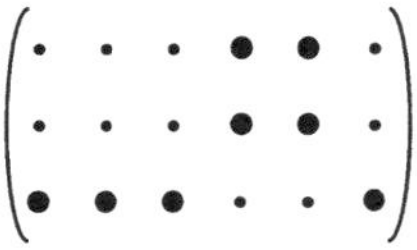

Class : m, C_s — $m \perp z$ (10)

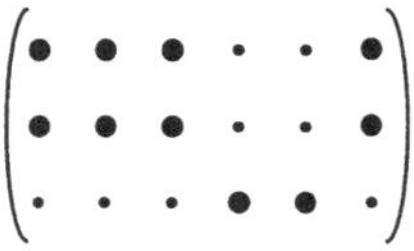

Orthorhombic

Class : 222, D_2 (3)

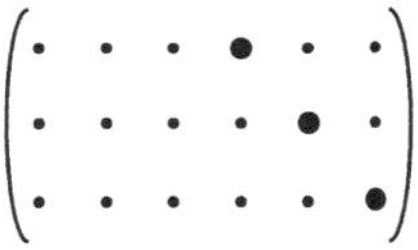

Class : mm2, C_{2v} (5)

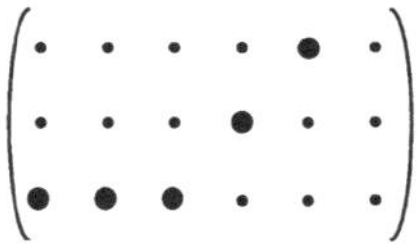

Tetragonal

Class : 4, C_4 (4)

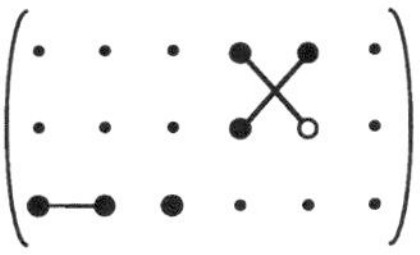

Class : $\bar{4}$, S_4 (4)

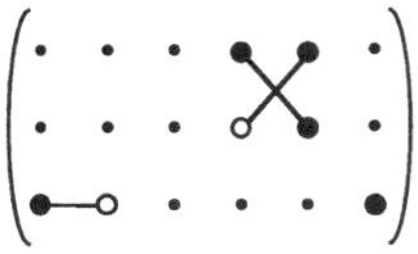

Class : 422, D_4 (1)

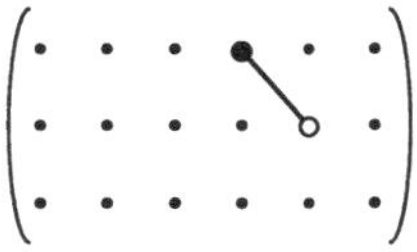

Class : $\bar{4}2m$, D_{2d} (2)

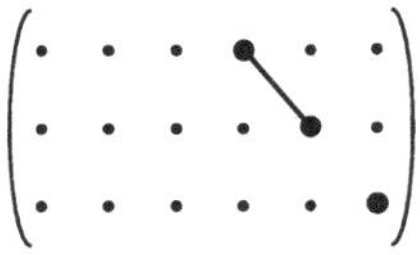

Class : 4mm, C_{4v} (3)

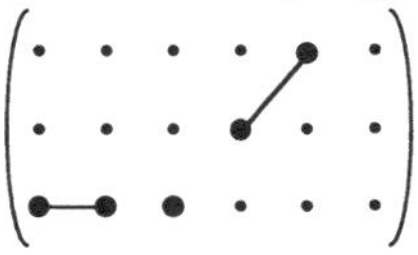

Trigonal

Class : 3, C_3 (6)

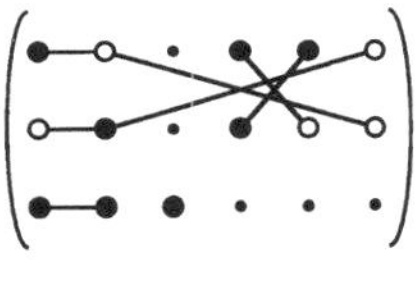

Class : 32, D_3 — $C_2 \| x$ (2)

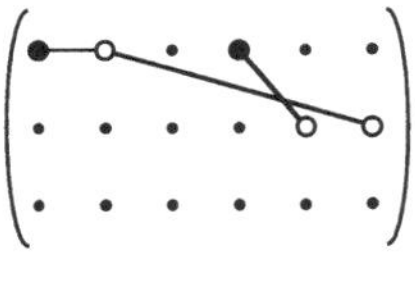

Class : 3m, C_{3v} — $m \perp x$ (4)

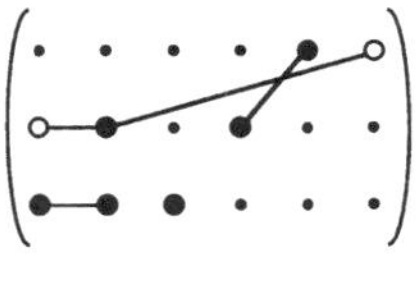

Hexagonal

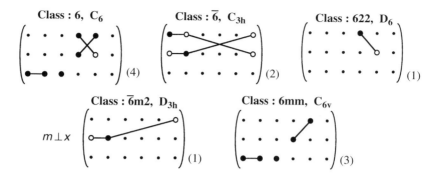

$m \perp x$

Cubic

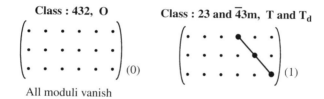

All moduli vanish

References

The above tensors have been established from the following two references:

1. J. F. Nye, *Propriétés physiques des cristaux*, Dunod, Paris (1961).
2. Kurtz/Jerphagnon/Choy, *Landolt-Börnstein*, Vol. 11 (1979).

Appendix 9
Irreducible Representations of the Group K_h

The spherical harmonics $Y_l^m(\Theta, \Phi)$ were introduced in Chapter 4. An essential property of these functions is that any function Y_l^m is transformed into $Y_l^{m'}$ by any proper rotation in the three-dimensional space. Consequently, the $(2l+1) Y_l^m (m \in [-l, \ldots, +l])$ functions form a basis of the irreducible representation D^l of the symmetry group K.

When spherical harmonics (Section 4.1.4) are referred to an orthogonal basis x, y, z (Figure 4.5), the application of a rotation $C_\alpha(z)$ around the z axis transforms the angle Φ into $\Phi + \alpha$. The Y_l^m function becomes

$$C_\alpha(z)[Y_l^m(\Theta, \Phi)] = Y_l^m(\Theta, \Phi - \alpha) = \exp(-im\alpha) Y_l^m(\Theta, \Phi)$$

In the equation only the exponential coefficient is modified (see Table 4.3). A matrix representation of the rotation $C_\alpha(z)$ is then given by the $(2l+1)$ diagonal terms $\exp(-im\alpha)$.

$$C_\alpha(z) \longrightarrow \begin{bmatrix} \exp(-il\alpha) & & & & \\ & \ddots & & 0 & \\ & & 1 & & \\ & 0 & & \ddots & \\ & & & & \exp(il\alpha) \end{bmatrix} \begin{matrix} Y_l^{m=+1} \\ \vdots \\ Y_l^0 \\ \vdots \\ Y_l^{m=-1} \end{matrix}$$

The character of any rotation of angle α for the D^l irreducible representation does not depend on the orientation of the axis. It is given by the sum

$$\chi_{(\alpha)}^l = \sum_{m=-l}^{m=+l} \exp(-im\alpha)$$

Table A9.1 Table of characters of the group of symmetry K_h. C_Φ means a rotation of an angle Φ (or α in the text) around all axes with an infinity of orientations

K_h	E	C_Φ	$\cdots C_\Phi^n$	i	iC_Φ	iC_Φ^n	
D_g^0	1	1	$\cdots$	1	1	$\cdots$	Scalar
D_u^0	1	1	$\cdots$	-1	-1	$\cdots$	Pseudo-scalar
D_g^1	3	$1 + 2\cos\Phi$	$\cdots$	3	$1 + 2\cos\Phi$	$\cdots$	Pseudo-vector
D_u^1	3	$1 + 2\cos\Phi$	$\cdots$	-3	$-(1 + 2\cos\Phi)$	$\cdots$	Vector
D_g^2	5	$1 + 2\cos\Phi + 2\cos 2\Phi$	$\cdots$	5	$1 + 2\cos\Phi + 2\cos 2\Phi$	$\cdots$	Deviator
D_u^2	5	$1 + 2\cos\Phi + 2\cos 2\Phi$	$\cdots$	-5	$-(1 + 2\cos\Phi + 2\cos 2\Phi)$	$\cdots$	Pseudo-deviator
D_g^3	7	$1 + 2\cos\Phi + \cdots + 2\cos 3\Phi$	$\cdots$	7	$1 + 2\cos\Phi + \cdots + 2\cos 3\Phi$	$\cdots$	Pseudo-septor
D_u^3	7	$1 + 2\cos\Phi + \cdots + 2\cos 3\Phi$	$\cdots$	-7	$-(1 + 2\cos\Phi + \cdots + 2\cos 3\Phi)$	$\cdots$	Septor
D_g^4	9	$1 + 2\cos\Phi + \cdots + 2\cos 4\Phi$	$\cdots$	9	$1 + 2\cos\Phi + \cdots + 2\cos 4\Phi$	$\cdots$	Nonor
D_u^4	9	$1 + 2\cos\Phi + \cdots + 2\cos 4\Phi$	$\cdots$	-9	$-(1 + 2\cos\Phi + \cdots + 2\cos 4\Phi)$	$\cdots$	Pseudo-nonor
$\cdots$	$\cdots$	$\cdots$		$\cdots$	$\cdots$		
D_g^l	$2l + 1$	$\dfrac{\sin(2l+1)\Phi/2}{\sin\Phi/2}$		$2l + 1$	$\dfrac{\sin(2l+1)\Phi/2}{\sin\Phi/2}$	$\cdots$	
D_u^l	$2l + 1$	$\dfrac{\sin(2l+1)\Phi/2}{\sin\Phi/2}$		$-(2l + 1)$	$-\dfrac{\sin(2l+1)\Phi/2}{\sin\Phi/2}$	$\cdots$	
$\cdots$						$\cdots$	

It can be shown that this sum is

$$\chi^l_{(\alpha)} = \frac{\sin(2l+1)\alpha/2}{\sin \alpha/2} = 1 + 2\cos\alpha + 2\cos 2\alpha + \cdots + 2\cos l\alpha$$

where $\chi^l_{(\alpha)}$ gives all the characters of the irreducible representations of the group K. In order to obtain all the symmetry operations of the group K_h we must add to the proper rotations C_α their product by the inversion centre: iC_α. Each D^l representation of K may be divided into gerade and ungerade representations for the group K_h:

(a) D^l_g where C_α and iC_α have the same character $\chi^l_{(\alpha)}$;
(b) D^l_u with $\chi^l_{(\alpha)}$ for C_α and $-\chi^l_{(\alpha)}$ for iC_α.

Multipolar moments are described in the group K_h by D^l_g if l is even and by D^l_u for odd values of l. The irreducible representations of K_h (see Table A9.1) correspond to a name (scalar, pseudo-scalar, vector, etc.) [1, 2].

The direct product of two $D^l_{g(u)}$ representations can be decomposed into a sum of irreducible representations known as the Clebsh–Gordan series:

$$D^i \otimes D^j = \sum_{k=|i-j|}^{k=i+j} D^k = D^{|i-j|} \oplus D^{|i-j+1|} \oplus \cdots \oplus D^{|i+j|}$$

with the following relations for gerade and ungerade products:

$$D^i_g \otimes D^j_g = D^i_u \otimes D^j_u = \sum_k D^k_g$$

$$D^i_g \otimes D^j_u = \sum_k D^k_u$$

A molecule or a molecular assembly could afford some anisotropic physical property only if its symmetry group G leaves unchanged at least one component of the basis of $D^l_{g(u)}$, which describes the physical property considered. The decomposition of $D^l_{g(u)}$ into irreducible representations of G must contain the totally symmetric representation (A_1 or A):

$$D^l_{g(u)} = a_1 A_1 \oplus a_2 A_2 \oplus \cdots \oplus a_i \Gamma_i + \cdots$$

The number of invariant components is given by

$$a_1 = \frac{1}{h} \sum_R \chi_{A_1}(R)\chi^l_{D_{g(u)}}(R)$$

where
  h = order of the point symmetry group G
  R = symmetry operation of G
  $\chi(R)$ = character of R

Since $\chi_{A_1}(R) = 1$, the formula reduces to

$$a_l = \frac{1}{h} \sum_R \chi^l_{D_{g(u)}}(R)$$

References

1. J. Jerphagnon, D. Chemla, R. Bonneville, *Adv. Phys.*, **27**, 609 (1978).
2. M. W. Evans, *Spectrochim. Acta.*, **46A**, 1475 (1990).

Appendix 10
The Main Dyes and Pigments

References

1. W. Herbst and K. Hunger, *Industrial Organic Pigments*, VCH, Weinheim (1993).
2. P. F. Gordon and P. Gregory, *Organic Chemistry in Colour*, Springer-Verlag, Berlin (1983).

Table A10.1 Some important natural dyes.

Color	Class	Typical dyes	Structure (name)	Source
Yellow	Flavone	Weld	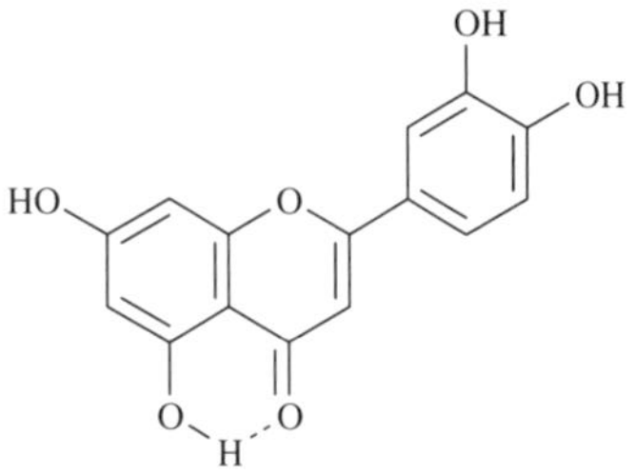 (Luteolin)	Seeds, stems and leaves of the *Reseda luteola* L plant
	Flavanol	Quercitron	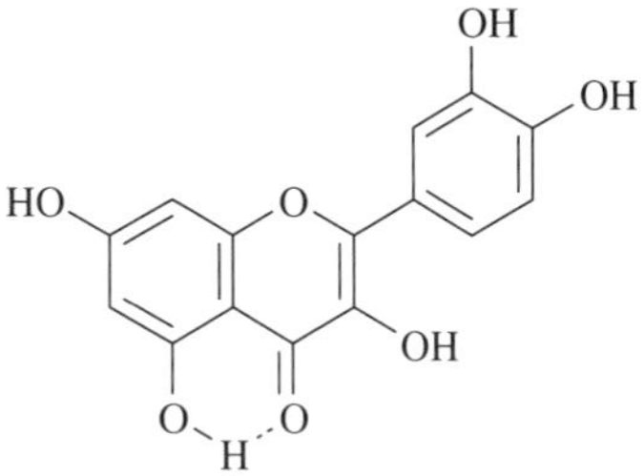 (Quercetin)	Bark of North American oak, *Quercus tinctoria nigra*
	Chalcone	Safflower	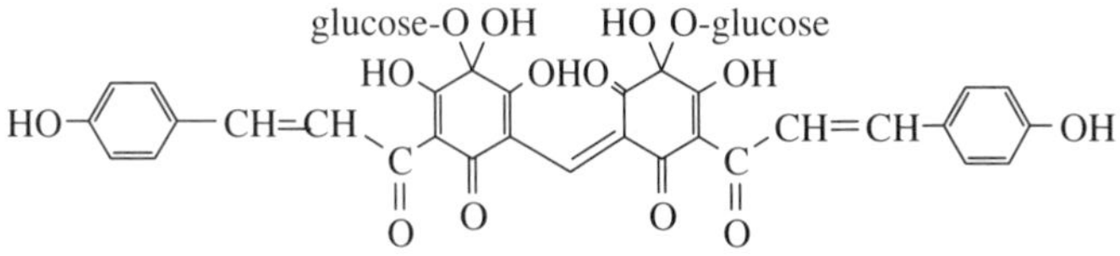 (Carthamin)	Dried petals of *Carthamus tinctorius*.

Polyene	Saffron	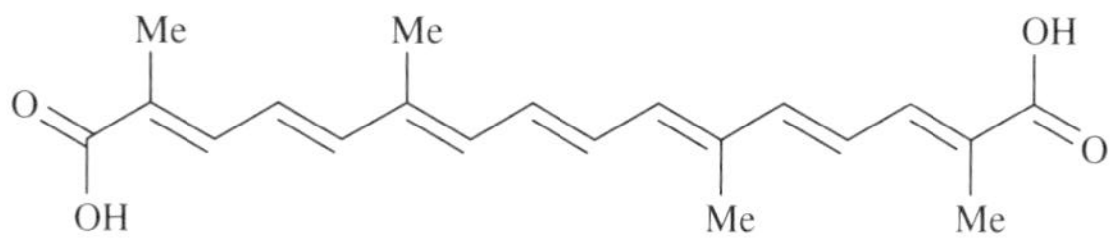 Crocetin structure (HOOC-C(Me)=CH-CH=CH-C(Me)=CH-CH=CH-C(Me)=CH-CH=CH-C(Me)=CH-COOH)	Stigmas of *Crocus sativus* L.
Red			
Anthraquinone	Kermes	Kermesic acid	Female insects, *Coccus ilicis*
Anthraquinone	Cochineal	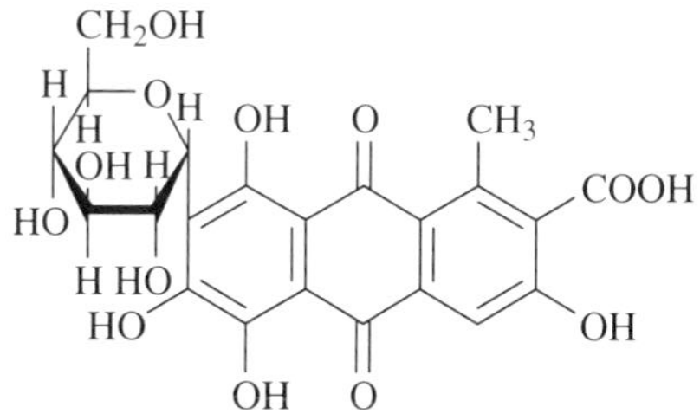 Carminic acid	Female insect, *Coccus cacti*

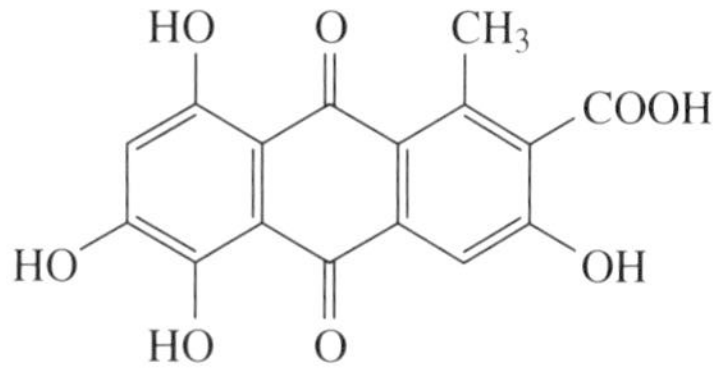

(continued overleaf)

Table A10.1 (continued)

Color	Class	Typical dyes	Structure (name)	Source
	Anthraquinone	Madder or alizarin	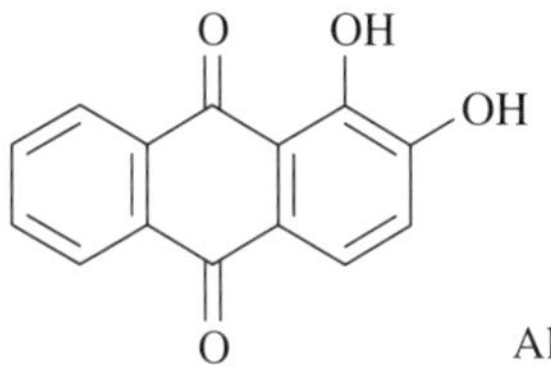 Alizarin (1,2-dihydroxyanthraquinone)	Roots of the *Rubia tinctorum* plant;
	Anthraquinone	Turkey red	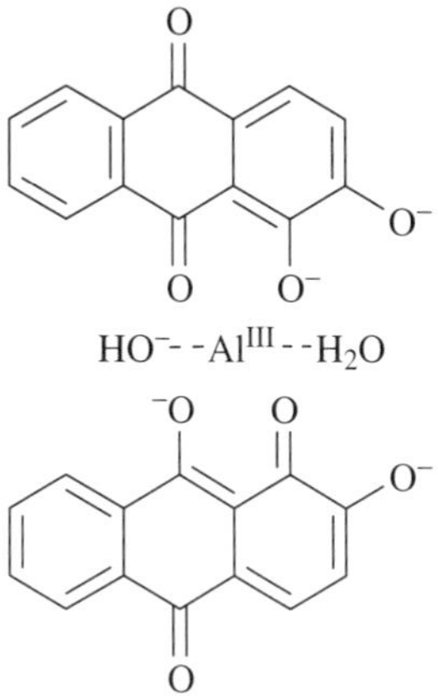 Turkey red, Ca^{2+}, $2H_2O$ (Al^{III} complex of alizarin)	
Purple	Indigoid	Tyrian purple	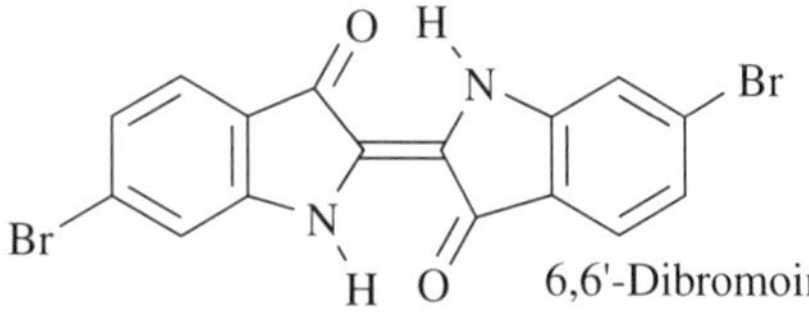 6,6'-Dibromoindigo	Mollusc usually *Murex brandaris*

Blue	Indigoid	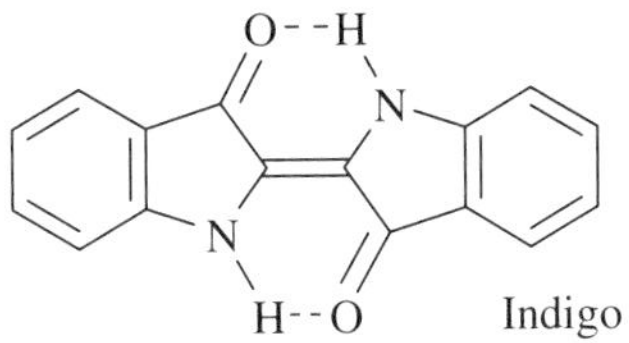 Woad; indigo	Leaves of indigo plant, *Indigofera tinctoria* L.
Black	Chroman	Logwood	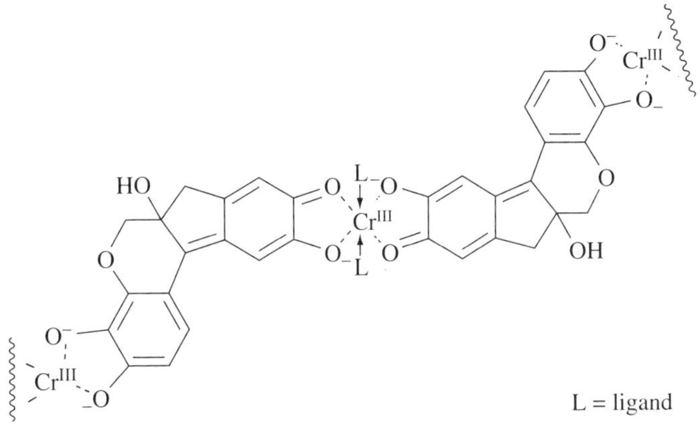 Heartwood of the tree *Haematoxylon campechiancum* L

L = ligand

Table A10.2 Commercially available copper (and metal-free) phthalocyanines (P.Bl. = Pigment Blue)

Phthalo blue

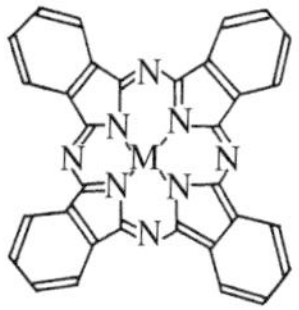

C.I. name	C.I. constitution number	Stabilized towards change of crystal modification	Crystal modification	Range of shades	Number of halogen atoms	Comments
P.Bl. 15	74160	No	α	Reddish-blue	—	
P.Bl. 15	74160:1	Yes	α	Greener than P.Bl. 15	0.5–1 Cl	
P.Bl. 15	74160:2	Yes	α	Reddish-blue	0.5–1 Cl	Non-flocculating
P.Bl. 15	74160:3	—	β	Greenish-blue	0[a]	
P.Bl. 15	74160:4	—	β		0[a]	
P.Bl. 15	74160:6	Yes	ε	Very reddish-blue	0[a]	Non-flocculating
P.Bl. 16	74100	Metal-free phthalocyanine				

[a] Depending on the synthetic route, small amounts of chlorine may be present.

Table A10.3 Phthalo green pigments (copper complex); P.Gr.: Pigment Green.

Phthalo green

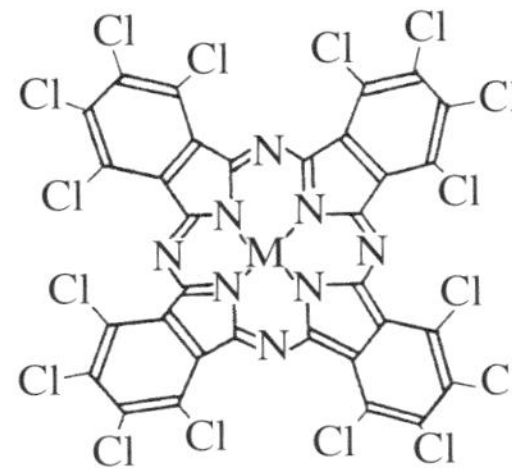

C.I. name	C.I. constitution number	Range of shades	Number of halogen atoms
P.Gr. 7	74260	bluish-green	14–15 Cl
P.Gr. 36	74265	yellowish-green	4–9 Br, 8–2 Cl

Table A10.4 Some of the commercially available diarylide yellow pigments (P.Gr.: Pigment Yellow)

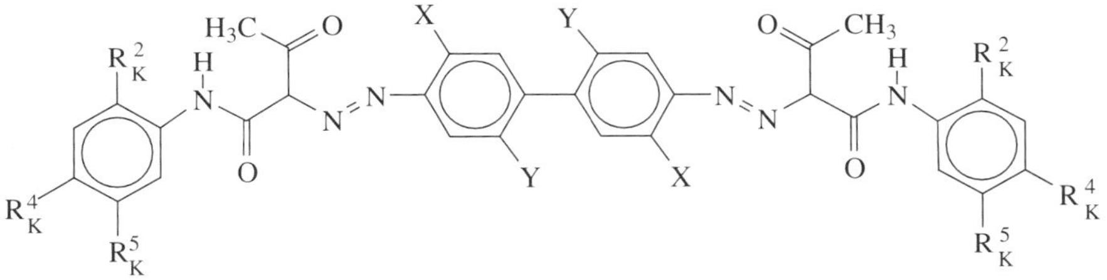

C.I. name	C.I. constitution number	X	Y	R_K^2	R_K^4	R_K^5	Shade
P.Y. 12	21090	Cl	H	H	H	H	Yellow
P.Y. 13	21100	Cl	H	CH_3	CH_3	H	Yellow
P.Y. 14	21095	Cl	H	CH_3	H	H	Yellow
P.Y. 17	21105	Cl	H	OCH_3	H	H	Greenish-yellow
P.Y. 55	21096	Cl	H	H	CH_3	H	Reddish-yellow
P.Y. 63	21091	Cl	H	Cl	H	H	Yellow
P.Y. 81	21127	Cl	Cl	CH_3	CH_3	H	Very greenish-yellow
P.Y. 83	21108	Cl	H	OCH_3	Cl	OCH_3	Reddish-yellow
P.Y. 87	21107:1	Cl	H	OCH_3	H	OCH_3	Reddish-yellow
P.Y. 113	21126	Cl	Cl	CH_3	Cl	H	Very greenish-yellow
P.O. 15	21130	CH_3	H	H	H	H	Yellowish-orange
P.O. 16	21160	OCH_3	H	H	H	H	Yellowish-orange

Table A10.5 Lithol rubine (P.R.: Pigment Red).

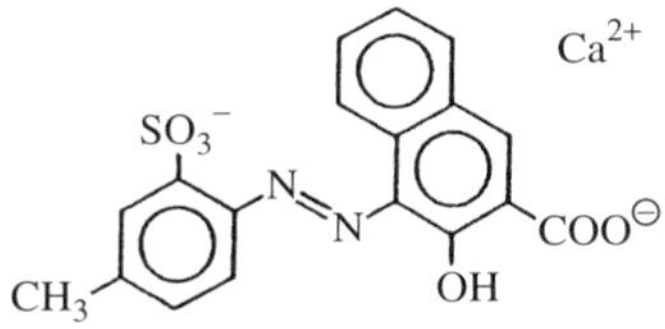

C.I. name	C.I. constitution number	Shade
P.R. 57:1	15850:1	Bluish-red (magenta, ruby)

Table A10.6 Some of the commercially available BONA (β-oxynaphthoic acid) pigment lakes [15] (P.Br.: Pigment Brown).

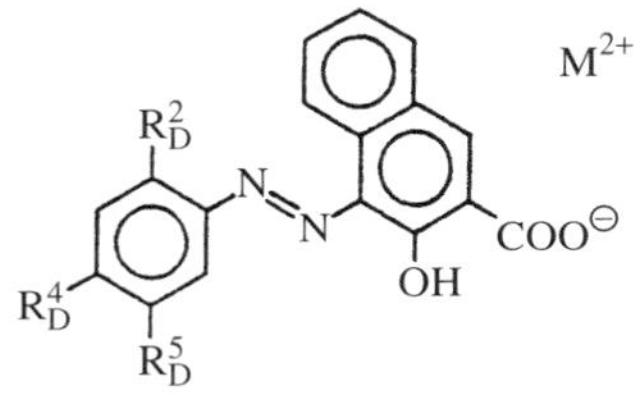

C.I. name	C.I. constitution number	R_D^2	R_D^4	R_D^5	M	Shade
P.R. 48:1	15865:1	SO_3^-	CH_3	Cl	Ba	Red
P.R. 48:4	15865:4	SO_3^-	CH_3	Cl	Mn	Bluish-red
P.R. 52:1	15860:1	SO_3^-	Cl	CH_3	Ca	Ruby
P.R. 52:2	15860:2	SO_3^-	Cl	CH_3	Mn	Maroon
P.R. 58:2	15825:2	H	Cl	SO_3^-	Ca	Bluish-red
P.R. 58:4	15825:4	H	Cl	SO_3^-	Mn	Medium-red
P.Br. 5	15800:2	H	H	H	Cu/2	Brown

Table A10.7 Hansa yellow

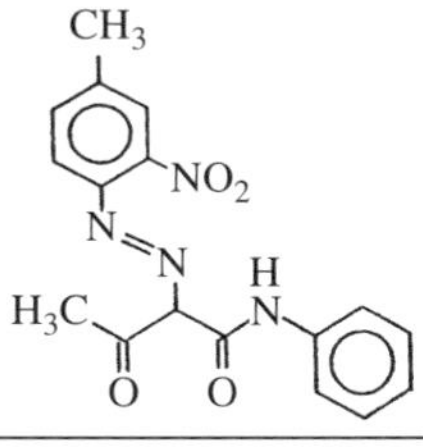

C.I. name	C.I. constitution number	Shade
P.Y. 1	11680	Yellow

APPENDIXES

Table A10.8 Some of the non-laked monoazo yellow and orange pigments (P.O.: Pigment Orange).

C.I. name	C.I. constitution number	R_D^2	R_D^4	R_D^5	R_K^2	R_K^4	R_K^5	Shade
P.Y. 2	11730	NO_2	Cl	H	CH_3	CH_3	H	Reddish-yellow
P.Y. 3	11710	NO_2	Cl	H	Cl	H	H	Very greenish-yellow
P.Y. 5	11660	NO_2	H	H	H	H	H	Very greenish-yellow
P.Y. 74	11741	OCH_3	NO_2	H	OCH_3	H	H	Greenish-yellow
P.Y. 111	11745	OCH_3	NO_2	H	OCH_3	H	Cl	Greenish-yellow
P.Y. 116	11790	Cl	$CONH_2$	H	H	$NHCOCH_3$	H	Yellow
P.O. 1	11725	NO_2	OCH_3	H	CH_3	H	H	Very reddish-yellow

Table A10.9 Some of the commercially available naphthol AS pigments (P.V.: Pigment Violet).

C.I. name	C.I. constitution number	R_D^2	R_D^4	R_D^5	R_K^2	R_K^4	R_K^5	Shade
P.R. 2	12310	Cl	H	Cl	H	H	H	Red
P.R. 7	12420	CH_3	Cl	H	CH_3	Cl	H	Bluish-red
P.R. 11	12430	CH_3	H	Cl	CH_3	H	Cl	Ruby
P.R. 12	12385	CH_3	NO_2	H	CH_3	H	H	Bordeaux
P.R. 13	12395	NO_2	CH_3	H	CH_3	H	H	Bluish-red
P.R. 14	12380	NO_2	Cl	H	CH_3	H	H	Bordeaux
P.R. 17	12390	CH_3	H	NO_2	CH_3	H	H	Red
P.R. 21	12300	Cl	H	H	H	H	H	Yellowish-red
P.R. 22	12315	CH_3	H	NO_2	H	H	H	Yellowish-red
P.R. 23	12355	OCH_3	H	NO_2	H	H	NO_2	Bluish-red
P.R. 112	12370	Cl	Cl	Cl	CH_3	H	H	Red
P.O. 24	12305	H	H	Cl	H	H	H	Orange
P.Br. 1	12480	Cl	H	Cl	OCH_3	H	OCH_3	Brown
P.R. 5	12490	OCH_3	H	$SO_2N(C_2H_5)_2$	OCH_3	OCH_3	Cl	Carmine
P.R. 31	12360	OCH_3	H	$CONHC_6H_5$	H	H	NO_2	Bluish-red
P.R. 32	12320	OCH_3	H	$CONHC_6H_5$	H	H	H	Red
P.R. 146	12485	OCH_3	H	$CONHC_6H_5$	OCH_3	Cl	OCH_3	Carmine
P.R. 245	12317	OCH_3	H	$CONH_2$	H	H	H	Bluish-red
P.R. 253	12375	Cl	SO_2NHCH_3	Cl	CH_3	H	H	Red
P.V. 25	12321	OCH_3	$NHCOC_6H_5$	OCH_3	H	H	H	Violet

Table A10.10 Toluidine red

[Structure: 2-methyl-4-nitrophenyl azo coupled to 2-naphthol]

C.I. name	C.I. constitution number	Common name
P.R. 3	12120	Toluidine red

Table A10.11 Commercially available β-naphthol pigments

[Structure: substituted phenyl azo coupled to 2-naphthol, with R² and R⁴ substituents]

C.I. name	C.I. constitution number	R^2	R^4	Common name
P.O. 2	12060	NO_2	H	Orthonitraniline orange
P.O. 5	12075	NO_2	NO_2	Dinitraniline orange
P.R. 1	12070	H	NO_2	Parachlor red
P.R. 4	12085	Cl	NO_2	Chlorinated parachlor red
P.R. 6	12090	NO_2	Cl	Parachlor red

Table A10.12 Some of the commercially available β-naphthol pigments lakes. Lake Red C is Pigment Red 53:1

[Structure: Diazo component (DC) azo coupled to 2-naphthol]

DC = Diazo component group

C.I. name	C.I. constitution number	DC	M/2	Shade
P.R. 49	15630	[2-methyl-1-naphthyl with SO_3M]	2 Na	Yellowish-red
P.R. 49:1	15630:1		Ba	Yellowish-red
P.R. 49:2	15630:2		Ca	Bluish-red or maroon

APPENDIXES

Table A10.12 (continued)

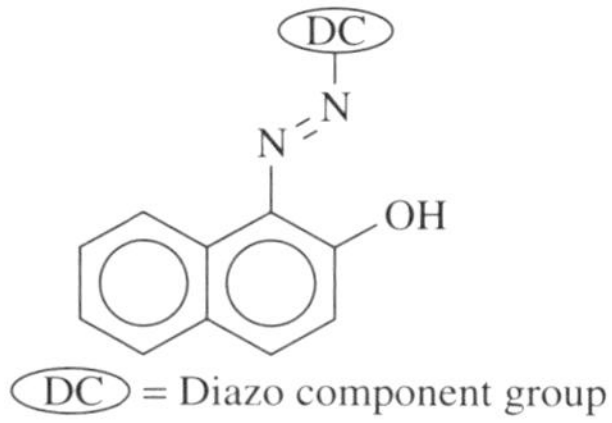

DC = Diazo component group

C.I. name	C.I. constitution number	DC	M/2	Shade
P.R. 51	15580	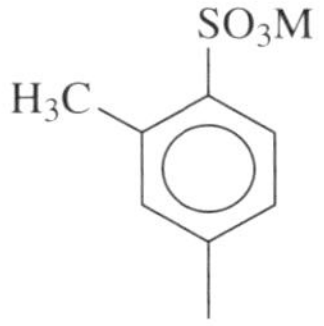 (2-methyl-5-sulfonate phenyl, with SO_3M, H_3C)	Ba	Scarlet
P.R. 53:1	15585:1	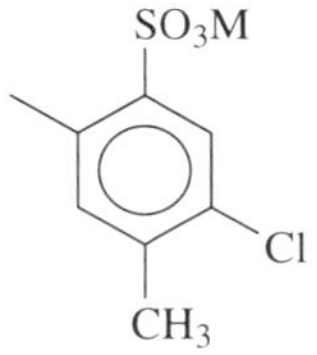 (methyl, SO_3M, Cl, CH_3 substituted phenyl)	Ba	Scarlet
P.R. 68	15525	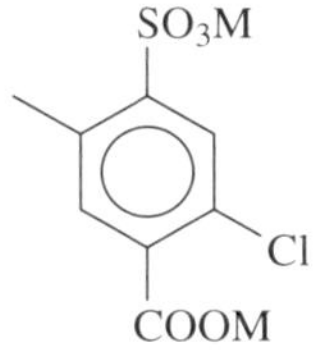 (methyl, SO_3M, Cl, COOM substituted phenyl)	2 Ca	Yellowish-red

Table A10.13 Some of the commercially available diazopyrazolone pigments (pyrazolones)

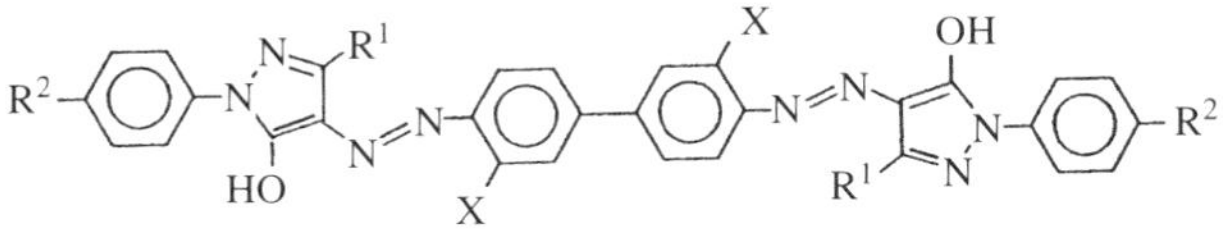

C.I. name	C.I. constitution number	X	R^1	R^2	Shade
P.O. 13	21110	Cl	CH_3	H	Yellowish-orange
P.R. 37	21205	OCH_3	CH_3	CH_3	Yellowish-red
P.R. 38	21120	Cl	$COOC_2H_5$	H	Red

Table A10.14 Triaryl carbonium pigments (alkali blue)

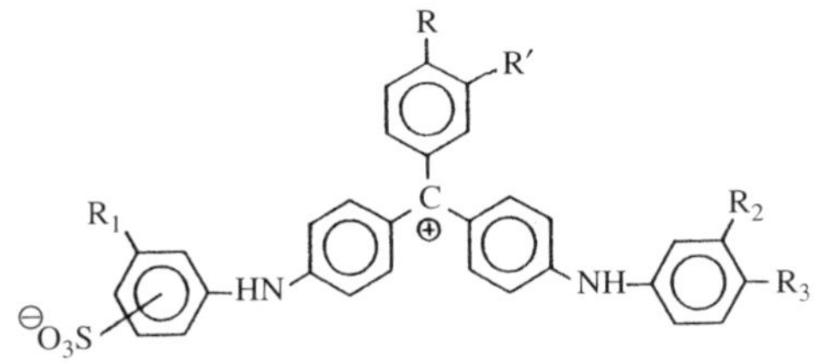

C.I. name	C.I. constitution number	R	R′	R_1	R_2	R_3
P.Bl. 18	42770:1	C_6H_5NH	H	H	H	H
P.Bl. 19	42750	NH_2	CH_3	H	H	H
P.Bl. 56	42800	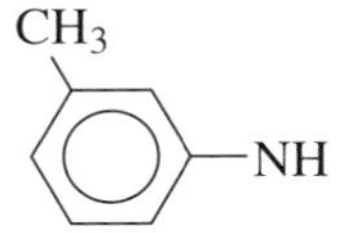 (3-methylphenyl)NH	H	CH_3	CH_3	H

Index

Note: abbreviations in the index follow the usage in text. For example, Pc indicates phthalocyanine, PcCu copper phthalocyanine etc.

absorption, light   319, 321–4, 415
absorption spectra   221, 230, 367, 401, 402, 403
acceptors
  electron   165, 328–30, 338, 393, 399
  p-n junctions   243, 244
activation energy   225, 226, 227–8, 231
adsorption   256–7, 261–2
air, PcM and   245–8, 252–5, 262, 276–8
  see also oxygen
amorphous materials   9, 219
  silicon   270, 271, 282, 283, 410
amphiphilic molecules and complexes   xiii, 15–18, 346–50
  metal ion   23–4, 29–30, 31, 32
    tiling   108–11
  soaps   16, 19–23, 346
anions   159–64, 170–1
aromatic derivatives   172–6, 182–3, 198
  and hyperpolarizability   328–31, 334–7
  for SHG   338–9
  see also benzene; naphthalene; perylene; phthalocyanines; triptycene
asymmetric unit definition   430
atomic orbitals   199–200, 202–5
azo derivatives   365, 394, 396, 398, 416, 483

band gap   216–17
band structures   202–12
  metallophthalocyanines   233–9
band theory   214–17
  applicability   197, 233–5
  for classical FETs   269–70
bands (symmetry)   71–9
bandwidth   207, 220, 233
  phthalocyanines   229, 235, 236–7
benzene   173, 175, 200

bicolor
  space groups   84, 85
  stripes   77–9
bisphthalocyanines   38, 220, 222, 264–8
  see also Pc_2Lu; Pc_2Sc; Pc_2Tm
Bloch functions   203–11, 216–17
bonds
  hydrogen   158, 159, 190–3, 303–4
  and polarizability   169–70
    hyperpolarizability   328, 329–30
  see also interactions
Bravais lattices   6, 7
Brillouin zone   206
broad-band semiconductors   229–33

cations   24, 25, 159–67, 170–1, 277
cause/effect relationships   140–55, 300–1, 312
cells
  LCD   406–11
  solar cells   242, 245, 249–50
  see also unit cells
charge carriers
  concentration   218, 219
  FETs   276–7
  generation   198, 217–18, 226–8, 269, 276
    photogeneration   386–9, 393, 397, 400–1
    and redox properties   219, 259–60
  injection   276, 406
  and intrinsic conductivity   230–2
  minority   249–50
  p-n junctions   243–4
  in van der Waals crystals   170–1
  see also mobility
charge transfer
  in molecular Schottky devices   250–2
  in xerographic processes   391–401

charge/charge interactions (coulomb forces) 157, 158, 159–67
charge/dipole interactions 157–9
  in van der Waals crystals 170–1
chemical potential 208–9, 227
chirality 32, 51, 66, 126, 141–3
  smectic C phases 304–7, 311–12
cholesteric mesophases 13, 405, 407
close packing 8, 88, 103, 115, 116
  Kitaigorodskii's approach 86–90, 119–20
  see also packing
coatings 367, 368–85
cohesive energy 157, 171–2
colloidal dispersions 369, 373
color 360–2
  electrochromism 220–1
  see also bicolor; monocolor
colorants 359–85, 475–86
  phthalocyanines 32, 364–5, 366, 394, 402–3
  see also dyes, pigments
columnar mesophases 4–5, 14, 32–5, 38–43, 186, 187, 240–2
complexes xiii, 23–43, 163–7
  charge transfer in 394, 399–401
  HSAB principle 167
conductivity, electrical 153, 197–9
  and charge carriers 198, 218, 219
  phthalocyanines 38, 39–40, 224–7, 230–2, 241–2, 261–8, 276–80
  and redox potentials 224–5
  see also charge carriers; insulators; photoconduction; semiconductors
conductivity, thermal 153
conductivity-based gas sensors 256–68
coordination number 86, 87–9, 95
  and topological class 93, 100, 101, 103
coulomb forces (charge/charge interactions) 157, 158, 159–67
crown-ether macrocycles xiii, 24–7, 104–6
  conductivity 263, 266
crystal class 72–3, 77
  and physical properties 144–5
  three-dimensional 432
  two-dimensional 82, 83
crystal growth
  crystallization 70–1
  epitaxial 9, 224, 258
  Kitaigorodskii's approach 84–90
crystal structure 70

nitrobenzene 180
Pc_2Lu 221–4
PcCu and PcPb 261, 262
  see also packing; structure
crystals 1–5
  hydrogen bonding in 191, 192
  ionic 159–62
  metallophthalocyanines 224, 225, 226, 231, 232–5, 258, 259
  noncentrosymmetric 181–2
  plastic 3, 14–15, 111–19
  second harmonic generation in 338–40
  soap 19–21
  see also lattices; liquid crystals; molecular crystals; polycrystalline materials; symmetry
cubic lattices 7, 113–15, 116, 126–8, 161
  soap 22–3
cubic system 7, 440, 469
Curie Principle 140–53, 308
current-voltage characteristics
  FETs 272–4, 281, 282
  p-n junctions 244, 245, 246–7
  solar cells 245, 250, 253
cylindrical phases (middle soap) 21–2, 357–8

d orbitals 23, 136, 137
dark response (solar cells) 245–8, 404, 405, 410
delocalised electrons 198–9, 228
density of states (DOS) 207–8
dichroism 405, 413–19
dielectrics, molecular 297–342
diffraction 3–5
diffusion 256, 257–8
diode effect *see* rectification
dioxygen (or oxygen)
  in cobalt complexes 26–8
  and phthalocyanines 232–3, 245–8, 252–5, 262–3, 276–8, 403
  and the siccative process 371, 372
dipole moments 66–7, 115–16, 328–31
  and ferroelectricity 301
  and piezoelectricity 314, 317
  and pyroelectricity 309
  see also dipoles; polarity; polarizability; polarization; transition moments
dipole/charge interactions *see* charge/dipole interactions
dipole/dipole interactions 157–9

INDEX

see also induced dipole/induced dipole interactions
dipoles 133–4, 324
  see also dipole/dipole interactions; induced dipoles; molecular dipoles
directionality 156–9, 180–3, 188, 190
discogens see columnar mesophases
disproportionation 217–18, 221, 230
domains 167–8, 299–300, 420–2
donors
  electron 165, 328–30, 336, 338, 393
  p-n junctions 243, 244
dopants 260, 261, 264–5, 406, 407
dyes 359–85, 475–86
  in charge generation layers 394
  dichroic 405, 413, 416–19
dynamic scattering displays 406–7

electrical fields 167–8, 171, 298–9
  Curie Principle 140–3, 146–53
  and dichroism 413–14
  in liquid crystal displays 420–1
  on nematics 405, 407–8
  see also piezoelectricity; polarizability
electrochromism 220–1
electromagnetic waves 3–5, 318–24
  polarized 141–3
  refraction 168–9, 318–21, 324–8
  see also light
electron acceptors 165, 328–30, 338
  in charge transfer 393, 399
electron donors 165, 328–30, 336, 338, 393
electrons
  in band theory 198, 215–17
  delocalised 198–9, 228
  energy levels 139
  see also charge carriers; charge transfer; orbitals
electrostatic interactions 157, 159–60
  see also charge/charge interactions
elementary tiles 96–104
  mesophases 104–11, 116–17, 118
  see also isohedral tilings
energy
  activation 225, 226, 227–8, 231
  charge carrier generation 198
  cohesive 157, 171–2
  E_k 207–8, 216–17
  electron 139, 215–17
  hydrogen bonds 190, 191
  lattice 157, 159–62, 171–2
  light exposure energy 403–5

packing 180–1
polarization 170–1
solar cells 248, 250, 255–6
solvation 227
transfer 251, 386–8
  see also free energy; interaction energy
energy gap 198, 216–17
energy levels, electron 139
  see also orbitals
enthalpy 153, 171–4, 183–4
entropy 185, 186, 208–9
excitation 318–21, 322
  in molecular Schottky devices 250–2
  photoconductive 386–8

f orbitals 136, 137
fatty acids 16, 18, 19, 346–7, 370–1, 373
Fermi level 208–9, 243
ferroelectricity 299, 301–7
  and LCDs 415–24, plate 12
field-effect transistors (FETs) 269–91, 409–11, 424
fields 140–3
  see also electrical fields
forces 156–67, 180–1
  directionality 156–9, 180–3, 188, 190
  see also interactions
free energy 138, 139–40, 162–3, 184, 208–9
frequency doubling 337–8
  see also second harmonic generation

gas sensors 256–68, 291
gel phase, soap 20–1, 22, 352–3
generalized shape 91, 93, 96–103
globular molecules 14–15, 112–13, 115, 118
groups 54, 57–60, 62
  molecular orbitals 200–2
  point symmetry groups 57, 58–60, 82, 135, 429–42
  subgroup relationships 66–7, 443
  for tensor coefficients 312–14
  see also site symmetry; space groups
growth see crystal growth

hexagonal lattice 79, 83
holes 399–400, 406
HOMO 212, 320, 322
  phthalocyanines 230, 235, 237
hopping 233–5, 275, 284–6, 289, 392, 393

HSAB principle 164–7
hydrogen bonds 158, 159, 190–3, 303–4
hydrophilic moieties 15–17, 18
  in soaps 19–23, 346
  see also polar groups
hydrophobic moieties 15–17, 18
  in soaps 19–23, 346
  see also lipophilic parts
hyperpolarizability 150–3, 328–37
hysteresis effects 300, 303
  see also nonlinear effects

icosahedral symmetry 117, 441
induced dipole/induced dipole interactions 6, 171–9, 180–1
  directionality 159
  and electrical properties 199
  and lattice energy 162
  and segregation 185
induced dipoles 167–8, 298–9
  from photons 318–19, 321
induced symmetry *see* site symmetry
industrial applications 343–427
  thin-film transistors 270–4
industrial production 345, 375–6
insulators 197–8, 199, 219, 244
  molecular 38, 40, 233–5, 259
interaction energy 1, 158, 183–5
  in molecular materials 198–9, 214, 217
  phthalocyanines 222, 231–2
interaction moments 330–1
interactions 15, 18, 70, 157–60
  directionality 156–9, 180–3, 188, 190
  electromagnetic waves 318–21
  electrostatic 157, 159–60
  free energy of 138, 139–40
  gases and matter 256–8
  see also intermolecular (forces and interactions)
intermolecular forces 156–9
intermolecular interactions 157, 191, 198–9, 214, 217, 231–2
  see also directionality
intramolecular interactions 198–9, 214, 217
intrinsic semiconductors 38, 220, 230–2, 239, 276, 278–9
ionic crystals, lattice energy 159–62
ions 159–65, 167, 168
  as charge carriers 170–1
  see also anions; cations

irreducible representations 62–6, 146–55, 471–4
isohedral tilings 91, 96, 128, 450–67

junctions 242–8, 249–50

Kitaigorodskii's approach 84–90, 119–21

lattice energy 157, 159–62, 171–2
lattices 5–9, 79, 161, 429
Laüe parallelepipeds 6, 7
Laves tiles 91–3, 96–7
  see also topological class
LCAO 199–200
LCDs *see* liquid crystal displays
ligands 23, 28, 29, 163–7, 334–7
light
  absorption of 319, 321–4, 415
  refraction 168–9, 318–21, 324–8
  see also photoconduction
lipophilic parts, soap 346, 352–8
liquid crystal displays 405–24
liquid crystals 9–15, 28–9, 32–41, 405–24
  columnar *see* columnar mesophases
  ferroelectric properties 301
  molecular semiconductors 239–42
  triptycene 106, 107, 108
lithium phthalocyanine *see* PcLi
luminance 360–1
LUMO 212, 237, 320, 322
lutetium bisphthalocyanine *see* Pc$_2$Lu
lyotropic mesophases 15–23, 28–9, 351

magnetic fields 140, 142
magnetic properties 229, 232–3
mesomeric moments 328–30
mesophases 2–5
  cholesteric 13, 405, 407
  lyotropic 15–23, 28–9, 351
  nematic 3, 12–13, 37, 38, 405, 406, 407–13
  organization 104–11, 179, 183–90, 191, 193
  see also smectic mesophases
metallophthalocyanines 213
  absorption spectra 367
  band structures 233–9
  charge carriers 226–7, 230–2, 276–7
  conduction 224–7, 230–2, 261–8, 276–80
  see rectification
  electrochromism 220–1

FETs 276–91
  as gas sensors 264–8
  liquid crystal 240–2
  magnetic properties 229, 232–3
  orbitals 237, 238
  supramolecular 235–6, 239, plates 4, 5, 6
  as photoconductors 402–3, 404, 405
  redox potentials 224–5, 228
  redox reactions 220–1, 230, 264–5, 267, 276
  Schottky contacts 245–8
  semiconductors 38, 220–33, 239, 276, 278–9
  as solar cells 251–6
  structure 8–9, 121–2, 123, 221–4, 258, 259, 261, 262
  symmetry 213–14
  unit cells 213, 259
  see also Pc$_2$Lu; Pc$_2$Sc; Pc$_2$Tm; PcAlF thin films; PcCu; PcLi; PcMg photoreceptors; PcNi; PcPb; PcZn; phthalocyanines; rare-earth phthalocyanines; thin films metals 197–8, 219
  complexes xiii, 23, 163–4
  see also cations; metallophthalocyanines
micelles 19, 20, 24–5, 27–8, 32, 352
mobility, charge carrier 218–19
  in charge transfer 392–3, 399–400
  and conductivity, perylene 233–4
  FETs 270, 271, 275
    molecular 287–9, 290, 291
    Pc$_2$Lu 226–7
molecular assemblies 1–46
  lyotropic 15–23
  see also supramolecular assemblies
molecular crystals 171–2, 198–9, 233, 301
  mobility of charge carriers 219
  pyroelectric 309–10
molecular dielectrics 297–342
molecular dipoles 181–2
molecular electronics xiii
molecular engineering xii
molecular orbitals 199–205, 212
  and energy transfer 386
  HOMO 212, 230, 235, 237, 320, 322
  LUMO 212, 237, 320, 322
  PcLi 237, 238
  supramolecular orbitals from 212–14
molecular polarizability 167–71, 321–4
molecular semiconductors 217–33

liquid crystals 239–42
solar cells 248, 250–6
molecular subunits 2
molecular units xii
  charge transport 393, 395–6
  in chiral smectic C phases 306–7
  electromagnetic wave interaction 318–24
  hyperpolarizability 332–7
  organization 100, 101, 103–11, 183–7
    three-dimensional 84–9, 116–17, 121–8
  shape 93–111
  symmetry of 54–60
  tiling 90, 91, 93, 100
  and unit cells 82, 83
  see also intermolecular interactions
molecular volume 118, 119
moments
  electrical 132–3
  see also dipole moments
monoclinic system 7, 326, 469
monocolor
  layers 80–1, 82, 83
  space groups 82–3, 444–9
  stripes 71–8
monophthalocyanines 38, 259–64
  see also PcCu; PcLi; PcMg photoreceptors; PcNi; PcPb; PcZn
multipoles 137, 138, 139–40

naphthalene 173, 175, 176, 234, 329
narrow-band semiconductors 220–9
nematic liquid crystals 3, 12–13, 37, 38
  LCDs 405, 406, 407–15
nitrogen dioxide 261–8, 291
noncentrosymmetric crystals 181–2
nonlinear effects xiv, 299–301

oblique system 79, 83, 87, 88, 444–5
one-dimensional bands 202–5
one-dimensional space groups 71–9, 82–3
one-sided bands 71–8
optical tensor coefficients 468–70
orbitals 136–7, 199–205
  d orbitals 23, 136, 137
  molecular 199–205, 212, 230, 235, 237, 238, 320, 322, 386
  p orbitals 65–6, 136, 137, 139, 207
  s orbitals 63–4, 136, 137, 200–1, 207, 208
  supramolecular 212–14, 235–6, 239, plates 4, 5, 6

order of crystal class (or crystallineclass) 73
order of groups 57–8
order of site symmetry group 73
orthorhombic system 7, 326, 469
oxygen 183, 264
 dioxygen *see* dioxygen

p orbitals 65–6, 136, 137, 139, 207
p-n junctions 242–8, 249
packing 172–9
 chains 174, 176–9, 314–16
 effect on segregation 187–90
paints 367, 369–76
paraffinic chains 172, 174, 353–4
 packing 174, 176–9, 314–16
 phthalocyanines 122, 240–1
 in soaps 346, 352–8
 triptycene 106–8, 109
paramagnetism 229, 233, 299
Pc_2Lu (lutetium bisphthalocyanine) 219
 semiconductors 220–9, 240–2
 FETs 275, 278–82, 283
 gas sensors 264–8
 see also phthalocyanines
Pc_2Sc 282–3
Pc_2Tm 278–80
PcAlF thin films 262–3, 265
PcCu (copper phthalocyanine) 8, 9, 225, 402, 480
 in FETs 288, 289
 in gas sensors 261–2
PcH_2 225, 235–7, 238, 252, 367
 as photoreceptors 402–3
PcLi (lithium phthalocyanine) 219, 229–33, 237–9, 240–2
PcM *see* metallophthalocyanines
PcMg photoreceptors 402
PcNi 225, 275, 277, 278, 282–3
PcPb 261–2, 263, 264
PcZn 85, 275, 276, 278, 280–2, 403
permanent dipoles 335
permanent multipole/permanent multipoles 180–1
perylene 173, 175, 233–4, 394
 colorants 366, 401, 402, 419
photoconduction 385–405
photocopying 385–6, 388–91, 403–5
photovoltaic effect 249–56
phthalocyanines 23–43
 absorption spectra 367, 402, 403
 adsorption 261–2
 bandwidth 235, 236–7

colorants 32, 364–5, 366, 371, 480, 481
 in charge generation layers 394, 396
 as photoreceptors 402–3
 conduction 38, 39–40, 224–7, 230–2, 241–2, 261–8, 276–80
 crown-ether-substitution 26, 104–6
 conductivity 263, 266
 molecular orbitals, HOMO 230, 235, 237
 structure 14, 32–43, 237, plate 1
 micellar phases 27–8
 paraffinic chains 122, 240–1
 see also PcH_2
pi bonds 169–70, 328, 329–30
pi orbitals 200, 239
piezoelectrical coefficients 312–13, 317, 468–70
piezoelectricity 143–5, 153, 312–17
pigments 359–85, 394, 475–86
 early use 362–3, plate 10
 particles in 367–8, 369, 373, plate 11
pixels 409–10, 411, 423–4
plastic crystals 3, 14–15, 111–19
pleochroism 405, 413
point group of crystal 429
 see also crystal class
point symmetry groups 57, 58–60, 429–42
 of molecules 82
 and spherical harmonics 135
polar groups 16, 18, 348–9
 in soaps 19–23, 346, 353–5
polarity 180–2, 405–6
 and ferroelectric LCDs 420–2
 and group-subgroup relations 66–7
 in piezoelectricity 314, 317
 in pyroelectricity 308
polarizability 317–40
 hyperpolarizability 150–3, 328–37
 molecular 167–71, 321–4
polarizability tensor coefficients 146–55, 321, 324, 327–8
polarization 153, 157, 158, 170–1
 in complexation 167
 by photons 318–21
 and pyroelectricity 308
polarization of light, LCDs and 407, 409
polarized waves 141–3
polycrystalline materials 8–9, 219, 245–8
 silicon 270, 271
polymerization, emulsion 373–5

INDEX

polymers 15, 198, 219, 225
  FETs from 275
  in photoconduction 393, 394, 396
  phthalocyanines 40–2
potentials 131–9
  on electrons, in band theory 215–17
  see also redox potentials
Principe des affinités électives xiii, 162–4, 183–90, 346
pyroelectric coefficients 308, 309, 310, 311–12
pyroelectricity 144–5, 153, 308–12

radicals 219, 220, 229, 239
radii
  ions 160, 161, 163–5, 167, 168
  molecular units 184, 188–9
rare-earth phthalocyanines 32, 38–9, 40, 219, 220
  magnetic properties 229, 233
  semiconductors 220–9, 240–2
    FETs 275, 278–82, 283
    gas sensors 264–8
  see also Pc_2Lu
rectangular system 79, 83, 88, 445–6
rectification (effect) 244, 245, 246, 247, 252, 253, 278
redox potentials 217–18, 219, 260
  and activation energy 225, 227–8
  and conduction 224–5, 231
  and p-n junctions 243–4
  Pc_2Lu 221, 224
  PcLi disproportionation 230
redox reactions 38–9, 260, 261
  charge carriers 217–18, 259–60, 276
  Pc_2Lu 220–1, 264–5, 267
  PcLi 230
reducible representations 63–4
refraction 168–9, 318–21, 324–8
refractive indexes 318, 324–8
rosin (colophony) 379–82
rotation 4–5, 15, 49–51, 61, 111
  and spherical harmonics 134, 471

s orbitals 63–4, 136, 137, 200–1
  band structure 207, 208
Schlieren texture 13
Schoenflies symbols 57, 432
Schottky devices 242, 245–56, 275–6
Schrödinger equation 215
second harmonic generation (SHG) 144–5, 328–32, 338–40
segregation 183–90, 346

semiconductors 38, 197–8, 219, 220, 230–2, 239, 276, 278–9
  for FETs 269–74
  molecular 217–33, 239–42
    solar cells 248, 250–6
  Schottky devices 242, 245–56, 275–6
shape, molecular 93–111
Shubnikov symbols 432
sigma bonds 169–70, 328, 329–30
silicon 270, 271
  FETs 269–70, 271, 282, 283, 410
  solar cells 248
site symmetry 72, 73, 74, 429
  examples 109, 117, 122
    phthalocyanines 104–6, 121
  in tiling 97, 103–4, 107, 466–7
smectic mesophases 3, 10–12
  chirality 304–7, 311–12
  for LCDs 405–6, 415, 420–3
soaps 16, 19–23, 344–59, 382
solar cells 242, 245, 249–50
solid state 5–9
space filling 111
  see also tessellation
space groups 6, 97, 429
  examples 114, 115, 116, 117, 122, 124, 126
  frequency of occurence 119–21
  metallophthalocyanines 121
  monocolor 82–3, 444–9
  one-dimensional 71–9, 82–3
  in organic crystals 181–2
  three-dimensional 119–21
  two-dimensional 79–93, 104, 444–9
spherical harmonics 134–9, 471
square system 79, 82, 83, 88, 101–2, 447
stereographic projections 48, 49, 50
stress 154–5, 312–14
structure 1
  chiral smectic C phases 311
  and ferroelectricity 303–4
  liquid crystals 10–15, 32–41
  phthalocyanines 8–9, 14, 32–43, 121–2, 123, 221–4, 237, 261, 262, plate 1
    biphenyl cores 186–7
    crown-ether macrocycles 24–6
    micellar phases 27–8
    paraffinic chains 122, 240–1
    thin films 258–9, 264–8
    poly(vinylidene fluoride) 314–17
    soap 16–23, 347, 350–8

structure (continued)
    thin films   258–9, 264–8
    tourmaline   308
supermolecules   xiii
supramolecular assemblies   23–43
supramolecular engineering   xii–xiii
    symmetry aspects   69–129
supramolecular orbitals   212–14
    PcH_2 235–6, plates 4, 5, 6
    PcLi   239
symmetry   47–68
    cause/effect relationships   140–55
    and ferroelectric materials   302
        smectic C phases   304, 306
    of molecular units   54–60
    and piezoelectricity   144–5, 312–14
    and pyroelectricity   308
    and SHG   338–9
    and supramolecular engineering   69–129
    and transition moments   323
    see also point symmetry groups; site symmetry
symmetry axes   49, 55, 56–7, 431
symmetry elements   47–51, 52–4, 72–7
symmetry operations   50, 54–8, 73
    matrix notation for   60–2
symmetry of orientation see crystal class
systemic chemistry   xi–xiv
systems   xii, xiii–xiv

tables of characters   433–42
tensors
    hyperpolarizability   150–3, 328, 468–70
    optical   468–70
    piezoelectrical   312–13, 468–70
    polarizability   146–55, 321, 324, 327–8
tessellation   93–6
    see also tiling
tetragonal system   7, 326, 469
texture, mesophase   10–12, 13
thermal conductivity   153
thermotropic liquid crystals   9–15, 28
thin films, phthalocyanine
    electrical properties   224, 225–6, 231, 233–5, 261–8
        charge carrier mobility   219, 226–7
    FETs   276–80
    as gas sensors   264–8
    growth   85
    morphology   258–9, 264–7

Pc_2Lu 224, 226–7, 264–8, 278–80
Pc_2Tm 278–80
PcLi   231
thin-film transistors   270–4, 276–80
three-dimensional band structures   209–12
three-dimensional structures   84–90, 103–4, 111–28
tiles   90–3
    see also elementary tiles; Laves tiles; tilings
tilings   90–6, 122, 128, 450–67
topological approach   96–104
topological class   91–3, 96–7, 99–103
    isohedral tilings   458–65
tourmaline   308, 312, plate 7, plate 8
transistors   269–91, 409–11, 424
transition metals   xiii, 23, 163–4
    HSAB principle   164–7
transition moments   321, 322–4, 331
    in pigments   367–8, 413
traps   231, 233–5, 275, 287
triclinic system   7, 326, 468
trigonal system   7, 326, 469
triptycene   60, 106–8, 109, 174
twisted nematic liquid crystal displays   405, 407–13
two-dimensional band structures   209–12
two-dimensional molecular shape   93–111
two-dimensional space groups   79–93, 104, 444–9
two-sided bands   77–9

unicolor see monocolor
unit cells   6, 72, 82, 83, 102, 429
    metallophthalocyanines   213, 259

van der Waals interaction see induced dipole/induced dipole interaction
vesicles, soap   20

wavefunctions   215–17
    see also Bloch functions
waves
    electromagnetic see electromagnetic waves
    stationary, in band theory   217

xerography see photocopying

Yin–Yang principle   90, 91, 98, 99, 176, 188
    in phthalocyanines   105, 110, 122